Get the eBook FREE!

(PDF, ePub, Kindle, and liveBook all included)

We believe that once you buy a book from us, you should be able to read it in any format we have available. To get electronic versions of this book at no additional cost to you, purchase and then register this book at the Manning website.

Go to https://www.manning.com/freebook and follow the instructions to complete your pBook registration.

That's it!
Thanks from Manning!

OpenStack in Action

V. K. CODY BUMGARDNER

For online information and ordering of this and other Manning books, please visit www.manning.com. The publisher offers discounts on this book when ordered in quantity. For more information, please contact

Special Sales Department
Manning Publications Co.
20 Baldwin Road
PO Box 761
Shelter Island, NY 11964
Email: orders@manning.com

∞ Recognizing the importance of preserving what has been written, it is Manning's policy to have the books we publish printed on acid-free paper, and we exert our best efforts to that end. Recognizing also our responsibility to conserve the resources of our planet, Manning books are printed on paper that is at least 15 percent recycled and processed without the use of elemental chlorine.

Manning Publications Co.
20 Baldwin Road
PO Box 761
Shelter Island, NY 11964

Development editors: Susan Conant, Cynthia Kane
Technical development editors: Bill Bruns, Andy Hill, Michael Kidd, Jeff Lim, Fabrizio Soppelsa
Copyeditor: Andy Carroll
Proofreader: Katie Tennant
Technical proofreaders: Alain Couniot, David Pombal
Typesetter: Dottie Marsico
Cover designer: Marija Tudor

ISBN 9781617292163
Printed in the United States of America

brief contents

PART 1 GETTING STARTED 1

1 ■ Introducing OpenStack 3
2 ■ Taking an OpenStack test-drive 20
3 ■ Learning basic OpenStack operations 55
4 ■ Understanding private cloud building blocks 84

PART 2 WALKING THROUGH A MANUAL DEPLOYMENT.............111

5 ■ Walking through a Controller deployment 113
6 ■ Walking through a Networking deployment 161
7 ■ Walking through a Block Storage deployment 195
8 ■ Walking through a Compute deployment 216

PART 3 BUILDING A PRODUCTION ENVIRONMENT 239

9 ■ Architecting your OpenStack 241
10 ■ Deploying Ceph 259
11 ■ Automated HA OpenStack deployment with Fuel 277
12 ■ Cloud orchestration using OpenStack 303

contents

foreword *xiii*
preface *xv*
acknowledgements *xvii*
about this book *xviii*

PART 1 GETTING STARTED 1

1 ***Introducing OpenStack*** ***3***

1.1 What is OpenStack? 6
1.2 Understanding cloud computing and OpenStack 9
Abstraction and the OpenStack API *10*
1.3 Relating OpenStack to the computational resources it controls 11
OpenStack and hypervisors *11* ▪ *OpenStack and network services* *14* ▪ *OpenStack and storage* *15* ▪ *OpenStack and cloud terminology* *17*
1.4 Introducing OpenStack components 18
1.5 History of OpenStack 18
1.6 Summary 19

2 *Taking an OpenStack test-drive* 20

2.1 What is DevStack? 22

2.2 Deploying DevStack 23

Creating the server 25 ▪ *Preparing the server environment* 26
Preparing DevStack 28 ▪ *Executing DevStack* 29

2.3 Using the OpenStack Dashboard 36

Overview screen 38 ▪ *Access & Security screen* 38
Images & Snapshots screen 41 ▪ *Volumes screen* 44
Instances screen 47

2.4 Accessing your first private cloud server 51

Assigning a floating IP to an instance 53 ▪ *Permitting network traffic to your floating IP* 53

2.5 Summary 54

3 *Learning basic OpenStack operations* 55

3.1 Using the OpenStack CLI 56

3.2 Using the OpenStack APIs 58

3.3 Tenant model operations 59

The tenant model 61 ▪ *Creating tenants, users, and roles* 62 ▪ *Tenant networks* 65

3.4 Quotas 78

Tenant quotas 79 ▪ *Tenant-user quotas* 80
Additional quotas 82

3.5 Summary 83

4 *Understanding private cloud building blocks* 84

4.1 How are OpenStack components related? 85

Understanding component communication 85
Distributed computing model 91

4.2 How is OpenStack related to vendor technologies? 95

Using vendor storage systems with OpenStack 96
Using vendor network systems with OpenStack 101

4.3 Why walk through a manual deployment? 108

4.4 Summary 109

PART 2 WALKING THROUGH A MANUAL DEPLOYMENT 111

5 *Walking through a Controller deployment* *113*

5.1 Deploying controller prerequisites 116

Preparing the environment 116 ▪ Configuring the network interface 117 ▪ Updating packages 120 ▪ Installing software dependencies 121

5.2 Deploying shared services 124

Deploying the Identity Service (Keystone) 125 ▪ Deploying the Image Service (Glance) 135

5.3 Deploying the Block Storage (Cinder) service 142

Creating the Cinder data store 143 ▪ Configuring a Cinder Keystone user 144 ▪ Creating the Cinder service and endpoint 145 ▪ Installing Cinder 146

5.4 Deploying the Networking (Neutron) service 147

Creating the Neutron data store 148 ▪ Configuring a Neutron Keystone user 149 ▪ Installing Neutron 151

5.5 Deploying the Compute (Nova) service 152

Creating the Nova data store 153 ▪ Configuring a Nova Keystone user 154 ▪ Assigning a role to the nova user 154 Creating the Nova service and endpoint 155 ▪ Installing the Nova controller 156

5.6 Deploying the Dashboard (Horizon) service 158

Installing Horizon 158 ▪ Accessing Horizon 159 Debugging Horizon 160

5.7 Summary 160

6 *Walking through a Networking deployment* *161*

6.1 Deploying network prerequisites 163

Preparing the environment 164 ▪ Configuring the network interfaces 164 ▪ Updating packages 167 ▪ Software and configuration dependencies 168 ▪ Installing Open vSwitch 171 ▪ Configuring Open vSwitch 174

6.2 Installing Neutron 177

Installing Neutron components 177 ▪ Configuring Neutron 178 ▪ Configuring the Neutron ML2 plug-in 178 ▪ Configuring the Neutron L3 agent 179

Configuring the Neutron DHCP agent 180 ▪ Configuring the Neutron Metadata agent 180 ▪ Restarting and verifying Neutron agents 181 ▪ Creating Neutron networks 182 Relating Linux, OVS, and Neutron 191 ▪ Checking Horizon 193

6.3 Summary 194

7 *Walking through a Block Storage deployment 195*

7.1 Deploying Block Storage prerequisites 197

Preparing the environment 198 ▪ Configuring the network interface 198 ▪ Updating packages 201 ▪ Installing and configuring the Logical Volume Manager 202

7.2 Deploying Cinder 206

Installing Cinder 208 ▪ Configuring Cinder 209 Restarting and verifying the Cinder agents 210

7.3 Testing Cinder 211

Create a Cinder volume: command line 211 ▪ Create a Cinder volume: Dashboard 213

7.4 Summary 215

8 *Walking through a Compute deployment 216*

8.1 Deploying Compute prerequisites 219

Preparing the environment 219 ▪ Configuring the network interface 219 ▪ Updating packages 222 Software and configuration dependencies 222 Installing Open vSwitch 223 ▪ Configuring Open vSwitch 225

8.2 Installing a hypervisor 226

Verifying your host as a hypervisor platform 226 Using KVM 227

8.3 Installing Neutron on Compute nodes 229

Installing the Neutron software 230 ▪ Configuring Neutron 230 ▪ Configuring the Neutron ML2 plug-in 231

8.4 Installing Nova on compute nodes 231

Installing the Nova software 231 ▪ Configuring core Nova components 232 ▪ Checking Horizon 233

8.5 Testing Nova 234

Creating an instance (VM): command line 234

8.6 Summary 238

PART 3 BUILDING A PRODUCTION ENVIRONMENT........... 239

9 *Architecting your OpenStack 241*

9.1 Replacement of existing virtual server platforms 242

Making deployment choices 245 ▪ What kind of network are you? 246 ▪ What type of storage are you? 247 ▪ What kind of server are you? 250

9.2 Why build a private cloud? 251

Public cloud economy-of-scale myth 251 ▪ Global scale or tight control 252 ▪ Keeping data gravity private 252 Hybrid moments 253

9.3 Building a private cloud 254

OpenStack deployment tools 254 ▪ Networking in your private cloud 255 ▪ Storage in your private cloud 257

9.4 Summary 258

10 *Deploying Ceph 259*

10.1 Preparing Ceph nodes 260

Node authentication and authorization 261 ▪ Deploying Ceph software 264

10.2 Creating a Ceph cluster 265

Creating the initial configuration 265 ▪ Deploying Ceph software 266 ▪ Deploying the initial configuration 267

10.3 Adding OSD resources 268

Readying OSD devices 269 ▪ Creating OSDs 271

10.4 Basic Ceph operations 273

Ceph pools 273 ▪ Benchmarking a Ceph cluster 274

10.5 Summary 276

11 *Automated HA OpenStack deployment with Fuel 277*

11.1 Preparing your environment 279

Network hardware 279 ▪ Server hardware 282

11.2 Deploying Fuel 290

Installing Fuel 290

11.3 Web-based basic Fuel OpenStack deployment 293

Server discovery 294 ▪ Creating a Fuel deployment environment 295 ▪ Configuring the network for the

environment 296 ▪ Allocating hosts to your environment 298
Final settings and verification 301 ▪ Deploying changes 302

11.4 Summary 302

12 Cloud orchestration using OpenStack 303

12.1 OpenStack Heat 304
Heat templates 304 ▪ A Heat demonstration 307

12.2 Ubuntu Juju 312
Preparing OpenStack for Juju 312 ▪ Installing Juju 314
Deploying the charms CLI 317 ▪ Deploying the Juju GUI 319

12.3 Summary 325

appendix Installing Linux 326
index 347

foreword

It's difficult for me to believe that it's already been five years since I was looking over the original source code of the Nova project. The code had just been released by the Anso Labs team who created it for NASA. Rackspace, where I worked at the time, was seeking a new code base to act as the next generation of the Rackspace Cloud. A few months later, Rackspace open-sourced the code for its Rackspace Cloud Files platform as the Swift project. Nova and Swift became the first two pillars of the nascent OpenStack project.

Since that time, both projects have undergone substantial change. Swift's core team and code base have remained remarkably stable, though the project has added a number of features and seen numerous enhancements in performance and scalability. On the other hand, when compared with its humble beginnings, Nova's source code is nearly unrecognizable. New code bases like Glance, Cinder, Keystone, and Neutron were constructed to deliver functionality that originally was handled by Nova.

At the same time that this new source code emerged to handle functionality essential to managing large computing infrastructure, a new kind of open source community was beginning to take form. Open source developers and advocates with experience in operating system distribution and packaging, configuration management, database design, automation, networking, and storage systems flocked to contribute to OpenStack projects.

Our community grew (and continues to grow) at breakneck pace, quickly becoming one of the largest, most active and influential open source communities on the planet. The OpenStack Foundation was created to handle the governance challenges that accompany the growth of a successful community. Design summits and conferences have grown to host more than 3,500 contributors per year all around the planet.

A world-class continuous integration and build system was created by the community to support the massive growth in both source code and number of contributors. This automated build system's size and scope now rival or exceed those of much older open source communities like the Apache and Eclipse foundations.

The OpenStack ecosystem has been the fertile ground from which new companies like SwiftStack and Piston Cloud emerged. Existing companies like HP, Mirantis, and Red Hat found the OpenStack landscape to be similarly fruitful, and they continue to drive innovation across the now dozens of projects that comprise the big tent of the OpenStack community.

This expansion of the OpenStack community has brought with it a bewildering complexity to how distributed software components are deployed and how they interoperate. Those who deploy OpenStack "in the wild" need to understand a broad set of concepts from networking and storage to virtualization and configuration management. Obtaining the necessary knowledge and skills has been and remains one of the key challenges facing those who wish to build cloud platforms using OpenStack. This book, *OpenStack in Action,* will provide readers with the knowledge they need to deploy and run OpenStack.

The author leads the audience through the complexity of an OpenStack deployment, demonstrating three ways to deploy the software: via a scripted tool called DevStack, via manual installation of operating system packages, and via the Fuel OpenStack installer. In each section, concepts around networking and storage setup are thoroughly explained, allowing readers to gradually dip their toes into the cloud computing waters and, by the end of the book, feel comfortable diving into the deep end of the pool.

Besides excellent coverage of OpenStack technology, the author also explains how to go about evaluating if and how your organization will benefit from cloud computing. The cloud does not magically solve the manual and time-consuming human-based process problems that exist in many organizations. But, when implemented smartly and for the right reasons, the cloud can transform an IT organization and dramatically improve the services they provide. In chapter 9, Mr. Bumgardner leads a discussion that should be required reading for any IT director who is considering replacing existing virtualized IT infrastructure with OpenStack or constructing a new private cloud offering for their internal customers.

In short, *OpenStack in Action* serves as an excellent primer on the complex world of cloud computing and the OpenStack software ecosystem. Read it. Absorb it. And become a "Stacker" at heart!

JAY PIPES
MEMBER, OPENSTACK TECHNICAL COMMITTEE
DIRECTOR OF ENGINEERING
MIRANTIS, INC.

preface

My first exposure to OpenStack came in the summer of 2011 while I was working at the University of Kentucky. My coworker and friend, Brent Salisbury, and I were invited to meet with a Fortune 50 technology company to discuss a product development project. During our meeting, the project's executive sponsor gave us the option to work with existing commercial tools or investigate the use of a community project called OpenStack. Naturally, we chose to work with the framework we knew nothing about, and so began our OpenStack journey. Nothing came of the product development project, but the OpenStack encounter, as it turned out, became a turning point in our professional, and in my case academic, careers. Brent left the university and cofounded a startup that was acquired by Docker, where he currently works. I, on the other hand, transferred from a master's to a doctoral program and wrote this book.

By early 2013, the Grizzly release of OpenStack somewhat resembled current versions, but instabilities due to rapid feature inclusion prevented us from considering OpenStack production-ready for our enterprise environment. But although I was not ready to put my neck on the line with OpenStack for the enterprise, research computing was another story. As part of a graduate independent study class, I documented the use cases, architecture, and strategy around using OpenStack in research computing. In addition, I described the process and eventual adoption of the platform as a private cloud for our enterprise.

I used figure 1 in my original academic report to represent the component-level distribution of OpenStack. I suspect cooking an elephant, much like eating one, must be done a piece at a time. Far too often in technology, we accept technological isolation as an organizationally sound practice—"I am a storage guy," or "I am a network girl"—but this is paramount to someone only eating one part of the elephant. In this book I've

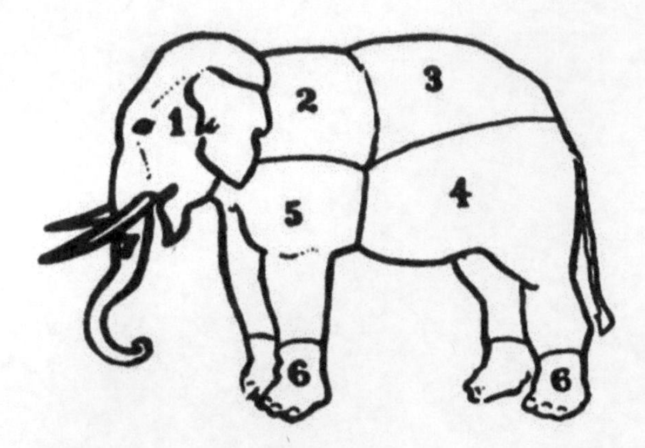

Figure 1 This image is from a sixteenth-century edition of *Libro de Arte Coquinaria (Book on the Art of Cookery)* by Maestro Martino.

tried to mix recognizable morsels with new concepts for easier digestion. Although you might not want to taste elephant feet, you'd better know, at least in principle, how they work if you are going to be successful in your adoption of cloud computing.

I'm writing this preface exactly two years to the day after I first spoke with a Manning acquisitions editor. When I started this project, there were fewer than 500 OpenStack contributors, and now there are thousands. Not only has OpenStack become one of the fastest-growing open source communities ever, it has been adopted by the biggest organizations in the world. More importantly, at least for you, OpenStack is now mature and ready to serve as a foundation for your organization's private cloud.

acknowledgments

This book would have never existed without the encouragement of my doctoral advisor and friend, Professor Victor W. Marek. The pushing was always welcome, the confidence was always needed, and the responsibility is mine to pay forward.

I would never have imagined the effort that it takes to produce a book if I hadn't experienced it myself. Whether or not this effort produced the desired result will be decided by my readers, but there should be no doubt that an army of reviewers, editors, and other contributors spent many hours in the pursuit of a high-quality book. Having contributed to and reviewed other books from other publishers as I wrote this book, I can honestly say that Manning does everything they can to produce the very best work possible. I especially want to thank Susan Conant, my development editor for most of this book, for her tireless work, continually pushing for improvements. Thanks also to publisher Marjan Bace and everyone on the editorial and production teams, including Mary Piergies, Cynthia Kane, Andy Carroll, Katie Tennant, and many others who worked behind the scenes. Many thanks to Bill Bruns, Andy Hill, Michael Kidd, Jeff Lim, and Fabrizio Soppelsa for their technical editing during the book's development. Finally, I want to thank the following individuals, who read drafts of this book and provided suggestions: Andy Kirsch, Chris Snow, Fernando Rodrigues, Hafizur Rahman, Kosmas Chatzimichalis, Matt Harting, Mayur Patil, Michael Hamrah, Peeyush Maharshi, and Toby Lazar.

Special thanks go to you, Sarah, my wife, who between caring for our two small children and supporting my work travel, graduate work, this book, and other work, took on far more than your share. Although the papers, presentations, and books have my first name on them, they will also always carry the name we share. Sarah, Sydney, and Jack, I'm sorry for the lost time and energy. I hope you can be as proud of me as I am of you. I love you all.

about this book

The primary topic of this book is deploying enterprise private clouds using OpenStack. In this context, I discuss private clouds as pools of infrastructure resources, or *infrastructure as a service* (IaaS), that are owned and managed by the organizations they serve. In contrast, public cloud IaaS resources are owned and operated by third-party service providers.

Financially, one can think of private clouds as primarily a capital expense, whereas public clouds are typically operational ones. The distinction is easy to understand, given that in private cloud deployments, organizations typically purchase or lease a fixed infrastructure for the duration of its serviceable life, regardless of actual usage. In public cloud deployments, cost is typically directly related to hourly occupancy (on or off) and communication costs.

The organizational adoption of private and public clouds is often related to the size and scope of those organizations' IT responsibilities. Enterprise IT departments, whose responsibility it is to centrally provide technical architectures and resources for the rest of the organization, have a vested interest in using a private cloud. A multi-tenant, fully orchestrated, private cloud provides great resource-management efficiency to enterprise IT. In this regard, enterprise IT becomes a cloud broker. In contrast, departmental IT units often lack the data center facilities and personnel to cost-effectively deploy private clouds. Often, due to their relatively small resource requirements, departments can take advantage of public cloud resources. If they're available, departments can also take advantage of private cloud resources managed by their enterprise IT units. Using both private and public clouds based on workload results in a hybrid cloud.

Despite the differences in clouds and the types of organizations that are best positioned to take advantage of them, the clouds themselves can be built using the same technologies. Although the ingredients that make up cloud resources might be the same, the recipes and methods of consumption can be very different.

OpenStack is a powerful framework for constructing both private and public clouds. Fundamentally, OpenStack abstracts and provides a common API for the hardware and software used in building clouds. The framework provides two very important things:

- Abstraction of hardware and software resources, which prevents vendor lock-in of any particular component
- A common API across resources, which allows for complete orchestration of connected components

The first aspect is nice from a financial perspective, but the second is the key to a modern IT transformation. For enterprise IT, OpenStack brings the same level of transformational efficiency to cloud deployment.

Why OpenStack in Action?

This book is intended as a step-by-step, bottom-up guide for constructing computational clouds of resources. My intended audience includes researchers, administrators, and students interested in the deployment of an OpenStack environment. There are no technical prerequisites beyond a basic operational knowledge of Linux, and the material is suited to people with very different backgrounds and technical abilities. Similarly, OpenStack is suited for many use cases.

Despite separate use cases utilizing the same OpenStack framework, the requirements and design of private clouds can vary greatly from those of service providers. Enterprises are interested in providing private resource clouds for their organizations. These private clouds don't just represent additional services; they can represent a transformation in the way organizations provide computational resources.

This book comprises

- An introduction to OpenStack through the automated deployment of a single-node development environment
- A deeper understanding of OpenStack through a step-by-step manual deployment of a multi-node environment
- The impacts of private cloud technologies (OpenStack, Ceph, Juju, and the like) from the perspective of IT operations
- The deployment of a production OpenStack environment using vendor-provided automated deployment and management tools

The architecture covered in the book is appropriate for small (5-node) to large enterprise (100-node) private cloud deployments. In addition, chapter 12 walks you through the use of application orchestration tools like OpenStack Heat and Ubuntu Juju on your newly constructed private cloud.

This book is about building an understanding of private cloud technologies, the deployment and operation of those technologies, and the long-term impact of cloud orchestration on traditional IT roles. This book will provide you with the ability to develop a convincing argument for the deployment of an OpenStack private cloud in your enterprise and will help you develop the technical knowledge to deploy your private cloud. Configurations and operational scripts demonstrated in the book are also available through the GitHub repository for the book at https://github.com/codybum/OpenStackInAction.

The most important thing for you to understand is that an OpenStack private cloud is not simply another virtualization tool. OpenStack is a framework that utilizes existing virtualization tools to construct and manage clouds. You'll learn how to think about cloud construction, deployment, and organization. On the technical side, you'll gain an understanding of the components of OpenStack and supporting technologies—specifically, OpenStack Compute, Networking, Block Storage, Dashboard, and API components.

Roadmap

The book is divided into three parts, where part 1 (chapters 1–4) gets you started, part 2 (chapters 5–8) provides a deep dive into the ecosystem, and part 3 (chapters 9–12) prepares you for using OpenStack in a production environment.

Chapter 1 introduces the OpenStack cloud operating system, the motivations for developing the framework, and what OpenStack can do for your organization.

In chapter 2 you'll jump right in and take a test drive with OpenStack, using a rapid deployment tool and minimal infrastructure. This test drive is not just to demonstrate the OpenStack Dashboard user experience; it will also provide you with a known working model to use while you learn the OpenStack framework. By the end of chapter 2, you'll be provisioning virtual machines under your own OpenStack environment.

Chapter 3 makes use of the environment you constructed in chapter 2 and introduces the OpenStack command-line interface (CLI). In this chapter, you'll walk through basic OpenStack operations as you create new tenants (projects), users, roles, and internal networks.

In chapter 4 you'll progress from using OpenStack to understanding the component-level functions and their interactions with each other in the overall OpenStack framework. You'll learn about several cloud design methodologies, which will prepare you for your own multi-node deployment. This chapter covers how OpenStack components work together and the relationship between OpenStack components and vendor resources.

Chapters 5–8 cover deep dives related to specific OpenStack projects, devoting a chapter to each of the major projects. These chapters will walk you through a manual deployment of OpenStack in a multi-node environment. Through these chapters, you'll gain a great understanding of how and why things work as they do in the OpenStack ecosystem. In addition, this manual deployment will give you valuable troubleshooting experience.

Chapter 9 covers architectural, organizational, and strategic decisions relating to a production OpenStack deployment. Chapter 10 covers the basic deployment and operation of Ceph storage. Chapter 11 will walk you through the automated HA deployment of OpenStack using Fuel. Finally, chapter 12 covers cloud orchestration using OpenStack Heat and Ubuntu Juju.

Who should read this book?

The book is suited for infrastructure specialists, engineers, architects, and support personnel interested in deploying a private cloud environment using OpenStack. Although the book has strategic value for those in executive and strategic roles, the message is tailored for a technical reader. There are no technical prerequisites beyond a basic operational knowledge of Linux.

Code conventions and downloads

All code in the book is presented in a `fixed-width font like this` to separate it from ordinary text. Code annotations accompany many of the listings, highlighting important concepts. In some cases, numbered bullets link to explanations that follow the listing.

You can download the code for the examples in the book from the publisher's website at www.manning.com/books/openstack-in-action and also from https://github.com/codybum/OpenStackInAction.

Author Online

The purchase of *OpenStack in Action* includes free access to a private web forum run by Manning Publications, where you can make comments about the book, ask technical questions, and receive help from the author and from other users. To access the forum and subscribe to it, point your web browser to www.manning.com/books/openstack-in-action. This page provides information on how to get on the forum once you are registered, what kind of help is available, and the rules of conduct on the forum.

Manning's commitment to our readers is to provide a venue where a meaningful dialogue between individual readers and between readers and the author can take place. It is not a commitment to any specific amount of participation on the part of the author, whose contribution to the forum remains voluntary (and unpaid). We suggest you try asking the author some challenging questions lest his interest stray! The Author Online forum and the archives of previous discussions will be accessible from the publisher's website as long as the book is in print.

About the author

Cody Bumgardner (http://codybum.com) has been in the IT industry for over 20 years, during which he has worked in technical, managerial, and sales roles in the areas of IT architecture, software development, networking, research, systems, and security. Over the last several years, he has focused on researching, implementing, and speaking about cloud computing and computational economics. He is also currently a PhD candidate in Computer Science at the University of Kentucky (UK), focusing on computational economics and distributed resource management. Cody currently serves as the Chief Technology Architect (CTA) of a large public land-grant university. As CTA, he developed a five-year university strategy and roadmap for cloud computing. This roadmap outlined the introduction of disruptive cloud technologies along with the related transformations of the IT workforce. The plan centered around the deployment of an enterprise OpenStack private cloud, supporting over 40,000 users in academic, research, and health care (academic) divisions. Cody is responsible for the architecture, financial model, deployment, and long-term strategy of the OpenStack private cloud, research computing, and other cloud computing initiatives.

About the cover

The figure on the cover of *OpenStack in Action* is captioned "Milkmaid from Coutances." The illustration is taken from a collection of works by many artists, edited by Louis Curmer and published in Paris in 1841. The title of the collection is *Les Français peints par eux-mêmes*, which translates as *The French People Painted by Themselves*. Each illustration is finely drawn and colored by hand, and the rich variety of drawings in the collection reminds us vividly of how culturally apart the world's regions, towns, villages, and neighborhoods were just 200 years ago. Isolated from each other, people spoke different dialects and languages. In the streets or in the countryside, it was easy to identify where they lived and what their trade or station in life was just by their dress.

Dress codes have changed since then, and the diversity by region, so rich at the time, has faded away. It's now hard to tell apart the inhabitants of different continents, let alone different towns or regions. Perhaps we have traded cultural diversity for a more varied personal life—certainly for a more varied and fast-paced technological life.

At a time when it's hard to tell one computer book from another, Manning celebrates the inventiveness and initiative of the computer business with book covers based on the rich diversity of regional life of two centuries ago, brought back to life by pictures from collections such as this one.

Part 1

Getting started

The first part of this book is an introduction to the OpenStack framework: how and why you'll want to use it. OpenStack components are decomposed, and relationships to underlying resources (compute, storage, network, and so on) are explained. You'll deploy OpenStack on a single node using the DevStack deployment tool. Along the way, this part will help you start thinking about how OpenStack could be used in your environment and develop your interest in the framework enough to gain a deeper understanding of how things work under the covers.

1 Introducing OpenStack

This chapter covers

- OpenStack and the cloud ecosystem
- Reasons to choose OpenStack
- What OpenStack can do for you and your organization
- Key components of OpenStack

Only a few decades ago, many large computer hardware companies had their own fabrication facilities and maintained competitive advantage by making specialty processors, but as costs rose, fewer companies produced the volume of chips needed to remain profitable. Merchant chip fabricators emerged, able to produce general-purpose processors at scale, and drove down costs significantly. Having just a few computer chip manufacturers encouraged standardized desktop and server platforms around the Intel x86 instruction set, and eventually led to commodity hardware in the client-server market.

The rapid growth of the World Wide Web during the dot-com years of the early 2000s created huge data centers filled with this commodity hardware. But although the commodity hardware was powerful and inexpensive, its architecture was often like that found in desktop computing, which was not designed with centralized management in mind. No tools existed to manage commodity hardware as a collection of resources. To make matters worse, servers during this period generally lacked hardware management capabilities (secondary management cards), just like their desktop cousins. Unlike mainframes and large symmetric multiprocessing (SMP) machines, these commodity servers, like desktops, required layers of management software to coordinate otherwise independent resources.

During this period, many management frameworks were developed internally by both public and private organizations to manage commodity resources. Figure 1.1 shows collections of interconnected resources spread across several data centers. With management frameworks, these common resources could be used interchangeably, based on availability or user requirements. While it's unclear exactly who coined the term, those able to harness the power of commodity computing through management frameworks would say they had a "cloud" of resources.

Out of many commercial and open source cloud management packages to be developed during this period, the OpenStack project was one of the most popular. OpenStack provides a common platform for controlling clouds of servers, storage, networks, and even application resources. OpenStack is managed through a web-based interface, a command-line interface (CLI), and an application programming interface (API). Not only does this platform control resources, it does so without requiring you to choose a specific hardware or software vendor. Vendor-specific

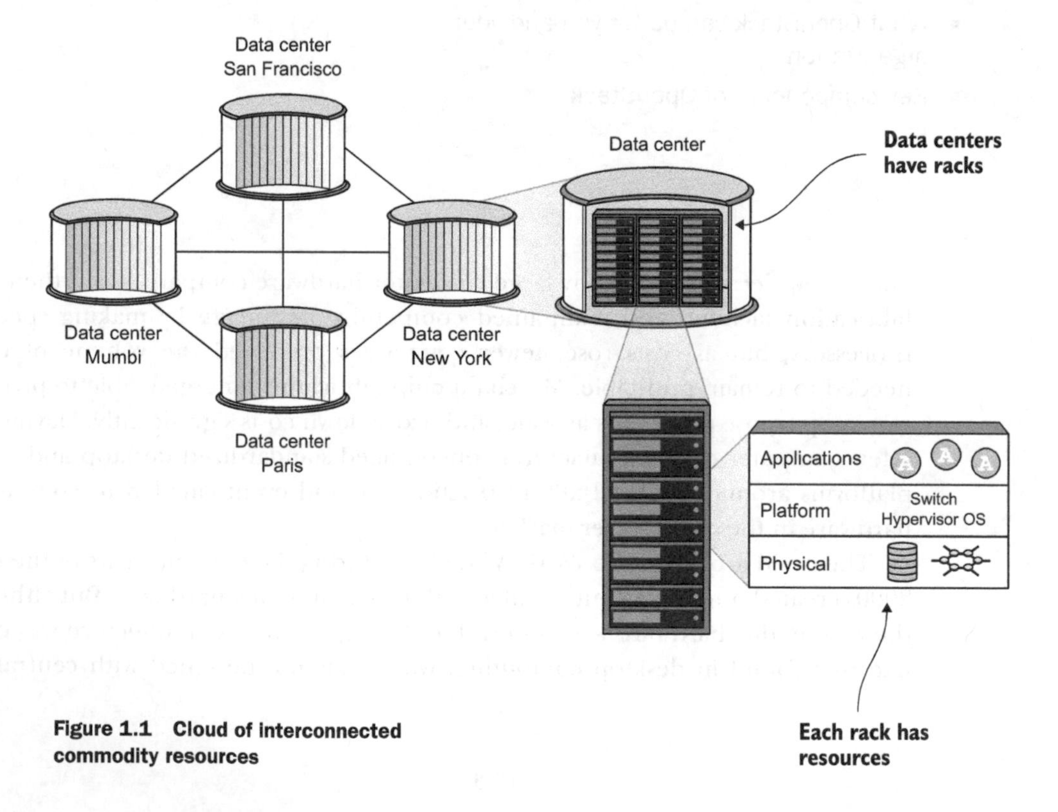

Figure 1.1 Cloud of interconnected commodity resources

components can be replaced with minimum effort. OpenStack provides value for a wide range of people in IT organizations.

One way to think about OpenStack is in the context of the Amazon buying experience. Users log in to Amazon and purchase products, and products are delivered. Behind the scenes, an orchestra of highly optimized steps are taken to get products to your door as quickly and inexpensively as possible. Twelve years after Amazon was founded, Amazon Web Services (AWS) was launched. AWS brought the Amazon experience to computing resource delivery. A server request that might take weeks from a local IT department could be fulfilled with a credit card and a few mouse clicks with AWS. OpenStack aims to provide the same level of orchestrated efficiency demonstrated by Amazon and other service providers to your organization.

What is OpenStack?

- *For cloud/system/storage/network administrators*—OpenStack controls many types of commercial and open source hardware and software, providing a cloud management layer on top of vendor-specific resources. Repetitive manual tasks like disk and network provisioning are automated with the OpenStack framework. In fact, the entire process of provisioning virtual machines and even applications can be automated using the OpenStack framework.
- *For the developer*—OpenStack is a platform that can be used not only as an Amazon-like service for procuring resources (virtual machines, storage, and so on) used in development environments, but also as a cloud orchestration platform for deploying extensible applications based on application templates. Imagine the ability to describe the infrastructure (*X* servers with *Y* RAM) and software dependencies (MySQL, Apache2, and so on) of your application, and having the OpenStack framework deploy those resources for you.
- *For the end user*—OpenStack is a self-service system for infrastructure and applications. Users can do everything from simply provisioning virtual machines (VMs) like with AWS, to constructing advanced virtual networks and applications, all within an isolated tenant (project) space. *Tenants,* also known as *projects,* are the way that OpenStack isolates assignments of resources. Tenant isolation includes storage, network, and VM isolation, so end users can be given much more freedom than in traditional virtual server environments. Imagine end users being assigned a quota of resources that they could easily provision how and when they want.

Virtual machines and tenants

Throughout the book, the term *virtual machine* (VM) will refer to an instance of an emulated physical machine (server). Virtual machines perform the same functions as physical machines and, from the perspective of the operating system, are intended to be indistinguishable from physical hardware. VMs are used for a variety of reasons, but most virtualization motivations boil down to the flexibility of controlling things through software outweighing the performance penalty. From a high-level view, you

> ***(continued)***
> can think of OpenStack as bringing the same level of operational efficiency to your data center that the software hypervisor brought to the server.
>
> As you'll learn in this book, the word *tenant* has a specific meaning in OpenStack. It's sufficient at this point to consider a tenant to be a quota-limited collection of resources used by VMs that are logically isolated from each other. For example, if a user misconfigures the network in tenant A, tenant B is unaffected.

The OpenStack foundation has hundreds of official corporate sponsors and a community of tens of thousands of people in over 130 countries. Like Linux, many people will be attracted to OpenStack as a community-supported alternative to commercial products. But what they'll soon learn is that when it comes to cloud frameworks, there are few that compare to OpenStack in terms of depth and breadth of services. Perhaps more importantly, there might not be another product, commercial or otherwise, that the average system administrator, developer, or architect can use on their own and that can provide a greater benefit to their organization.

1.1 What is OpenStack?

Let's expand on the definition of OpenStack as a framework for managing, defining, and utilizing cloud resources. The official OpenStack website (www.openstack.org) describes the framework as "open source software for creating private and public clouds." It goes on to say, "OpenStack Software delivers a massively scalable cloud operating system." If you have experience in server virtualization, you may quickly, yet incorrectly, conclude that OpenStack is just another way to provide virtual machines. Although this is a service enabled by the OpenStack framework, it's by no means OpenStack's definitive function.

Figure 1.2 shows several of the resource components that OpenStack coordinates to create public and private clouds. As the figure illustrates, OpenStack doesn't replace these resource providers; it simply manages them, through control points built into the framework.

An experienced systems administrator might take the description of OpenStack as a "cloud operating system" with great skepticism. It's not like administrators run around to hundreds of servers with a boot disk, and load OpenStack on bare metal, like a traditional operating system. Nevertheless, through its management of resources, OpenStack shares operating systems characteristics, but in the context of cloud computing.

With an OpenStack cloud you can

- Harness the resources of physical and virtual servers, networks, and storage systems
- Efficiently manage clouds of resources through tenants, quotas, and user roles
- Provide a common interface to control resources regardless of the underlying vendor subsystem

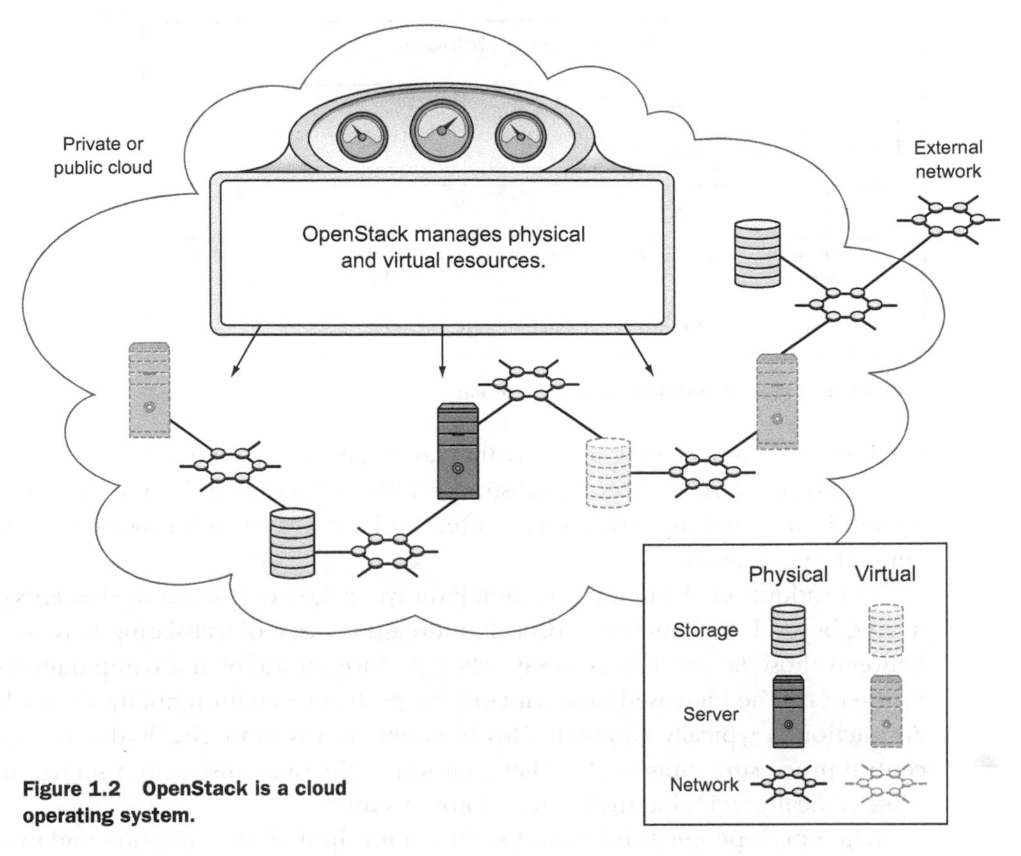

Figure 1.2 OpenStack is a cloud operating system.

At first glance, OpenStack doesn't look like a traditional operating system, but then again, the "cloud" doesn't look like a normal computer. One must take a step back and consider the fundamental benefits of an operating system.

Before there were operating systems or even hardware-level abstraction languages (assembly), programs were written in the language (binary machine code) of a specific computer. Then traditional operating systems came along and, among other things, allowed users to standardize not just application code, but also the management functions of the hardware. Administrators could now manage hardware instances using a common interface, developers could write code for a common system, and users only had to learn a single user interface. This held true regardless of underlying hardware, as long as the operating system remained the same. In the evolution of computers, the development and proliferation of operating systems gave rise to the field of systems engineering and administration.

Figure 1.3 shows the many layers of abstraction in modern computational systems.

No doubt, in the past there were developers who didn't like the idea of losing direct control over hardware through the use of an operating system, just as some systems administrators don't like the idea of losing control over the underlying hardware and operating systems through server virtualization. In each of these transitions, from

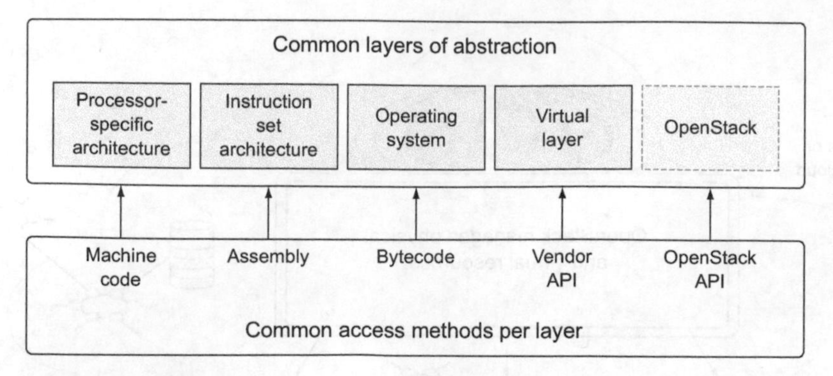

Figure 1.3 Layers of computational abstraction

machine code to assembly to the virtual layer, we didn't lose the underlying layer; it was simply standardized through abstraction. We still have highly optimized hardware, we still have operating systems, and often we have a layer of hardware virtualization between those layers.

Wide adoption of a new abstraction layer typically occurs when the benefits of optimizing beyond the standard approach outweigh the cost of translating (virtualization) between those layers. This is to say, when the overall utility of a computational environment can be increased by sacrificing raw performance for usability, then a layer of abstraction is typically adopted. This phenomenon is most clearly demonstrated by central processing units (CPUs) that conform to the same instruction set for decades, while radically changing their internal architectures.

When most people think of CPUs, they don't think of virtualization and execution variability on the hardware level, but this is the case. Many of the instructions on x86 processors are virtualized inside the processors themselves, where some complex legacy instruction is executed through a series of simpler, yet faster, instructions. The complexity of instruction-level optimization is beyond the scope of this book, but it's important to understand that even when using bare metal, some form of virtualization is at play, even on the processor level. Now, instead of focusing on losing control, imagine taking it through the use of a common framework for managing, monitoring, and deploying private and public clouds of infrastructure and applications. Once you take this evolutionary step forward, you have OpenStack.

Decades of CPU abstraction and virtualization

The Intel x86 instruction set was introduced with the Intel 8086 CPU in 1978, as a backward-compatible alternative to the Intel 8080. The x86 instruction set defined assembly instructions that would remain available regardless of processor changes. Since then, new "processor extensions" were added and clock cycles were increased, but the existing instructions remained.

As the demand for faster processors increased, so did the desire to ensure software interoperability between processor generations. CPU designers needed the flexibility to optimize at lower levels of abstraction, while maintaining instruction-level compatibility (standardization). Designers didn't worry about keeping the underlying hardware the same, which allowed them to greatly increase the clock speed of processors between generations. In 1995, the Pentium Pro introduced the idea of micro-op decoding. Instead of a specific instruction taking a single clock cycle, it might be translated into several simple instructions, which could take many cycles.

In addition to micro-ops, the Pentium Pro processor introduced optimizations through out-of-order execution of instructions and memory virtualization (addressing 36 bits of memory with a 32-bit bus). But this was all completely abstracted from the developer, allowing the same applications to run on several generations of processors from multiple vendors. This method of maintaining instruction-level compatibility continues to this day with the current generation of x86_64 processors.

1.2 Understanding cloud computing and OpenStack

This book focuses on deploying private enterprise clouds using OpenStack. In this context, I'll describe private clouds as pools of infrastructure resources (VMs, storage, and so on), also known as *infrastructure as a service* (IaaS), owned and managed by the organizations they serve. In contrast, public cloud IaaS resources are owned and operated by third-party service providers, like Amazon AWS, Microsoft Azure, and the like. The goal of this book is to help you bring the ease and efficiencies of public cloud offerings to your enterprise.

The economics of private vs. public clouds

Financially, one can think of private clouds as primarily capital expenses, whereas public clouds are typically operational expenses. The distinction is easy to understand given that in private cloud deployments, organizations typically purchase or lease a fixed infrastructure for the duration of its serviceable life, regardless of actual usage. In public clouds, the cost is typically directly related to hourly occupancy (if it's on and provisioned, you pay; if it's off and destroyed, you don't) and communication costs.

Whether organizations opt for private or public clouds is often related to the size and scope of those organizations' IT responsibilities. Enterprise IT departments, with the responsibility to centrally provide the technical architecture and resources for the rest of the organization, have a vested interest in leveraging a private cloud. A multi-tenant (where data, configuration, and user management are logically isolated per tenant), fully orchestrated private cloud provides enterprise IT the ability to become a private cloud broker.

Multi-tenancy and full orchestration

Multi-tenant refers to the ability of a cloud platform to manage computational resources on a departmental level. For instance, a marketing department could be allocated a portion of shared resources (*X* VMs, *Y* storage, and so on), and from this resource allocation the department could provision resources without interacting with the central organization (think of the Amazon purchase model). Likewise, *fully orchestrated* describes the ability to allocate resources in relation to application dependencies. For instance, an accounting application along with its web and database server dependencies could be programmatically deployed in this environment. So not only could the marketing department manage its own resources, platform orchestration could be used to deploy both the infrastructure (VMs) and applications (MySQL, Apache2, custom application, and so on) inside a dedicated tenant.

In contrast, departmental IT units often lack the data center facilities and personnel to cost-effectively deploy private clouds. Due to their relatively small resource requirements, departmental IT units can often take advantage of public cloud resources, or also take advantage of private cloud resources managed by their enterprise IT units, if any are available.

If you use both private and public clouds based on workload, that combination is referred to as a *hybrid cloud*. Both public and private clouds are built using the same technologies, but although the building blocks might be the same, the motivations for using private and public clouds can be very different. For instance, users often use a private cloud for security compliance reasons. It's common for a public cloud to be used for workloads that are cyclic in nature or that require a global scale that would be very costly for an enterprise to provide.

Although this book focuses on using OpenStack for private clouds, many of the lessons learned translate directly to public cloud provider services that are based on the OpenStack API.

1.2.1 Abstraction and the OpenStack API

Fundamentally, OpenStack abstracts and provides a common API for controlling hardware and software resources provided by a wide range of vendors. The framework provides two very important things:

- Abstraction of hardware and software resources, which prevents vendor lock-in for any particular component. This is accomplished by managing resources through OpenStack, not directly using the vendor component. The drawback is that not all vendor features are supported by OpenStack, but common required features are.
- A common API across resources, which allows for complete orchestration of connected components.

OK, what's the catch?

OpenStack provides support for a wide range of functions that are both scalable and abstracted from the underlying hardware. What OpenStack (or any other cloud framework) can't do is conform to all of your current practices. To take advantage of the power of cloud computing, you might have to change some of your own business and architectural practices.

If your architectural standards are based on using <insert propriety vendor feature> to do <some function> for all servers in your data center, this approach may be in conflict with a vendor-abstracted cloud deployment. If your business practice is to create virtual machines for users on request, you're missing the point of cloud self-service. If end-user requests can be effectively automated, or end users can provision resources themselves, then you're harnessing the power of cloud computing.

While the first point is nice from a financial perspective, the second is the key to a modern IT transformation.

The next section will relate OpenStack to other technologies that you might be familiar with.

1.3 Relating OpenStack to the computational resources it controls

You've read all about the great things that OpenStack provides, but how does it work? Perhaps the easiest way to grasp how OpenStack works is to relate the framework to common technologies found in an enterprise environment.

In this section, you'll learn how OpenStack is related to the foundational resources (compute, storage, network, and so on) that it controls. As you'll see, OpenStack generally doesn't provide the actual resource; it simply controls lower-level resources. Figure 1.4 shows how OpenStack manages providers of resources, which in turn are used by virtual machines.

In the following subsections, you'll see the details related to specific resource components: server virtualization, via the control of hypervisors; networking, via the control of vendor-provided hardware and OpenStack services; block and object storage, via the control of vendor and OpenStack services. Finally, we'll look at OpenStack services in relation to common cloud terms. As you'll see, OpenStack is a framework that coordinates resources and services, regardless of the underlying technology vendor.

1.3.1 OpenStack and hypervisors

A *hypervisor* or *virtual machine monitor* (VMM) is software that manages the emulation of physical hardware for virtual machines. OpenStack is not a hypervisor, but it does control hypervisor operations. Many hypervisors are supported under the OpenStack framework, including XenServer/XCP, KVM, QEMU, LXC, ESXi, Hyper-V, BareMetal,

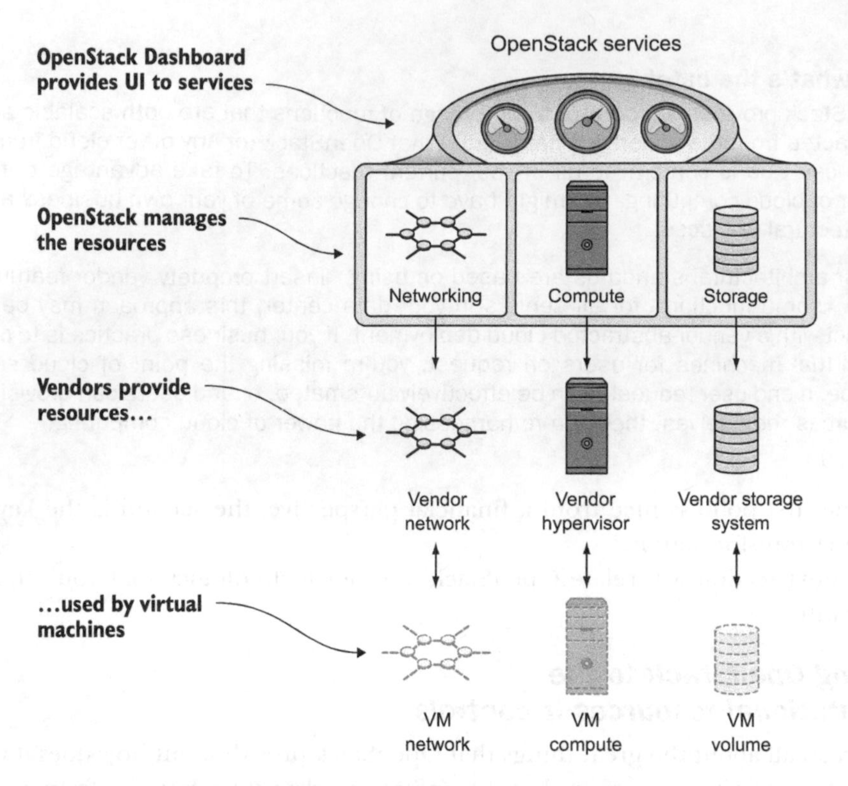

Figure 1.4 OpenStack resource management model

and others (see the hypervisor support matrix: https://wiki.openstack.org/wiki/HypervisorSupportMatrix). You're likely familiar with VMware ESX, VMware ESXi, and Microsoft Hyper-V, which at the current time are the dominant hypervisors in enterprise virtualization space. Because of licensing restrictions, cost, and other factors, there has been less OpenStack community support for commercial hypervisors than for open source alternatives.

Figure 1.5 shows OpenStack managing resources that are virtualized by a hypervisor on physical hardware. OpenStack coordinates the management of many hypervisor resources and virtual machines in an OpenStack cluster.

The majority of people and organizations, regardless of deployment size, use either XenServer or KVM hypervisors, which currently support the widest range of features. XenServer, a Citrix product, is technically an open source hypervisor, but commercial support is available through Citrix. KVM is included as part of the Linux kernel, so many Linux distribution maintainers provide commercial KVM support, including Red Hat, Ubuntu, and SUSE, to name a few.

ARE YOU CERTIFIED? As large providers started designing public IaaS offerings based on the OpenStack framework, they soon realized that their customers would require Microsoft certification for Windows hosts running

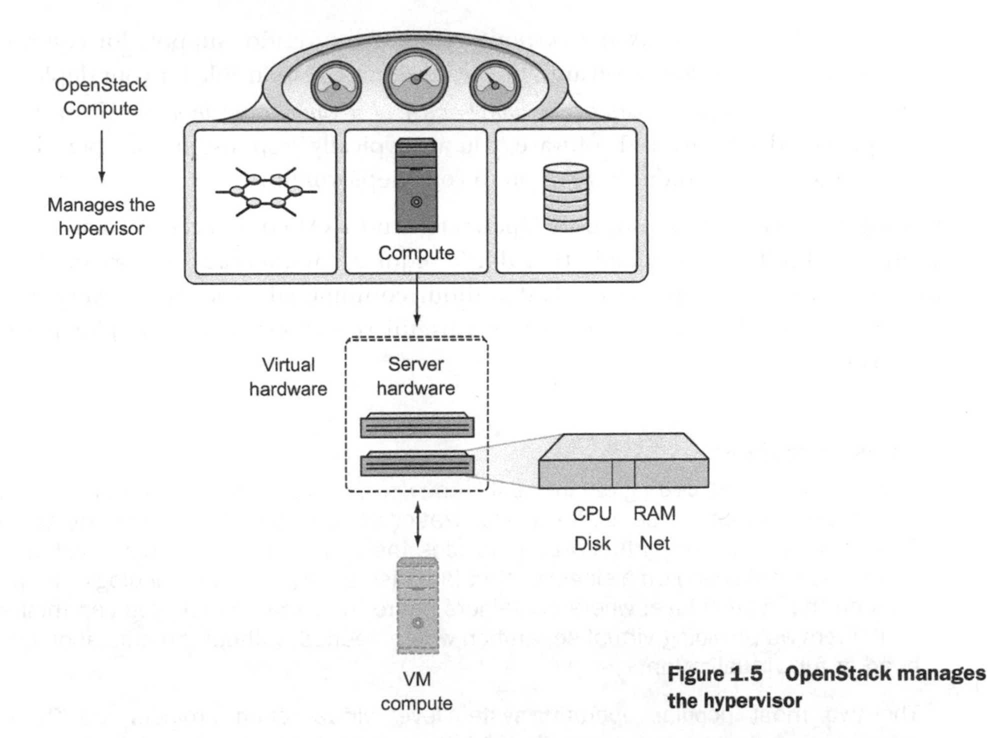

Figure 1.5 OpenStack manages the hypervisor

under their hypervisors. At the time, Citrix had gone through the Microsoft certification process with XenServer and fulfilled those requirements. But despite Citrix having a competing platform in the form of CloudStack, many organizations used XenServer as their OpenStack hypervisor. Since that time several Linux distribution providers have gone through the Microsoft certification process and now fully support Windows guests on the KVM hypervisor, including hypervisors controlled by OpenStack.

Throughout this book, the Kernel-based Virtual Machine (KVM) will be used. KVM has been part of the Linux kernel since the 2.6.20 release in early 2007, and it's fully supported by OpenStack. KVM also provides paravirtualization, which must be either supported natively by the operating system or added through the use of hypervisor-specific drivers installed on the virtualized operating system image. The traditional problem with open source hypervisors is that there's a steep learning curve for deploying and maintaining them, often requiring experience in system, network, and application administration. In organizations lucky enough to have centrally supported virtualized resources, the resource request process must pass through the organization's network, systems, security, and financial elements in the provisioning process. This typically leaves users with three choices:

- *Self-support using community code*—Using community support for community-maintained software, you take responsibility for the design, development, and operation of your deployment.

- *Commercial support using community code*—Using vendor support for community-maintained software, you and/or the vendor are responsible for your deployment.
- *Commercial support using a vendor branch of a community project*—Using vendor-provided support and software, you are typically responsible for operation and vendor management in relation to your deployment.

Although several vendors support OpenStack and KVM commercially, many internal clouds are built for workloads that don't require commercial support or certification, so OpenStack support of KVM without commercial support is a very popular option. This book covers material that's useful regardless of your deployment and support path.

Linux containers

Recently, there has been great interest in the use of operating-system-level virtualization in place of infrastructure-level visualization for providing OpenStack instances. Operating-system-level virtualization provides the ability to run multiple isolated OS instances (*containers*) on a single server. But it isn't a hypervisor technology—it operates on the system level where containers share the same kernel. You can think of containers as providing virtual separation where needed, without the emulation overhead of full virtualization.

The two most popular operating-system-level virtualization projects are Docker (https://www.docker.com/) and Rocket (https://github.com/coreos/rkt). Although it can be argued that containers are better suited for application runtime delivery than OS-level instances, technologies based on containers will undoubtedly be widely adopted for building clouds.

1.3.2 OpenStack and network services

OpenStack isn't a virtual switch, but it does manage several physical and virtual network devices and virtual overlay networks. Unlike the OpenStack control of hypervisors, which is limited to the services provided by the hypervisor alone, OpenStack directly provides network services like DHCP, routing, and so on. But much like hypervisor management, OpenStack is agnostic to the underlying vendor technology, which could be a commercial or open source technology.

More importantly, a backend technology change like moving from one type of network or vendor to another doesn't necessitate client configuration changes. Given the great deal of proprietary hardware, software, and user interfaces involved in networking, it's often not trivial to switch from one vendor or technology to another. With OpenStack, those interfaces are abstracted by the OpenStack API, as shown in figure 1.6.

OpenStack can manage many types of network technology (mechanisms), including those provided by Arista Networks, Cisco Nexus, Linux bridging, Open vSwitch (OVS), and others. Throughout this book we use networking services provided by

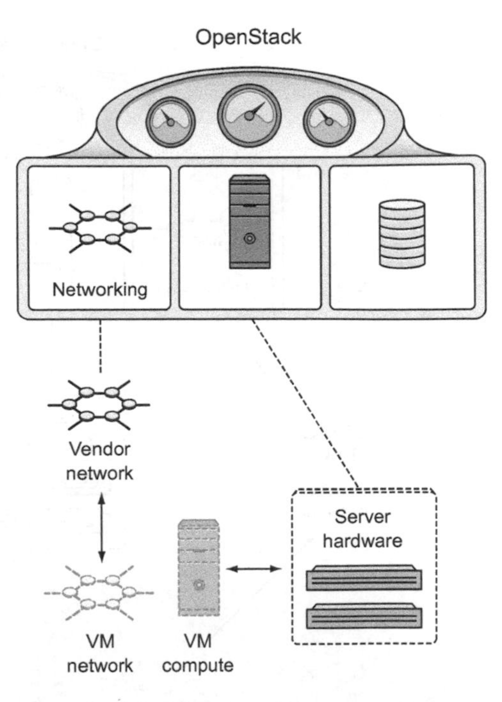

Figure 1.6 OpenStack manages the network

OpenStack and OVS. OVS is a common switch choice for OpenStack that you can easily obtain and replicate in your environment without specific hardware requirements. Along with the network mechanisms, there are several network types (VLANs, tunnels, and so on) supported by OpenStack, and those are covered in chapter 6.

1.3.3 OpenStack and storage

OpenStack isn't a storage array, at least not in the way you'd generally think of storage. OpenStack doesn't physically provide the storage used by virtual machines.

If you've ever used a file share (NFS, CIFS, and the like), you were using "file-based" storage. This type of storage is easy for humans to navigate and computers to access, but it's generally an abstraction of another type of storage: block storage. You can think of operating systems or filesystems as being the primary users of block storage.

There's also another type of storage that people in systems roles might not be familiar with: object-based storage. This type of storage is generally accessed through software APIs (for example, `GET /obj=xxx`). Object-based storage is a further abstraction of file and block storage, but without the restrictions of either. Object-based storage can be easily distributed and replicated between many participating nodes. Unlike block storage, which must quickly be accessed by a VM, distributed object storage is much more latent and wouldn't be used for things like VM volumes (which are actively attached to an instance). It's common to use object storage to store backups of volumes and images (containing operating systems) to be applied to volumes when they're created.

Let's first address how OpenStack works with block storage, and then we'll talk about object storage.

Block storage

OpenStack doesn't currently manage file-based storage for end users. In figure 1.7, you can see that OpenStack manages block (VM) storage much like it manages the hypervisor and network.

This figure shows a complete picture from a basic VM resource-management prospective. OpenStack can manage many vendor-provided storage solutions, including solutions from Ceph, Dell, EMC, HP, IBM, NetApp, and others. As it does with hypervisor

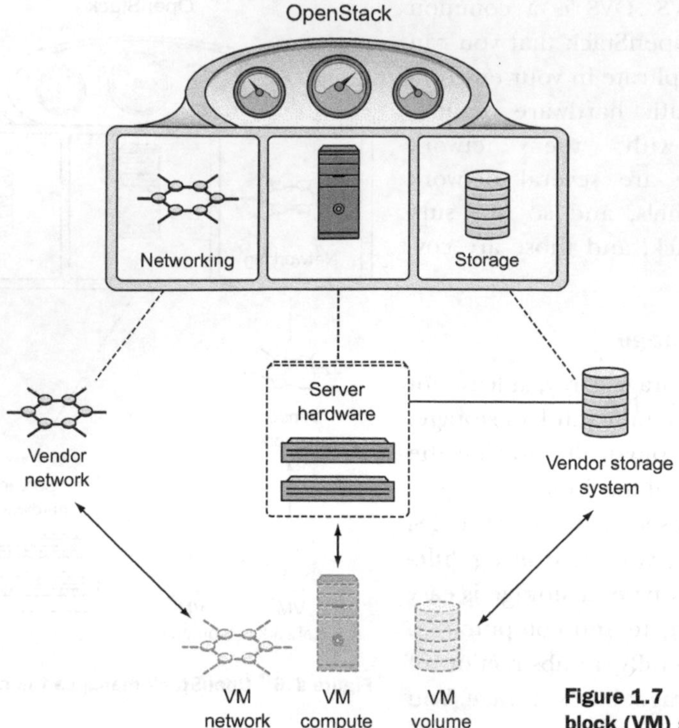

Figure 1.7 OpenStack manages block (VM) storage.

and network components, OpenStack provides you with the flexibility to switch between storage vendors and technologies without changing the client configuration.

OBJECT STORAGE

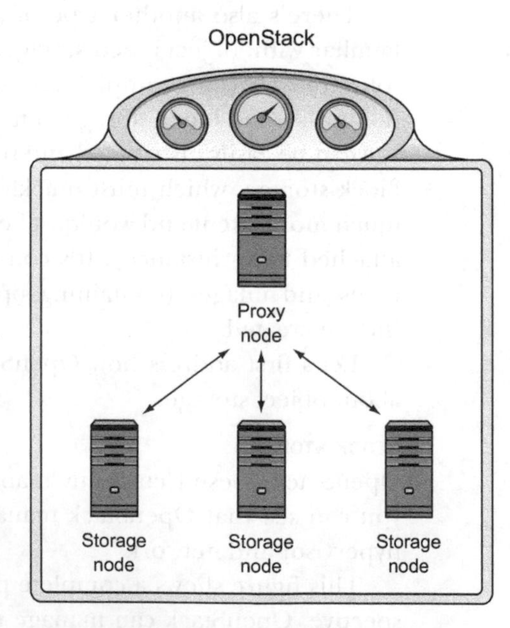

Figure 1.8 OpenStack provides object-based (API) storage.

Although OpenStack isn't a storage array for block storage (used to boot VMs), it does have the native ability to provide object storage. Other than physical hardware running a supported version of Linux, no other software is required for OpenStack to provide a distributed object storage cluster. This type of storage can be used to store volume backups, and it's also common to use object storage for large amounts of data that can be easily isolated into binary objects. Figure 1.8 shows a basic object server deployment that's all contained in the OpenStack environment.

Object storage doesn't have to be in a single location. In fact, nodes (proxy and storage) could be in several locations that replicate between each other.

The traditional use case for object storage is to store data that will be accessed by an application, such as a document or file that will be used by an application on a user's behalf. There are several use cases for object storage in OpenStack environments. For example, it's common to use this type of storage as a repository for VM images. This isn't to say that the VMs use this storage directly, but they're provisioned from data maintained in this system. This is reasonable because the provisioning process won't need low-latency access to random data. Object storage might also be used to back up a snapshot of an existing VM for a long-term backup.

1.3.4 OpenStack and cloud terminology

OpenStack is a framework for cloud construction and is used to construct both public and private clouds. Aside from the public and private cloud definitions, there are the "as-a-service" cloud destinations. Which as-a-service is OpenStack? OpenStack can be used as the foundation for several as-a-service clouds.

Suppose you're interested in providing your enterprise with an Amazon-like experience for acquiring VMs and storage resources. This would be considered *infrastructure as a service* (IaaS). In this context, users are given direct access to provision individual virtual machines, which users directly manage. Although the physical components that make up the cloud are hidden from the user, virtual components are directly accessible. The responsibility of OpenStack is to control resources that provide end users with the infrastructure.

Now suppose your cloud users are not given direct access to IaaS features and are only given access to application orchestration functions provided or supported by OpenStack. In this context, OpenStack could be considered the back end of a *platform as a service* (PaaS) offering. The underlying physical and virtual infrastructure components are hidden from the user. Consider the case where a development team requires an isolated application landscape (application-layer deployment on IaaS) for software testing. Through cloud orchestration, OpenStack could be used on the back end in the deployment of the described testing platform.

Now suppose your company provides a service to its customers using either IaaS or PaaS provided by OpenStack. In this context, OpenStack serves as the back end component of *software as a service* (SaaS). As you can see, OpenStack can be used as a fundamental component in many layers of cloud computing.

Now that you have a better idea of what OpenStack does and how it does it, it's time to introduce you to the components of OpenStack that do the work. The next section introduces the individual components of OpenStack and their role in the overall framework.

1.4 Introducing OpenStack components

I introduced the basic capabilities of OpenStack in section 1.1; in this section we'll look at the fundamental components that make up the framework.

Table 1.1 lists several of OpenStack's components or *core projects*. There are many more projects in various stages of development, but these are the foundational components of OpenStack. The most up-to-date roadmap of OpenStack's services can be found on the OpenStack Roadmap page: www.openstack.org/software/roadmap/.

Table 1.1 Core projects

Project	Code name	Description
Compute	Nova	Manages VM resources, including CPU, memory, disk, and network interfaces.
Networking	Neutron	Provides resources used by the VM network interface, including IP addressing, routing, and software-defined networking (SDN).
Object Storage	Swift	Provides object-level storage, accessible via a RESTful API.
Block Storage	Cinder	Provides block-level (traditional disk) storage to VMs.
Identity	Keystone	Manages role-based access control (RBAC) for OpenStack components. Provides authorization services.
Image Service	Glance	Manages VM disk images. Provides image delivery to VMs and snapshot (backup) services.
Dashboard	Horizon	Provides a web-based GUI for working with OpenStack.
Telemetry	Ceilometer	Provides collection for metering and monitoring OpenStack components.
Orchestration	Heat	Provides template-based cloud application orchestration for OpenStack environments.

Now that you know a bit about what OpenStack is and does, let's take a quick look at where it came from.

1.5 History of OpenStack

On his first day in office in 2009, US President Barack Obama signed a memorandum to all federal agencies directing them to break down barriers to transparency, participation, and collaboration between the federal government and the people it serves. The memorandum became known as the *Open Government Directive.*

One hundred and twenty days after the directive was issued, NASA announced its Open Government framework, which outlined the sharing of a tool called *Nebula.* Nebula was developed to speed the delivery of IaaS resources to NASA scientists and researchers. At the same time, the cloud computing provider Rackspace announced it would open-source its object storage platform, Swift.

In July 2010, Rackspace and NASA, along with 25 other companies, launched the OpenStack project. Over the past five years there have been ten releases. OpenStack releases are shown in table 1.2.

Table 1.2 OpenStack releases

Name	Date	Core components
Austin	October 2010	Nova, Swift
Bexar	February 2011	Nova, Glance, Swift
Cactus	April 2011	Nova, Glance, Swift
Diablo	September 2011	Nova, Glance, Swift
Essex	April 2012	Nova, Glance, Swift, Horizon, Keystone
Folsom	September 2012	Nova, Glance, Swift, Horizon, Keystone, Quantum, Cinder
Grizzly	April 2013	Nova, Glance, Swift, Horizon, Keystone, Quantum, Cinder
Havana	October 2013	Nova, Glance, Swift, Horizon, Keystone, Neutron, Cinder, Ceilometer, Heat
Icehouse	April 2014	Nova, Glance, Swift, Horizon, Keystone, Neutron, Cinder, Ceilometer, Heat, Trove
Juno	October 2014	Nova, Glance, Swift, Horizon, Keystone, Neutron, Cinder, Ceilometer, Heat, Trove, Sahara
Kilo	April 2015	Nova, Glance, Swift, Horizon, Keystone, Neutron, Cinder, Ceilometer, Heat, Trove, Sahara, Ironic
Liberty	October 2015	Nova, Glance, Swift, Horizon, Keystone, Neutron, Cinder, Ceilometer, Heat, Marconi, Trove, Sahara, Ironic, Searchlight, Designate, Zaqar, DBaaS, Barbican, Manila

OpenStack has maintained a six-month release cycle, which is coordinated with OpenStack Summits. The project has grown from 25 participating companies to over 200, with thousands of participating users in over 130 countries.

1.6 Summary

- IaaS clouds are collections of commodity resources, which are coordinated through management frameworks.
- OpenStack is a management framework that provides end-user self-service coordination of IaaS and application orchestration (PaaS/SaaS).
- OpenStack controls existing commercial and community technologies like hypervisors, storage systems, and networking hardware and software.
- OpenStack is composed of a collection of projects, each with a specific purpose.
- Each OpenStack project has a related code name.

Taking an OpenStack test-drive

This chapter covers

- Using DevStack to take a test-drive with OpenStack
- Preparing an environment for DevStack
- Configuring and deploying DevStack
- Interacting with the OpenStack Dashboard
- Understanding the OpenStack tenant (project) model
- Creating virtual machines with OpenStack

In chapter 1 you learned many of the benefits of OpenStack and how OpenStack fits into the cloud ecosystem. But now that you have an idea of what OpenStack can do for you, you may wonder what it looks like. What will the experience be like for your users? This chapter lets you test-drive OpenStack through the use of DevStack, a rapid OpenStack deployment tool, and answers these questions.

DevStack lets you interact with OpenStack on a small scale that's representative of a much larger deployment. You can quickly deploy or "stack" (as fellow Open-Stackers call it) components and evaluate them for production use cases. DevStack

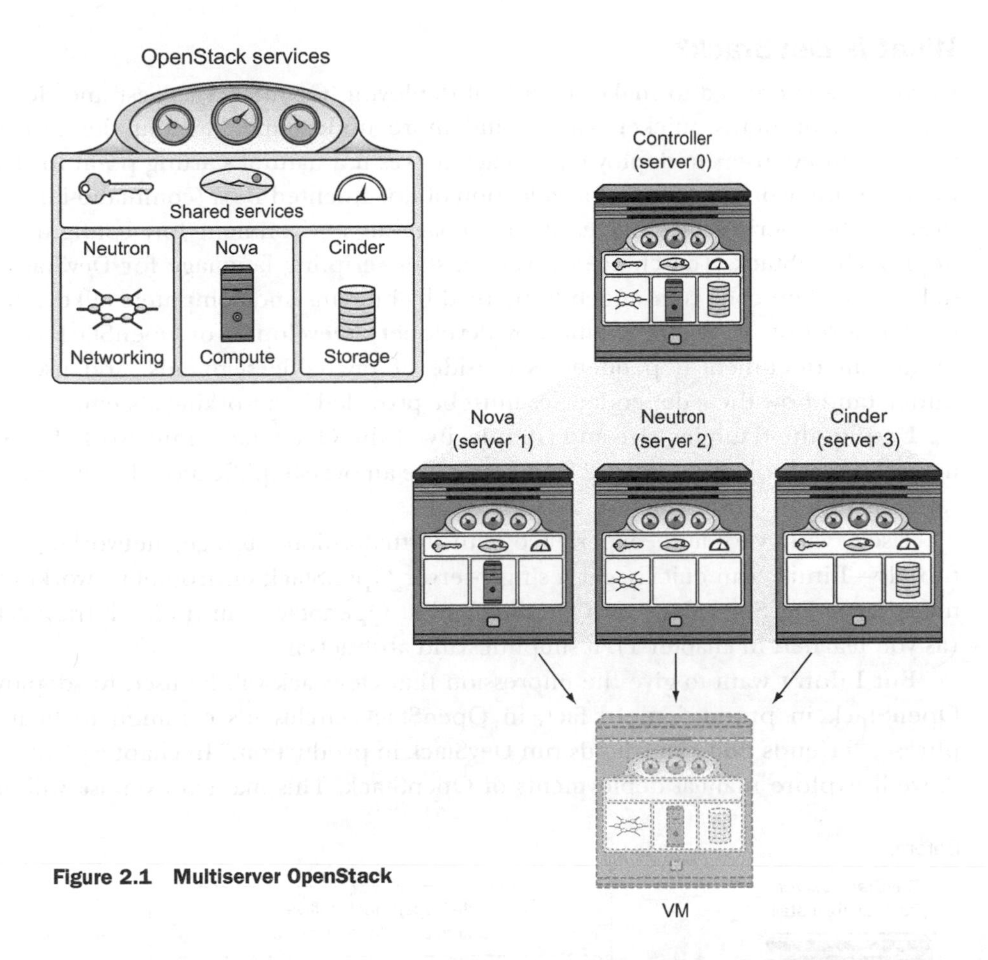

Figure 2.1 Multiserver OpenStack

helps you deploy the same OpenStack components found in large multiserver environments on a single server, as shown in figure 2.1. Without knowing a great deal about OpenStack and without the need for a bunch of hardware, you can use DevStack to get the OpenStack experience, just on a smaller scale.

The figure shows several components, including Cinder, Nova, and Neutron, deployed on an arbitrary number of nodes. OpenStack uses codenames for its components, so the codename *Cinder* refers to storage components, *Nova* to compute components, and *Neutron* to network components. At this point it's not important that you know the OpenStack components, the codenames, or what they do; this is explained in detail in chapter 4. What you need to know now is that OpenStack is made up of several core components that can be distributed among nodes (servers) based on the intended design. Information related to OpenStack design is also covered in chapter 9.

2.1 What is DevStack?

DevStack was created to make the job of deploying OpenStack in test and development environments quicker, easier, and more understandable, but the ease with which it allows users to deploy OpenStack makes it a natural starting point for learning the framework. DevStack is a collection of documented *Bash* (command-line interpreter) shell scripts that are used to prepare an environment for, configure, and deploy OpenStack. The choice of using a shell-scripting language for DevStack was deliberate. The code is intended to be read by humans and computers alike, and it's used as a source of documentation by developers. Developers of OpenStack components can document dependencies outside of raw code segments, and users can understand how these dependencies must be provided in a working system.

Despite the daunting size and complexity of the OpenStack framework, DevStack makes things look easy. Figure 2.2 may look like an oversimplification, but it's an accurate illustration of the function of DevStack.

Users with very little experience with virtualization, storage, networking, or—frankly—Linux, can quickly get a single-server OpenStack environment working. In many ways, DevStack does for OpenStack what OpenStack can do for infrastructure (as you learned in chapter 1): it simplifies and abstracts it.

But I don't want to give the impression that DevStack will be used for deploying OpenStack in production. In fact, in OpenStack circles it's common to hear the phrase, "Friends don't let friends run DevStack in production." In chapters 5 through 8, we'll explore manual deployments of OpenStack. This manual exercise will allow

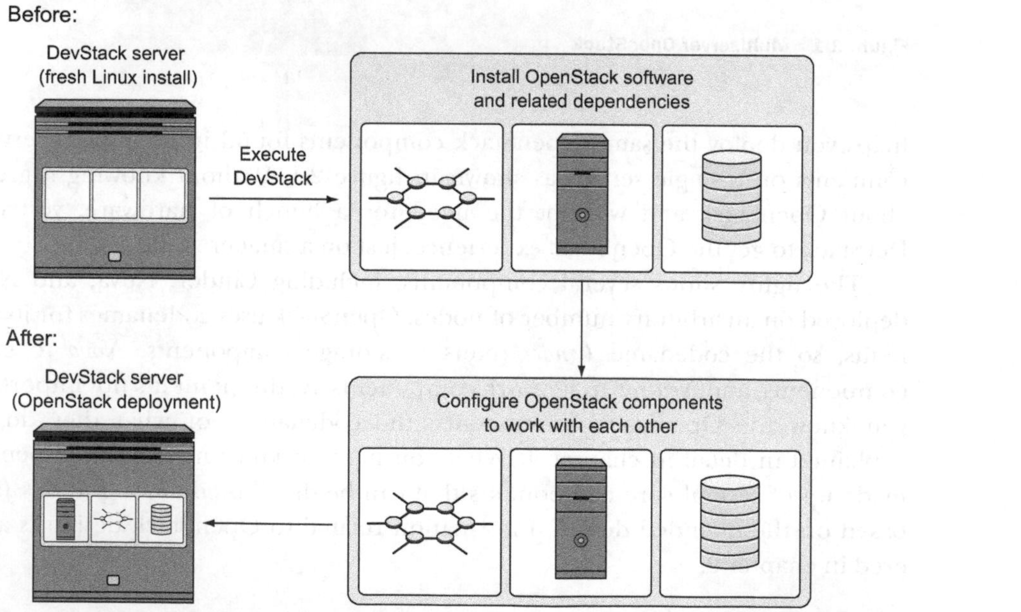

Figure 2.2 DevStack will install and configure OpenStack on a single node automatically.

you to learn about the configuration options and components of OpenStack and develop your ability to troubleshoot OpenStack deployments. In chapter 11, we'll cover an automated deployment of OpenStack that's intended to be used in a production environment.

In this chapter, you'll prepare an environment and deploy OpenStack using DevStack. You don't need to know much about Linux, storage, or networking to deploy a working single-server OpenStack environment. Using this deployment, we'll walk through ways you can interact with OpenStack, and this will give you some familiarity with both the components and the overall system. Then we'll discuss the OpenStack tenant model, which is how OpenStack logically separates, controls, and assigns resources to projects. In OpenStack terminology, *tenant* and *project* can be used interchangeably. Finally, you'll take what you've learned and create a virtual machine in a virtualization environment.

Let's get started stacking!

2.2 Deploying DevStack

As the name suggests, DevStack is a development tool, and its related OpenStack code is under constant development. The support packages used by DevStack to deploy OpenStack code are also under development. When DevStack works, it works beautifully, but when it fails, things get ugly, and first-time users get very frustrated. Although a large part of this chapter is dedicated to deploying OpenStack using DevStack, there's no way of knowing from day to day if a set of instructions will work with the latest versions of DevStack and OpenStack. The same DevStack instructions might fail on Monday but work on Friday.

To reduce reader frustration, a companion virtual machine (VM) that contains a DevStack-deployed OpenStack instance is provided for this book. This VM can be used to test-drive OpenStack with limited hardware resources and effort. In the event that DevStack doesn't work for you, simply make use of the companion VM. You can always try DevStack again when you have a better grasp of the overall OpenStack framework.

What version of OpenStack does the companion VM use?

The companion VM, along with examples in the first and second parts of this book, use the Icehouse release of OpenStack. Although Icehouse was already several revisions old at the time of writing, it was still the most widely deployed and arguably stable version of OpenStack. In addition, there are several Linux distributions and OpenStack production deployment tools that maintain the Icehouse release in long-term support. The third part of the book covers production deployment tools, which can be used to deploy several versions of OpenStack, including Icehouse or the most recent.

If you end up using the companion VM, follow the instructions found in the sidebar entitled "Instructions for using the companion VM," and then skip to section 2.3.

Instructions for using the companion VM

Follow these steps:

1. Go to http://manning.com/bumgardner/ and download the VM image under Links.
2. Make sure you have VirtualBox installed (this VM image was tested with version 4.3.30).
3. Unzip the file devstack_icehouse_openstackinaction.
4. Double click the dev_stack_icehouse_openstackinaction.vbox file (or use command-line arguments—see the VirtualBox docs for details).
5. VirtualBox should now be launched, and you should see the devstack_icehouse_openstackinaction VM.
6. Start the devstack_icehouse_openstackinaction VM.
7. In the VM configuration, several ports are forwarded from the VM to your local host (IP=127.0.0.1). These ports include 2222 for SSH access to the VM and 8080 for access to the OpenStack Dashboard.
8. Once the VM is started, log in to the VM using the `sysop` ID and password `u$osuser01` (for example, `ssh -u sysop@127.0.0.1 -p 2222`).
9. Once on the console, switch to the `stack` user: `sudo -i -u stack`.
10. Execute the rejoin script: `sudo /opt/devstack/rejoin-stack.sh`.
11. You'll now see screens related to the output of OpenStack components. To select a specific screen, hold and release Control-+, and then press " (double-quotes). You'll be presented with a list of screens.

Here are a couple of tips for accessing OpenStack:

- If you want to access the VM for working with the OpenStack CLI, use the instructions found in section 3.1, keeping in mind that the internal address of the OpenStack VM instance is 10.0.2.32.
- If you want to access the VM for working with the Dashboard, access the URL http://127.0.0.1:8080 with the username `admin` and the password `devstack`.

ARE YOU A VAGRANT USER? Although covering the use of Vagrant is outside the scope of this book, there are several community projects that use Vagrant to deploy DevStack on VirtualBox (such as devstack-vagrant at https://github.com/openstack-dev/devstack-vagrant and vagrant_devstack at https://github.com/bcwaldon/vagrant_devstack).

It's recommended that you attempt to deploy OpenStack using DevStack, as outlined in the following steps. This process will provide you with quick access to the framework, while you build a foundational understanding of OpenStack components. Although DevStack can be used to get started quickly, the documented scripts hide nothing from you, and each OpenStack component is configurable based on your needs. It's even possible to use DevStack to deploy a multiserver OpenStack environment.

In this chapter, we'll focus on deploying all components to a single server. This approach reduces the configuration problems you might experience before you fully understand the OpenStack component distribution model, which is discussed in chapter 3. Once you understand the interaction of components on a single server, deploying a multiserver configuration is far more understandable. In part 2 of this book, we'll discuss manual deployments, and in part 3 we'll cover automated deployments of multiserver OpenStack configurations.

To get started, you'll need a single physical or virtual server running a supported distribution of Linux.

2.2.1 Creating the server

For your DevStack deployment, you'll want to start with a fresh install of Linux. That will ensure that dependency conflicts are avoided altogether. I recommend using Ubuntu 14.04 (Trusty Tahr), which is one of the most widely documented and tested Linux distributions for working with OpenStack.

The examples presented in this chapter are based on Ubuntu 14.04, but users with experience in other distributions should be able to adapt the examples. Scripts and configuration files for this and other chapters can be found in the source code for this book: https://github.com/codybum/OpenStackInAction.

I recommend using physical hardware for the deployment if possible. Although it's possible to run OpenStack "nested" in a virtual environment, the VMs deployed in this nested OpenStack environment are notably slow. In this context, I define a *VM* as virtual hardware running a full operating system. I say that OpenStack (the hypervisor) is *nested* if you're trying to use a VM to virtualize another VM. If no hardware is available, the deployment process will be the same, with the exception of the noted performance issues. A walk-through guide for a basic install of Ubuntu 14.04 can be found in the appendix.

LINUX DISTRIBUTIONS Although Ubuntu is widely used, Fedora and CentOS/RHEL are also well documented. Additional Linux flavors such as OpenSUSE and Debian provide OpenStack packages and are known to work.

OpenStack on VMs (nested virtualization)

Hypervisors, or virtual machine monitors (VMMs), live between physical hardware and virtual machines. The hypervisor emulates the operations of the physical hardware, letting the operating system think that it has exclusive access to the underlying system. Hypervisors can take advantage of CPU virtualization extensions, allowing specific operations that would normally be emulated in software to be offloaded to the CPU directly. This greatly increases performance.

OpenStack manages the hypervisor to provide virtual infrastructure. When the hypervisor managed by OpenStack is run on a VM, CPU visualization extensions are generally not available. All instructions that would normally be offloaded to hardware are emulated (via QEMU, an open source hypervisor) in software. Pure software emulation of hardware is extremely slow and shouldn't be used in practice.

2.2.2 Preparing the server environment

As you might have gathered from figure 2.2, DevStack will install and configure the entire OpenStack suite for you. The process of deploying the OpenStack framework, regardless of the method, is called *stacking*. The stacking process will retrieve and configure OpenStack software and related package dependencies from online repositories. OpenStack dependencies will be satisfied by the Advanced Packaging Tool (APT), which along with packages is provided by the Linux distribution.

> **What user are you running as?**
>
> Right now, you should be running as a user with *sudo* privileges, but not as the *root* user. There are various permissions issues related to using the root user with a default Ubuntu 14.04 install and DevStack, so it shouldn't be used.
>
> Once the environment is prepared, you can create and then switch to the *stack* user for the deployment of DevStack. Specific user types used in this chapter will be explained later in this section.

You'll use the sudo command to execute commands with the security privileges of the *root* user. According to Wikipedia, the name *sudo* is a concatenation of "su" (substitute user) and "do." Sudo privileges give the user the ability to execute a command with the security privileges of another user. The other user is typically the root user, so sudo privileges are exactly what you need. The user you created when you installed your operating system would have the appropriate privileges.

In the following examples, the *sysop* user will be used as the normal user with sudo privileges. The first time you use the sudo command, you'll be prompted again for your password, but don't be confused: this is the same password you used for your normal account. Subsequent sudo commands executed within a timeout period (which, for Ubuntu 14.04, is 15 minutes) won't prompt for a password.

The APT database used to determine package availability and dependencies is maintained locally, so by the time you install your Linux distribution, it's already out of date. Your first step in preparing the environment is to update your APT package information from online sources. From a shell prompt, update your APT packages as shown in the following listing. This process won't update any packages, but any package installed after the update will be current.

Listing 2.1 Updating packages

```
sysop@devstack:~$ sudo apt-get -y update        <-- Updates local package information
[sudo] password for sysop:
Hit http://us.archive.ubuntu.com precise Release.gpg
...
Fetched 3,933 kB in 1s (2,143 kB/s)
Reading package lists... Done
```

Your local package database is now up to date, as the first line of code synchronizes the local package information with the newest online sources. After the update, it's recommended that you upgrade packages with the following command.

Listing 2.2 Upgrading packages

```
sudo apt-get -y upgrade
```

This upgrade step isn't necessary from the DevStack component standpoint, but a DevStack dependency could depend on a kernel update, which would require an update and reboot. If you proceed with the upgrade step, it's critical you *reboot* after the upgrade.

DevStack doesn't use the APT system to install OpenStack components, despite the packages being available in its repository. This choice is understandable given the need for component flexibility in a development and testing system. For example, you might run several OpenStack components from a stable release, along with a few others from a development branch. This level of modularity isn't generally possible with package management systems.

In place of using a package management system provided by the Linux distribution, DevStack retrieves OpenStack components directly from online OpenStack repositories. Git, a source revision control utility, is used to access OpenStack repositories, so your next step is to install the git client, as shown in the next listing. The client will be used both to retrieve the DevStack scripts and later by DevStack to retrieve OpenStack components.

Listing 2.3 Installing git

```
sysop@devstack:~$ sudo apt-get -y install git     <-- Installing git from the package management system
Reading package lists... Done
Building dependency tree
Reading state information... Done
The following extra packages will be installed:
  git-man liberror-perl
Suggested packages:
  git-daemon-run git-daemon-sysvinit git-doc git-el git-arch
  git-cvs git-svn git-email git-gui gitk gitweb
The following NEW packages will be installed:
  git git-man liberror-perl
0 upgraded, 3 newly installed, 0 to remove and 112 not upgraded.
...
Unpacking git (from .../git_1%3a1.7.9.5-1_amd64.deb) ...
Processing triggers for man-db ...
Setting up liberror-perl (0.17-1) ...
Setting up git-man (1:1.7.9.5-1) ...
Setting up git (1:1.7.9.5-1) ...
```

Now that the git client is installed, you can proceed with retrieving the DevStack scripts.

2.2.3 Preparing DevStack

The following examples describe the deployment of DevStack and OpenStack using the very latest versions. As previously mentioned, one can't be sure from day to day whether the latest DevStack code will properly deploy the latest version of OpenStack. If you experience trouble with your DevStack deployment, use the companion VM. You can always give DevStack another try at a later time.

Using git, retrieve the latest release of DevStack, as shown in the following listing.

Listing 2.4 Retrieving DevStack scripts

```
sysop@devstack:~$ sudo git clone \
                    https://github.com/openstack-dev/devstack.git \
                    /opt/devstack/
Cloning into '/opt/devstack'...
remote: Counting objects: 28734, done.
remote: Total 28734 (delta 0), reused 0 (delta 0), pack-reused 28734
Receiving objects: 100% (28734/28734), 9.86 MiB | 5.29 MiB/s, done.
Resolving deltas: 100% (19949/19949), done.
Checking connectivity... done.
```

Retrieving DevStack from the current branch

WANT TO USE A SPECIFIC DEVSTACK BRANCH? Use the -b <branch name> option with git to specify a specific DevStack branch. A list of current DevStack branches can be found on GitHub at https://github.com/openstack-dev/devstack/branches.

You should now have a clone (copy) of the DevStack scripts in the /opt/devstack directory.

DON'T STACK AS ROOT If you try to stack as the root user with DevStack, the process will fail with an error scolding you for running the script as the root user. In general, you only want to run as root or with root privileges when elevated administrative rights are required. One could argue that using root in a development setting is of little risk, but you do want to "practice how you expect to play," so it's a good idea to keep development as close as possible to production. At any rate, stacking as root isn't allowed by DevStack, so you need to prepare the environment to stack as another user.

The next step is to set appropriate directory permissions and create a new service account (under which services will run) for OpenStack, as shown in the next listing. This process will create the stack user and set ownership of all DevStack files to that user.

Listing 2.5 Preparing DevStack directory

```
sysop@devstack:~$ cd /opt/devstack/
sysop@devstack:/opt/devstack$ sudo \
    chmod u+x tools/create-stack-user.sh
sysop@devstack:/opt/devstack$ sudo \
    tools/create-stack-user.sh
```

Enters devstack directory

Makes create-stack-user.sh tool executable

Creates stack user

```
Creating a group called stack
Creating a user called stack
Giving stack user passwordless sudo privileges
sysop@devstack:/opt/devstack$ sudo \
    chown -R stack:stack /opt/devstack/
```

Makes stack user owner of all files in directory

Your directory has now been prepared with appropriate permissions, and a new user has been created. Your next steps are to switch to the stack account you just created, create a DevStack configuration file, and then stack (deploy) your configuration.

2.2.4 Executing DevStack

DevStack was designed for development and for testing OpenStack components, so there are many possible configurations. DevStack is controlled through the use of configuration options maintained in the local.conf file. You now must create a local.conf file in the devstack directory.

For the remainder of the installation, you need to switch to the stack user, as follows.

Listing 2.6 Switching to stack user

```
sysop@devstack:/opt/devstack$ sudo \
    -i -u stack
stack@devstack:~$ cd /opt/devstack/
```

Switches to stack user

At this point you should be in the /opt/devstack directory and running as the stack user. In a previous step, you assigned ownership of this directory to the stack user, so there shouldn't be any problems related to directory permissions.

WHAT USER ARE YOU RUNNING AS NOW? Regardless of the user you started out with, you should now be using the stack user. From this point forward, even if you need to log out or reboot (more on that in this section), you'll want to run as the stack user because DevStack can't be run as root. If you need to run as the stack user, follow the instructions in listing 2.6.

Using the stack user, you'll now create a local.conf file. I show file creation using the Vim text editor, which is a commonly used console-based editor. Given the large number of configuration and log files used in OpenStack, I recommend finding an editor you feel comfortable using.

Find a comfortable text editor

I can't overstate the importance of finding a text editor that you can use efficiently. You can configure OpenStack to do almost anything, but with that power comes configuration responsibility. You need to pick a text editor just as you'd choose shoes for a long walk. If it hurts when you start the journey, it will be painful; if you're comfortable, you won't even notice the effort.

There's a nice Stack Overflow post about Linux text editors here: http://stackoverflow.com/questions/ 2898/text-editor-for-linux-besides-vi.

CONFIGURING DEVSTACK OPTIONS

In this section, you'll build your local.conf file, which is used by DevStack to configure your deployment.

Using your favorite text editor, open your local.conf file as follows.

Listing 2.7 Creating local.conf

```
sysop@devstack:/opt/devstack$ vim local.conf    ← Using Vim to edit the local.conf file
```

Inside your editor, copy the contents of listing 2.8 to the local.conf file. Although it's a good idea to be familiar with local.conf options, these configurations are specific to DevStack and won't be directly used in a production environment.

Listing 2.8 Your local.conf

```
[[local|localrc]]                        ← local.conf header uses the format
                                           [[ <phase> | <config-file-name> ]].
# Credentials                            ← Assigns passwords for each supporting
ADMIN_PASSWORD=devstack                    service and service token and password.
MYSQL_PASSWORD=devstack
RABBIT_PASSWORD=devstack
SERVICE_PASSWORD=devstack
SERVICE_TOKEN=token

#Enable/Disable Services                 ← Disables Nova networking (n-net) and
disable_service n-net                      replaces it with Neutron networking services.
enable_service q-svc
enable_service q-agt
enable_service q-dhcp
enable_service q-l3
enable_service q-meta
enable_service neutron
enable_service tempest
HOST_IP=10.0.2.32                        ← IP address of the host running DevStack.
                                           Change this to your specific IP address.
#NEUTRON CONFIG
#Q_USE_DEBUG_COMMAND=True

#CINDER CONFIG
VOLUME_BACKING_FILE_SIZE=102400M         ← Default file used to store volumes in
                                           DevStack is very small. This line
#GENERAL CONFIG                            increases total volume size.
API_RATE_LIMIT=False

# Output                                 ← Consolidates logs and sets
LOGFILE=/opt/stack/logs/stack.sh.log       logging to verbose.
VERBOSE=True
LOG_COLOR=False
SCREEN_LOGDIR=/opt/stack/logs
```

> **CHECK YOUR DIRECTORY** Make sure that you create the local.conf file inside the devstack directory.

Want to use a specific OpenStack release with DevStack?

You can specify the release or branch of OpenStack that DevStack will use for each component in the local.conf file. For example, to specify the OpenStack branch for Nova, your local.conf file should contain the line `NOVA_BRANCH=<nova branch>`.

Current OpenStack Nova branches can be found here: https://github.com/openstack/nova/branches.

RUNNING THE STACK

At this point you're ready to run the DevStack build script, stack.sh. This script will read your local.conf configuration and deploy OpenStack components accordingly. The stacking process could take a long time, depending on the speed of the DevStack server and network connectivity. On a fast server with good network connectivity, stacking will take about 15 minutes.

Execute stack.sh as follows.

Listing 2.9 Stacking

```
./stack.sh          <-- Executes stack script
```

Stacking, with or without problems, will generate thousands of lines of output. The last line of output from a successful stack.sh execution will display "stack.sh completed in <*second count*> seconds," indicating the number of seconds elapsed during the process.

MY STACK DIDN'T FINISH Don't panic! There are several reasons for the DevStack process to fail, including updates that require a reboot or bad configuration. First, check your local.conf configuration and make sure there's nothing wrong there. Then follow the instructions in listing 2.14 to unstack and re-stack. Many problems can be corrected through the unstack and re-stack process.

Even if everything works as expected, it's hard to make heads or tails of the screen output during a stack. In local.conf you configured a central logging location: /opt/stack/logs. This directory is full of verbose logs, which capture the screen output of the entire process for each component. Fortunately, you're provided a summary log (stack.sh.log.summary) that shows each major step. A summary log for a stack using the local.conf configuration in listing 2.8 is shown next.

Listing 2.10 Stack summary log

```
Installing package prerequisites
Installing OpenStack project source
Installing Tempest
Starting RabbitMQ
Configuring and starting MySQL
```

```
Enabling MySQL query logging
Starting Keystone
Configuring and starting Horizon
Configuring Glance
Configuring Neutron
Configuring Cinder
Configuring Nova
Starting Glance
Uploading images
Starting Nova API
Starting Neutron
Creating initial neutron network elements
Starting Nova
Starting Cinder
Configuring Heat
Starting Heat
Initializing Tempest
stack.sh completed in 565 seconds.
```

> **DEVSTACK COMPONENTS DON'T RUN AS LINUX SERVICES** DevStack doesn't start OpenStack components as Linux services—it runs as screens. After successfully running stack.sh, if you want to restart any OpenStack service, access the screen console using the command `screen -r`. To restart the Nova network, go to the Nova network screen, which is screen 9, using the Ctrl-A command followed by 9. Then kill the Nova network using Ctrl-C, and restart it by pressing the up arrow and Enter.

TESTING THE STACK

Stop and take a deep breath. If all has gone well, you should have a fully functional OpenStack deployment at your fingertips. It may be tempting to skip over this testing section and jump right in, but wait! Skipping testing arouses contempt in the system's heart, and something will invariably break.

> **READ ME NOW** Upon completing your stack, you've just deployed literally hundreds of interoperating components, a huge tree of dependencies, and a web of integrations. From an engineering perspective, it's amazing that computers work at all (look up dynamic random-access memory—DRAM—as an example), much less a single computer running a fully orchestrated cloud platform. Save yourself the pain and agony of fighting a potentially broken stack, and work through the testing section now; it will be worth it.

The good news is that testing requires little configuration. The bad news is that it takes some time to go through all the tests. You'll run two test suites: the DevStack *exercises* and OpenStack *Tempest*.

DevStack exercises, as the name suggests, are specific to DevStack and have been included with DevStack since early releases. The exercises are intended to be run against a DevStack environment after the stacking process has completed, and they provide basic testing of primary functions.

In comparison to the DevStack exercises, OpenStack Tempest is an 800-pound gorilla ready to inflict all manner of torture on your OpenStack deployment. Tempest works on single-server DevStack deployments or 1000-node clouds. In this section, we'll focus on passing the DevStack exercises and then run some basic Tempest tests.

DEVSTACK EXERCISES FOR CHECKING AND TEMPEST FOR VALIDATION DevStack exercises do a sufficient job of checking core OpenStack services for a DevStack deployment. In fact, due to the constraints of a single-server deployment, it's likely that Tempest validation will fail on DevStack even when the exercises succeed (Tempest can't test multi-node operations on a single node). But for production deployments, Tempest can be a very powerful validation tool.

Go ahead and run the exercise suite found in the devstack directory, as follows.

Listing 2.11 Running DevStack exercises

```
stack@devstack:/opt/devstack$ ./exercise.sh
...
<lots of screen output>
...
*********************************************************************
SUCCESS: End DevStack Exercise:
*********************************************************************
=====================================================================
SKIP marconi
SKIP sahara
SKIP swift
SKIP trove
PASS aggregates
PASS boot_from_volume
PASS bundle
PASS client-args
PASS euca
PASS floating_ips
PASS horizon
PASS neutron-adv-test
PASS sec_groups
PASS volumes
FAILED client-env
=====================================================================
```

With a bit of luck, all of the exercises run against your system will PASS, with the exception of client-env. Because you haven't configured your shell variables yet, the client-env test will fail, but this is fine. You won't be working with the CLI commands until chapter 3.

If a particular test fails, the test can be run again from within the devstack/exercises directory. If the test continues to fail, follow the process shown in listing 2.14, and repeat the DevStack execution process, starting at the beginning of this section.

If all of your exercises pass, you have the option of running the Tempest test suite. The Tempest project page states "Tempest was originally designed to primarily run against a full OpenStack deployment. Due to that focus, some issues may occur when running Tempest against devstack." You don't have to run the full test suite; specific tests can be run individually.

Listing 2.12 Running single OpenStack Tempest test

```
cd /opt/stack/tempest
nosetests tempest/scenario/test_network_basic_ops.py
..
----------------------------------------------------------------------
Ran 2 tests in 247.376s

OK
```

The full Tempest suite contains thousands of tests and can take a very long time (20 minutes on a fast server) to run on virtual or low-resource machines. Listing 2.13 demonstrates how to run a full Tempest test. Please keep in mind that under DevStack, you can expect several failures. In the example, approximately 8% of the Tempest tests failed against a single-server DevStack deployment of OpenStack, despite passing the exercises tests.

Listing 2.13 Running the full OpenStack Tempest suite

```
devstack@devstack:~/devstack$ /opt/stack/tempest/run_tempest.sh
No virtual environment found...create one? (Y/n) Y
Creating venv... done.
Installing dependencies with pip (this can take a while)...
Downloading/unpacking pip>=1.4
  Downloading pip-1.5.2.tar.gz (1.1Mb): 1.1Mb downloaded
...
<loads of screen output>
...
setUpClass (tempest.api.compute.admin.test_fixed_ips_negative
    FixedIPsNegativeTestXml)
SKIP  0.00
    FixedIPsNegativeTestJson)
SKIP  0.00
tempest.api.compute.admin.test_availability_zone.AZAdminTestXML
    test_get_availability_zone_list[gate]
OK  1.93
    test_get_availability_zone_list_detail[gate]
OK  1.07
    test_get_availability_zone_list_with_non_admin_user[gate]
OK  1.94
...
<loads of screen output>
...
Ran 2376 tests in 1756.624s
FAILED (failures=19)
```

At this point you should be comfortable with the stacking and testing process.

If you experienced any problems or would like to experiment with some DevStack options, you can start the process over without reloading your operating system. To start over at any point in the process, follow the unstack and stack steps shown in listing 2.14. This process will take you back to where you were at the start of the subsection "Running the stack," just before listing 2.9.

Listing 2.14 Unstacking and stacking

```
./unstack.sh
./clean.sh
sudo rm -rf /opt/stack
sudo reboot
```

STACKING SUMMARY

Given the use of DevStack in development and testing, it stands to reason that you'd want a consistently configured demonstration environment. When developers are testing a specific feature, they don't want to have to manually create a sample user environment with each stack.

As mentioned in chapter 1, OpenStack distributes resources based on tenants or projects. OpenStack allows for multi-tenancy on the same deployment environment (think of a hotel or condominium). Multi-tenancy means that multiple users, departments, or even organizations can share the same OpenStack deployment without interfering with each other's configuration. I'll explain the OpenStack tenancy model in more depth in chapter 3, but for now it's sufficient to know that DevStack creates sample tenants/projects, roles, and user accounts for you. The *admin* account, as the name suggests, has administrative access to your newly stacked OpenStack deployment. The *demo* account has access to the Demo project, with access rights representative of a normal OpenStack user. Initially, both accounts use the default password of *devstack*.

You've now deployed OpenStack components on a single server using DevStack. Along with providing a test-drive of OpenStack services, you can also use this deployment as a working reference as you manually deploy OpenStack components in the second part of this book. You can now move on to interacting with OpenStack.

> **Rebooting DevStack**
>
> DevStack isn't intended for production, so features like services starting automatically after reboot don't work. If you reboot your system and want to continue with the same configuration you had before rebooting, you must take some manual steps.
>
> When the stack.sh script is run, it populates the stack-screenrc file with the commands used to start each OpenStack component at the time of stacking. After you reboot, you must run the rejoin-stack.sh script. The rejoin-stack script reads stack-screenrc and restarts the services.

> ***(continued)***
> Because you're using Cinder for volume (block storage) management, you must additionally set up the loopback volume using the `losetup` command. From the same directory that you ran ./stack.sh from, run the following commands:
>
> ```
> sudo losetup -f /opt/stack/data/stack-volumes-backing-file
> /opt/devstack/rejoin-stack.sh &
> ```

2.3 Using the OpenStack Dashboard

There are three primary ways to interface with OpenStack:

- *OpenStack Dashboard*—A web-based GUI, introduced in this section
- *OpenStack CLI*—Component-specific command-line interfaces, introduced in chapter 3
- *OpenStack APIs*—RESTful (web) services, briefly introduced in chapter 3

Regardless of interface method, all interactions will make their way back to the OpenStack APIs.

For most people, their first hands-on exposure to OpenStack will be through the Dashboard. In fact, the majority of end users will use the Dashboard exclusively, so that's the access method we'll discuss here. System administrators and programmers will need to understand how to access the CLI and APIs, and we'll cover those subjects in chapter 3.

Go ahead and access the Dashboard by entering the following URL into your browser: http://<*your host ip*>. You should be presented with the login screen shown in figure 2.3, where you can enter the following user name and password:

- *User name*—demo
- *Password*—devstack

The demo user simulates an unprivileged user. If you don't see the login screen, it's likely that an error occurred during the stacking process, and you should review section 2.2.4.

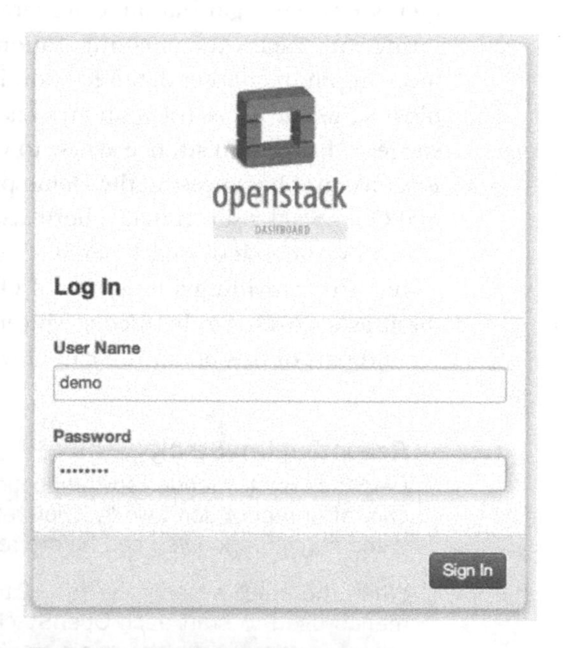

Figure 2.3 Dashboard login screen

> **Checking the port**
>
> You can save a great deal of time troubleshooting socket-based services (HTTP, SSH, and the like) by simply logging in to the server running the service, checking if the port is listening for a connection, and then working your way up from there.
>
> One widely distributed tool that checks if a port is listening is curl. To check if port 80 (the HTTP port) is listening for web requests on IP 10.0.2.32, you'd issue the following command: `curl 10.0.2.32:80`. If something is listening on that port, the output that would normally be sent to the browser will be returned to the console.
>
> Modern browsers often block malformed data, so the server could be failing on a specific error that you'd never see in the browser. If the connection is rejected, you'll know (assuming you don't have a local firewall problem) that the service that should be running on that port has not started. If the service for that port hasn't started, you should check the logs for the service.

The Dashboard is laid out in a two-column design, as shown in figure 2.4. The left column is fixed in size, and the right is dynamically sized based on your browser window. As you can see, the left column contains the Project tab, links to other management screens, project selection drop-downs, and the Admin tab if you're an admin. After you log in, you'll be taken directly to the Overview screen, which is the one shown in figure 2.4. We'll discuss many of the different screens in the following subsections.

MY DASHBOARD DOESN'T LOOK LIKE THAT If you're using the companion VM, the Dashboard should look the same as figure 2.4. But if you completed the DevStack process, you'll be using the very latest OpenStack version, and likely things will look a little different. Despite the different appearances, though, the examples in the next few chapters should work fine.

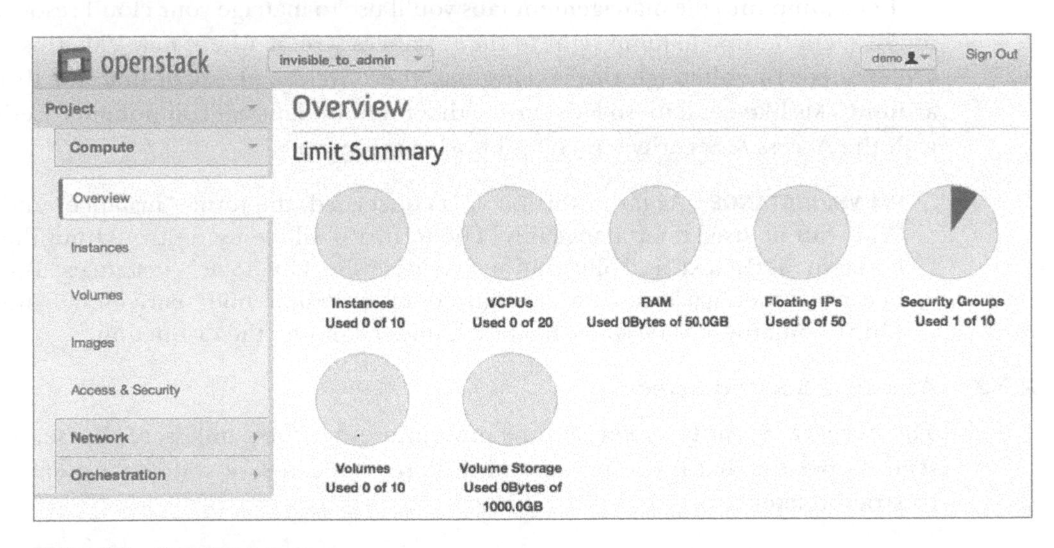

Figure 2.4 Overview screen

2.3.1 Overview screen

The Overview screen displays the current user's utilization based on their *current project* quota. A user might be part of several projects, with various quotas. The Management toolbar, shown in figure 2.5, lists all management screens available to the user in the current project.

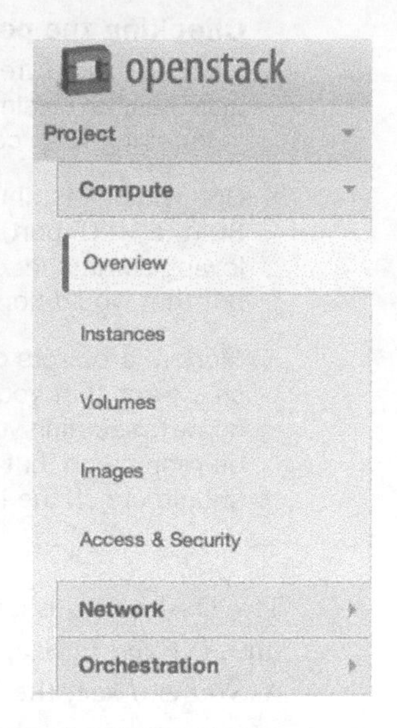

Figure 2.5 Management toolbar

Management screens are divided up into several sections, including Manage Compute, Manage Network, Manage Object Store, and Manage Orchestration. Tabs for each are included on the Management toolbar. The Object Store and Orchestration headings and screens are not shown or covered in this section, but we'll cover these topics in the third part of the book. Object storage, while very useful, isn't directly related to creating virtual machines. Orchestration, the automated combining of virtual hardware and software to deploy applications, is a very interesting topic. Cloud orchestration is so important that I devote all of chapter 12 to the topic.

In order to log in to the Dashboard, you must have a role in an existing project. When you log in to the Dashboard, one of your existing projects will be selected for you, and any project-level configuration will be related to that project. Your currently selected project is indicated by a drop-down menu on the left side of the top toolbar. DevStack will create two projects, *demo* and *invisible_to_admin.* To switch projects, click on the project drop-down menu.

Let's jump into the management tabs you'll use to manage your cloud resources. It might seem natural to jump right to the Instances screen, where new VMs (instances) are created. But although this is tempting, there will be plenty of time for throwing around VMs like confetti, so let's start with a few foundational components. Let's start with the Access & Security screen and work our way up.

> **VM VS. INSTANCE** As far as this book is concerned, the terms "instance" and "VM" can be used interchangeably. The term *VM* will be used throughout the book in both text and illustrations to describe OpenStack instances. But because OpenStack can be configured to provision both bare-metal and Linux containers as *instances,* it's worth understanding the distinction.

2.3.2 Access & Security screen

The Access & Security screen isn't the most interesting area, unless, of course, you're a security person. But if you pay attention here, you can save yourself a great deal of frustration later.

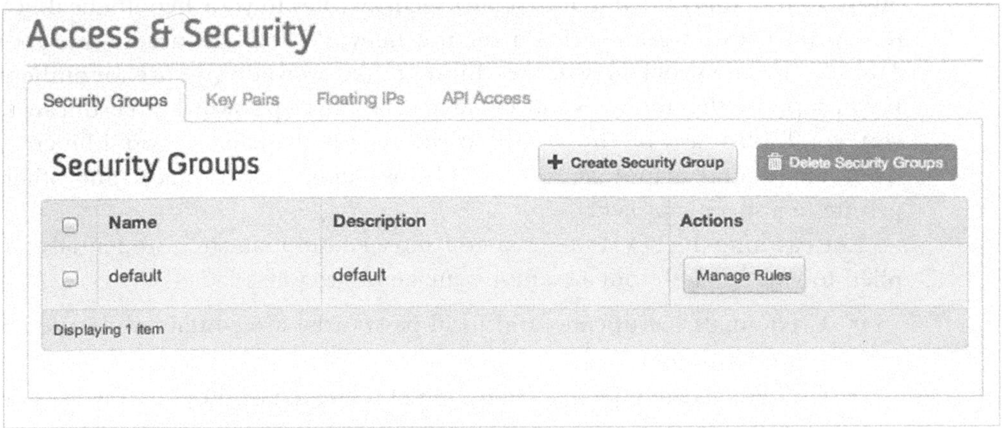

Access & Security

Security Groups | Key Pairs | Floating IPs | API Access

Security Groups

+ Create Security Group | Delete Security Groups

Name	Description	Actions
default	default	Manage Rules

Displaying 1 item

Figure 2.6 Access & Security screen

Take a look at the Access & Security screen shown in figure 2.6. The first three tabs along the top of the screen (Security Groups, Key Pairs, and Floating IPs) are related to how VMs are accessed. The API Access tab is also found on this page, but it isn't related to the other tabs for the most part.

Imagine that you have a VM instance that is *network policy inaccessible* (PI). In this context, PI refers to the inability to access an instance over the network based on some access-limiting network policy, such as a global rule that denies all network access by default. In OpenStack, security groups define rules (access lists) to describe access (both incoming and outgoing) on the network level. A security group can be created for an individual instance, or collections of instances can share the same security group.

DevStack creates a default security group for you. The default group contains rules that allow all IPv4 and IPv6 traffic in (ingress) and out (egress) of a virtual machine. If you applied this default configuration to the PI virtual machine, you'd have no VM-specific network restrictions. In short, security groups are like personal firewalls for specific groups or instances of VMs.

Per VM security groups

While at first it might seem like security groups are just a way of configuring a VM's local firewall (such as Iptables or Windows Firewall), this isn't the case. Security rules are generally enforced on the physical node running OpenStack Networking, and there are several options (drivers) for enforcing security groups, including offloading to physical firewalls.

The examples demonstrated in this book are based on a hybrid driver (OVSHybridIptablesFirewallDriver) that enforces security rules on the virtual switch level. For now it's sufficient to understand that this functionality is part of OpenStack Networking; we'll discuss it in more detail in chapter 6.

Suppose you apply the default (open) security rules to your hypothetical PI VM. Now that you're no longer restricted on the network, you can access the VM via SSH. There's just one problem: what certificate or password do you use for authentication? If you suppose that the VM's source image contains a password or certificate known to you, you'll obviously be fine, but that isn't generally realistic. You'll be creating VMs from images and snapshots; some will be available to all projects, and others will be private to a specific project.

Let's assume the PI VM was created from a common image. Credentials can be supplied to VMs created from a common image in two ways:

- Credentials (certificates and local passwords) are contained on the image and shared with users.
- Credentials are injected when the VM is created or they already exist on the VM image.

The security people are now clutching their chests at the idea of the first way. Sharing a root password or certificate for all VMs makes taping passwords to your monitor seem benign in comparison. For this reason, OpenStack provides the ability to inject credentials into VMs at the time of creation. The Key Pairs tab is used to create new or import existing certificates, to be used for user authentication on VMs.

Let's now assume you were able to access the PI VM using a certificate from the Key Pairs tab. Your VM was created on a network (subnet) that OpenStack Network manages, but this network is private (see Request For Comments (RFC) 1918 at www.ietf.org/rfc.html) and only accessible inside your organization. If you want to access the VM from outside of your organization, the VM must be related (either directly assigned or linked in some way) to a public (see RFC 791) network address. You could assign a public address directly to the VM, but aside from security concerns related to exposing a VM publicly, there are a limited number of IPv4 addresses (see the sidebar, "IPv4 exhaustion"). OpenStack provides the ability to expose VMs to external networks through the use of floating IPs, which can be allocated through the Floating IPs tab. *Floating IP* stands for *floating Internet Protocol* address, which means an address that can be assigned or *floated* between instances as needed. Floating addresses don't have to be public, but for the sake of this example, let's say you assigned a public floating IP to the PI VM. You now have the required access (security groups), credentials (key pairs), and connectivity (floating IPs) to connect to the VM.

IPv4 exhaustion

In ancient internet times, public Internet Protocol Version 4 (IPv4) addresses were assigned to all devices. By 1981, when the final RFC 791 (Internet Protocol) was ratified for the IPv4 specification, the exhaustion of the 2^32 (4,294,967,296) specified addresses was already being predicted. In 1996, RFC 1918 (Address Allocation for Private Internets) described additional address space that could be used for private networks.

If you compare IP addresses to telephone numbers, it's easy to draw the distinction between public and private addresses. Two companies can't have the same telephone number, just as they can't share the same public IP address space. But two companies can use the exact same private address space, just as two companies could use the exact same internal numbering scheme for phone extensions. In both cases, you can't remotely reach a private address or extension without first routing to a public address or phone number. In the private space, you might have thousands of private addresses or extensions behind a single public address.

In 1998, RFC 2460 (Internet Protocol Version 6 (IPv6)) was developed, and it specified the direct addressability of a mind-boggling 3.4×10^{1038} objects. There are far more IPv6 addresses than grains of sand on Earth (10×10^{24}), so it should have the "Internet of Things" covered. Most devices and operating systems now support IPv6, but its native deployment is still limited due in part to the use of private address ranges.

You should now have a good understanding of how access and security are handled in OpenStack, at least on the VM level. The examples in this section are based on a hypothetical VM, and we'll continue this exercise in the next subsection, where we'll talk about images and snapshots.

2.3.3 Images & Snapshots screen

If you're familiar with virtualization technologies such as XenServer, KVM, VMware, or Hyper-V, your idea of VM creation might be to start with creating the virtual hardware and then loading the software. But as was previously explained, OpenStack is on a higher level of abstraction than a traditional hypervisor. The VM (*Instance*) screen doesn't provide a way to attach virtual media, and based on the expected usage of OpenStack, it shouldn't. Users are expected to either import VM images pre-made for OpenStack (such as Ubuntu at https://cloud-images.ubuntu.com/ and CentOS at http://cloud.centos.org/centos/) or select an existing image. OpenStack *images* can be thought of as bundles of data that OpenStack applies to virtual hardware to provide a VM.

CLOUD IMAGES OpenStack can support any OS your underlying hypervisor supports. However, images typically used with OpenStack, as well as public cloud providers like Amazon EC2, contain additional tools used in the provisioning and operation of the underlying VM environment. One of these tools is cloud-init (https://help.ubuntu.com/community/CloudInit), which allows cloud frameworks to provide the operating system information related to resource assignments (host name, IP address, and so on).

This isn't to say that there's no way to boot a VM from an installation ISO (International Standards Organization 9660/13346) image and create a new image based on that installation. In fact, OpenStack has made specific provisions, like booting from ISO, to accommodate the use of commercial license–restricted operating systems, like Microsoft Windows. Many image formats, including RAW, VHD, VMDK, VDI, ISO,

OpenStack image formats

When dealing with image files, you might notice various file extensions, related to file formats. The following list describes image formats supported by OpenStack:

- *RAW*—An unstructured format. The extension could be "raw" or you could simply have an image with no extension.
- *VHD (Virtual Hard Disk)*—Originally a Microsoft virtual disk format, but the image specification has been licensed to other vendors.
- *VMDK (Virtual Machine Disk)*—Originally a VMware format, but it has since been placed in the public domain. This is a very common disk format.
- *VDI (Virtual Disk Image or VirtualBox Disk Image)*—An Oracle VirtualBox-specific image container.
- *ISO*—An archival format for optical images. It's mostly used in the creation of VMs from install disks.
- *QCOW (QEMU Copy On Write)*—A machine image format used by the hosted virtual machine monitor QEMU.
- *AKI*—An Amazon kernel image.
- *ARI*—An Amazon ramdisk image.
- *AMI*—An Amazon machine image.

There are also specifications for container formats for images, but OpenStack doesn't currently support containers.

QCOW, AKI, ARI, and AMI, are natively supported by OpenStack. But although many images are supported, there are several OpenStack-specific image requirements for full functionality, so it's recommended you start with pre-built images.

The name of the Images & Snapshots management screen would suggest that images and snapshots are technically different things, but there's very little technical difference between the two. It's common to think of *images* as "VMs waiting to happen," void of user data. You can think of *snapshots* as pictures or "snapshots in time" of existing VMs and their related data. You can also think of snapshots as backups, but given that we also create VMs (instances) from snapshots, the distinctions between images and snapshots are blurred.

There's a transitive relationship between images and snapshots: image => VM, VM => snapshot, so image => snapshot. Thus, in OpenStack, images = snapshots + metadata. Figure 2.7 illustrates how an image becomes a VM, a VM a snapshot, and a snapshot a VM.

On the Images & Snapshots management screen shown in figure 2.8 you have the ability to do the following:

- *Create images*—You can import an image by uploading a file or specifying a location on the web. You must specify an image format supported by OpenStack.
- *Create volumes*—You can create a volume (a bootable disk) from either an image or a snapshot. This action prepares the storage component for the VM (instance) creation process (it acquires resources, clones data, and consumes block storage), but it doesn't actually create the instance.

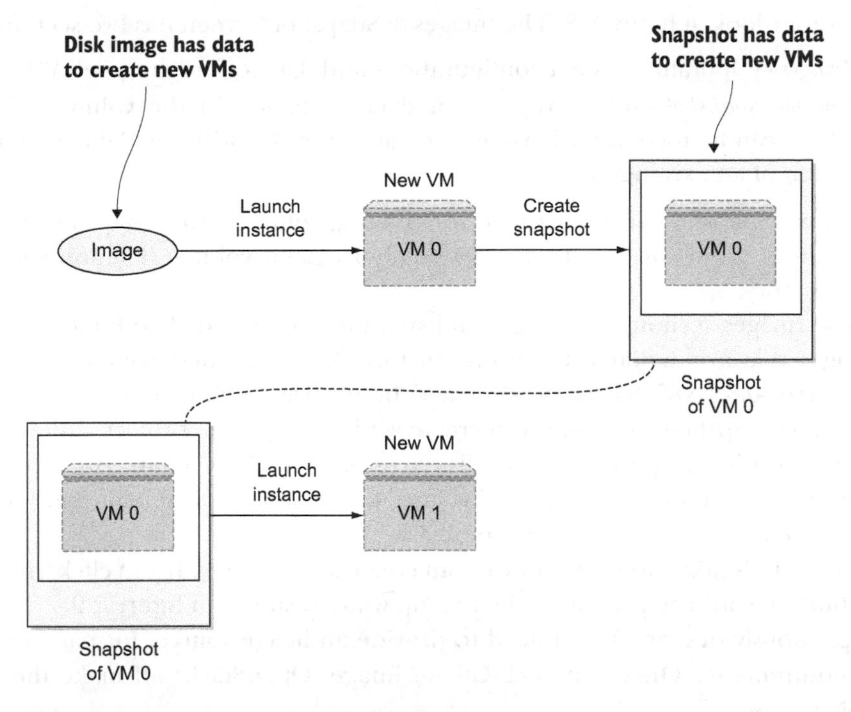

Figure 2.7 Image-snapshot relationships

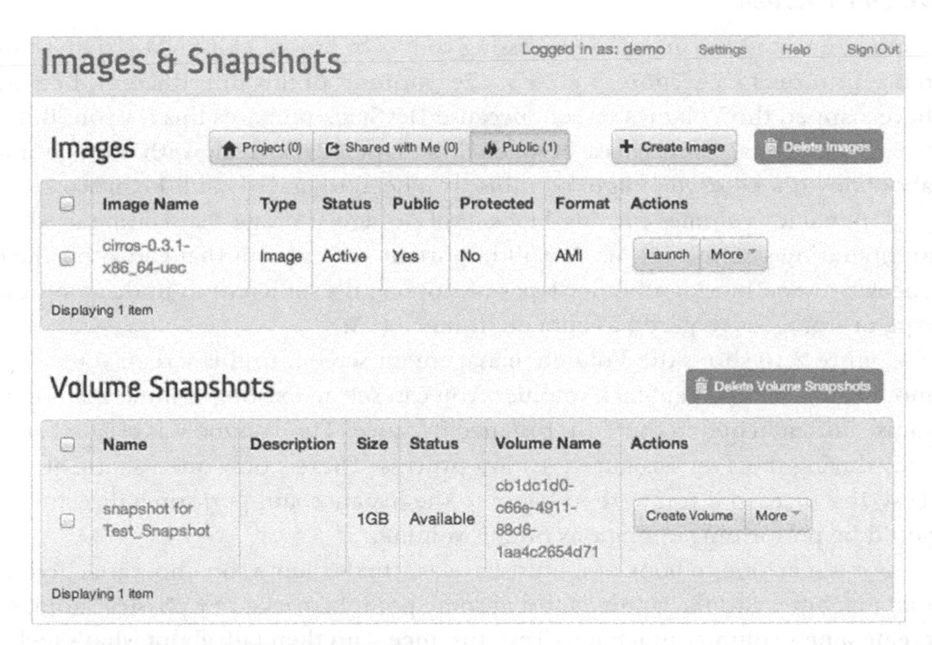

Figure 2.8 Images & Snapshots screen

Take another look at figure 2.8. The Images & Snapshots screen has two sections:

- *Images*—Operating system configurations and data for creating new VMs.
- *Volume Snapshots*—Exact replicas of data contained in the volume of a VM. These can be used as backups or to create a new VM with the data and configuration of an existing VM.

The Volume Snapshots section is used for creating volumes from snapshots or deleting snapshots. We'll cover volume creation shortly, and volume deletion should be self-explanatory.

In the Images section, the images and snapshots are listed. The Public tab shows the images that are available to anyone in this OpenStack deployment. The image named cirros-0.3.1-x86_64-uec was created by the DevStack process for testing. In contrast, any snapshots or images you create while in a specific project, unless they're explicitly specified as public, will be listed under the Project tab. By clicking the Launch button, you can create a new instance based on an available image. We'll discuss instance creation shortly, in section 2.3.5.

For now, let's take a look at how you can create a new image. If you click the Create Image button, you'll be presented the pop-up window shown in figure 2.9.

As previously described, you need to provide an image source, format, and minimum requirements. Once you click Create Image, OpenStack will make the image available for use.

Now let's take a look at the storage mechanism for OpenStack instances (VMs).

2.3.4 Volumes screen

You're almost to the fun part—accessing your own private cloud VM. There's only one more introductory section to go. For the purposes of this introduction, I could even have skipped the Volumes screen, because DevStack prepares this for you. But due to the fundamental differences between how OpenStack deals with storage and just about anything else you might be using, you need to understand the process.

OpenStack volumes provide block-level storage (storage that can be used to boot an operating system) to VMs. It isn't important at this point that you know the differences between block and other types of storage; it's sufficient to understand that this type of storage is required to boot an instance (VM).

Figure 2.10 shows the Volume management screen. In this screen you can create, modify, or delete OpenStack volumes. You can see an existing volume that's currently in use and attached to the Test_Instance instance. This volume was created automatically during the Test_Instance creation process. There's only one volume attached to Test_Instance, so if you were to log in to the instance and perform a directory listing, you'd be performing operations on this volume.

For a machine to boot, you must have a virtual volume somehow attached to a virtual machine, and the volume must at some point be backed by physical storage. Let's create a new volume, attach it to Test_Instance, and then talk about what's technically going on.

Create An Image

Name *
ubuntu_12_04_64

Description
Ubuntu 12.04 x86_64 Cloud Image

Image Source *
Image Location

Image Location
http://uec-images.ubuntu.com/precise/current/prec

Format *
QCOW2 - QEMU Emulator

Minimum Disk (GB)
5

Minimum Ram (MB)
512

Public

Protected

Description:
Specify an image to upload to the Image Service.

Currently only images available via an HTTP URL are supported. The image location must be accessible to the Image Service. Compressed image binaries are supported (.zip and .tar.gz.)

Please note: The Image Location field MUST be a valid and direct URL to the image binary. URLs that redirect or serve error pages will result in unusable images.

Cancel Create Image

Figure 2.9 Creating an image

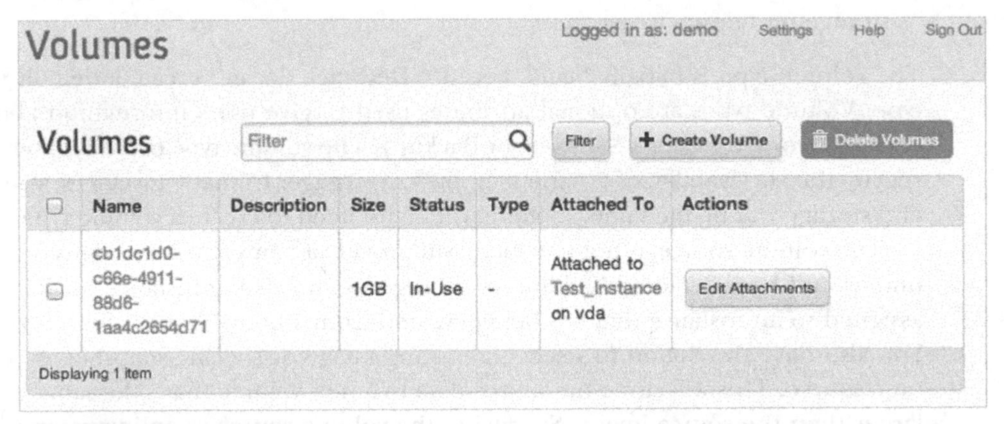

Figure 2.10 Volumes screen

To create a new volume, click Create Volume. You'll be presented with the pop-up window shown in figure 2.11. During volume creation, you specify volume name, type, size, and source.

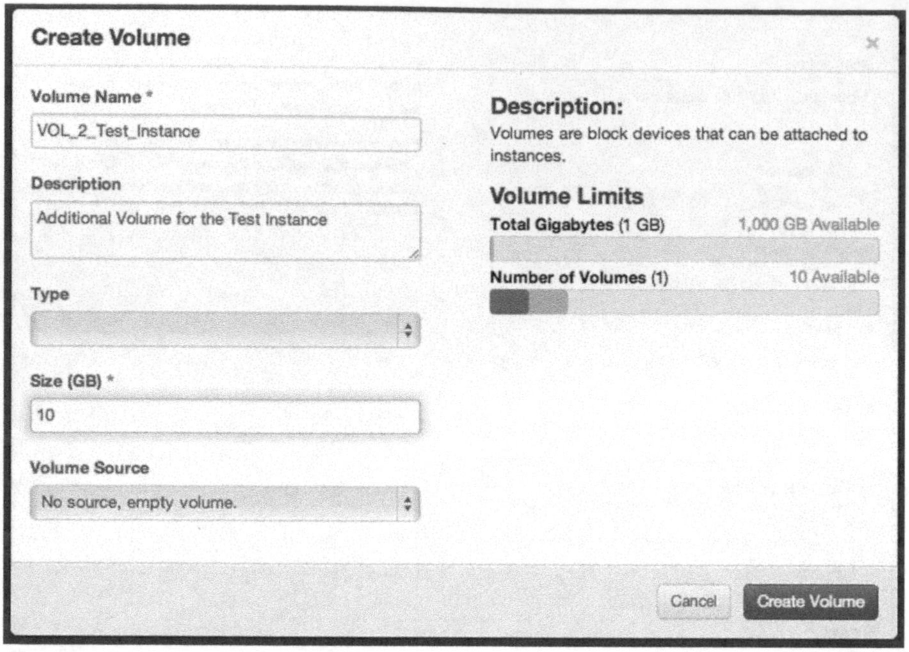

Figure 2.11 Create Volume screen

> **HOW MUCH SPACE DO YOU HAVE?** Way back in listing 2.8, which showed the local.conf file, you had the option to specify the `VOLUME_BACKING_FILE_SIZE`. This value, which is specific to DevStack, controls the total size of storage available to you. By default this value is set to `10240M` (10 GB), so unless you've increased this value, don't exceed 10 GB of total volume storage, including storage pertaining to instances.

The volume type is initially blank, because DevStack doesn't create a default volume type. Volume types are optional attributes used to give users information about the backend storage (such as SSD, SAS, or Backup). The volume type could also be used to specify the survivability of a particular class of storage. In many instances, your backend storage will be the same, so there will be no need to create a volume type.

The volume source, if not specified, will provide an empty volume. This type of volume would be used to add additional storage to an existing instance, or it could be assigned to an instance that was being created from a bootable installer (ISO) image. You also have the option to select an existing image or specific snapshot as your volume source. This will clone the source data to a new volume that's the same size as or larger than the source image. Specifying the volume source as an image or snapshot in the Create Volume screen (figure 2.11) is the same as creating a volume using the buttons in the Images & Snapshots screen (figure 2.8).

Next we'll walk through the process of creating an instance (VM).

> **Block vs. file vs. object storage**
>
> Long-term persistent storage can be provided by several devices, including spinning disk, solid state disk, or tape. Although there are several types of storage devices, typical storage access methods can be grouped into three categories:
>
> - Block—Abstracted at the memory level. For example, the computer fetches memory range 0–1000.
> - File—Abstracted at the network share level. For example, a user fetches a file from share nfs://somecomputer.testco.com/file.txt.
> - Object—Abstracted at the API level. For example, an application fetches an object/file from the API with `GET /bucket/0000`.

2.3.5 *Instances screen*

At this point you might not be an expert, but you should understand enough OpenStack terminology and technology to create an OpenStack instance. Before you start, keep in mind the warning in the previous subsection not to request a larger volume than you have capacity.

Figure 2.12 shows the Instances management screen. Each instance and its current state is listed, as are options to create new instances, reboot, snapshot, and terminate (delete) existing instances, and many other options.

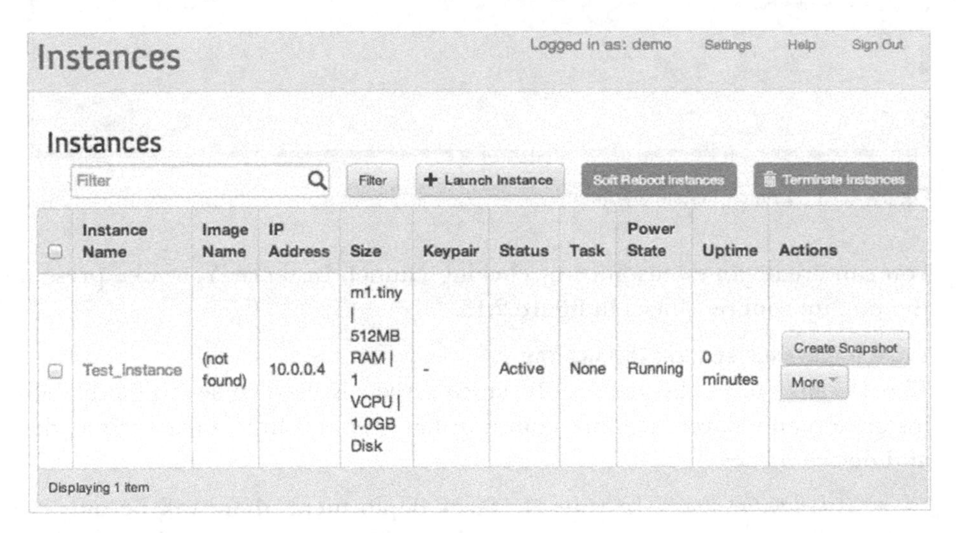

Figure 2.12 Instances screen

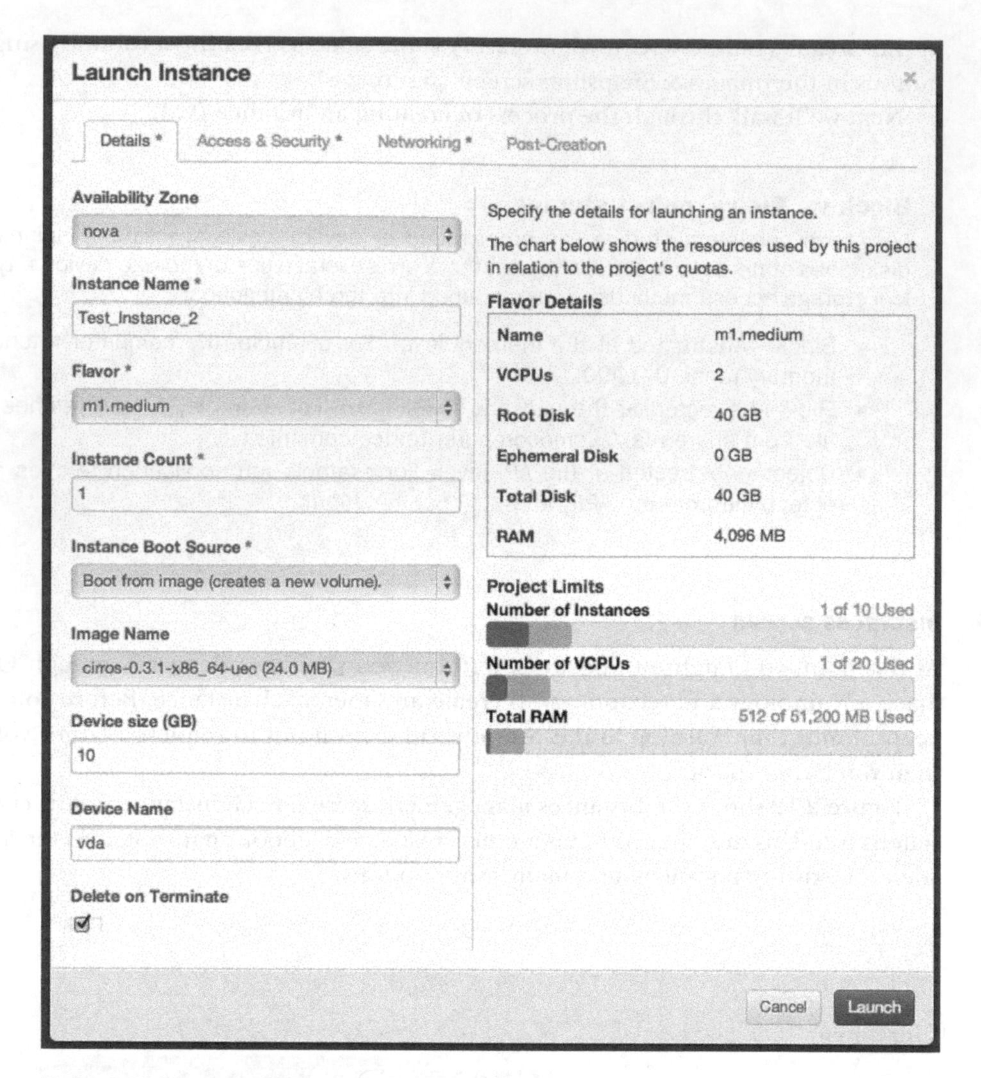

Figure 2.13 Launch Instance screen

You can create a new instance by clicking Launch Instance. You'll be presented with the pop-up window shown in figure 2.13.

LAUNCH INSTANCE SCREEN: DETAILS TAB

The Details tab in the Launch Instance screen is used to set the availability zone, instance name, flavor, instance count, instance boot source, image name, devise size, and device name:

- *Availability Zone*—In your DevStack deployment, there will be only one Availability Zone configured and available. However, in a production deployment you might have several zones, depending on your deployment. Zones are generally used to separate OpenStack deployments by data center or purpose.

- *Instance Name*—This is the name of your instance, which is both a reference for your instance in OpenStack and the host name of the instance.
- *Flavor*—OpenStack flavors specify the virtual resource size of the instance. DevStack will have created several flavors, but the values are configurable.
- *Instance Count*—It's possible to create more than one instance at a time by setting the Instance Count value.
- *Instance Boot Source*—There are several options to choose from under Instance Boot Source, but for this example you can choose "Boot from image," because it will create a volume for you based on an existing image.
- *Image Name*—As previously mentioned, DevStack will have made available at least one image under Image Name.
- *Device Size*—Specify your device size.
- *Device Name*—Specify the name of the boot device for your instance.

Once you've set these values, you can move on to the Access & Security tab.

Launch Instance screen: Access & Security tab

The Access & Security tab, shown in figure 2.14, is where you set the access and security options we first discussed in section 2.3.2. DevStack will have provided you with working defaults, as shown in the figure, so you can choose the Networking tab.

Launch Instance screen: Networking tab

We haven't really talked about networks, but fear not. OpenStack Networking is covered in later chapters, including an in-depth discussion in chapter 6.

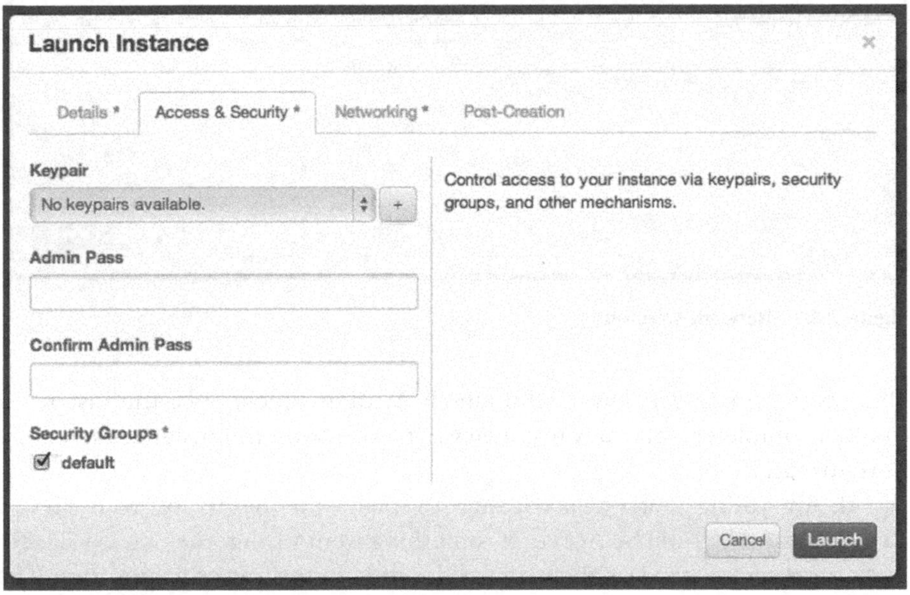

Figure 2.14 Access & Security tab

In the Networking tab (figure 2.15), click the + button on the right side of the *private* network, under Available Networks. Once you do so, the private network should move up to the Selected Networks box, as shown in figure 2.16.

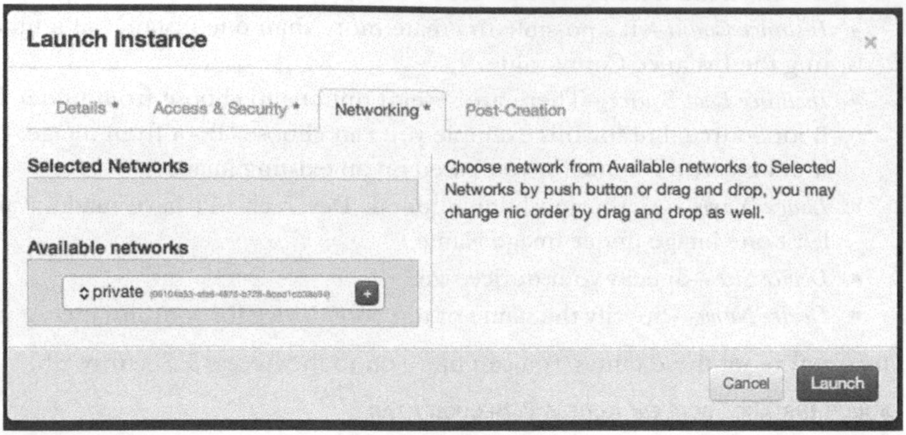

Figure 2.15 Networking tab

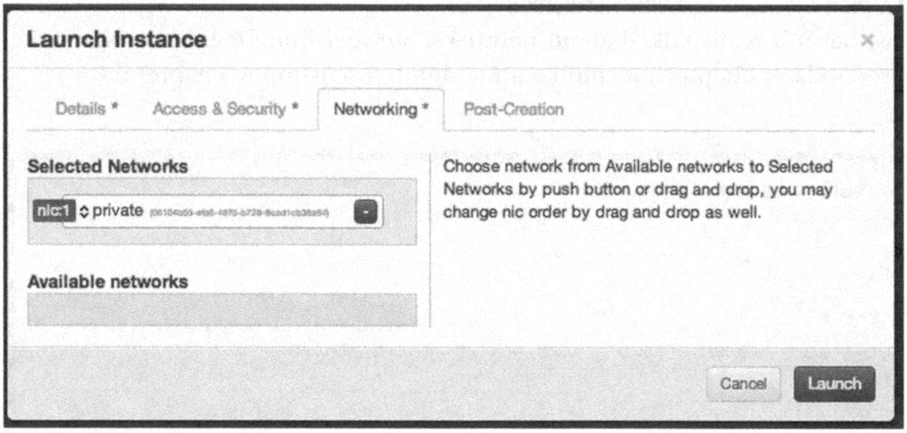

Figure 2.16 Network selected

OK. Cross your fingers and click Launch. With any luck, once the instance creation process completes, your new instance will be visible on the Instances screen, as shown in figure 2.17.

If everything worked according to plan, the Status of your new instance (Test_Instance_2) will be Active. If something went wrong, the Status will show Error, but unfortunately the Dashboard provides little in the way of diagnostics. If you experience an error, you can try reducing the size of your requested virtual machine; a

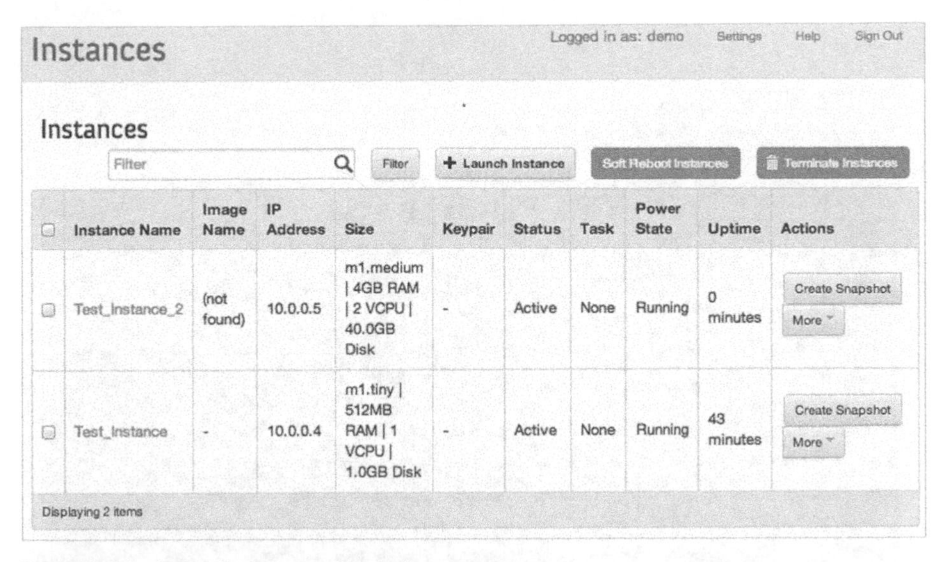

Figure 2.17 A new instance on the Instances screen

common mistake is to exceed the amount of storage space available for creating volumes. If problems persist, you can take a look at the screen logs for each service, located in /opt/stack/logs, or you can observe the screen logs directly using the command `screen -r`. Both of these methods of observing OpenStack component logs are described in section, "Running the stack."

No doubt you're ready to access the instance you just created. This will be especially true if you experienced an error and have spent time wading through logs to finally bring up an instance with an active status. In the next section, we'll walk through the process of accessing your new server.

2.4 Accessing your first private cloud server

We've reached the moment you've been waiting for—it's time to log into that VM. You can do that through the Instance Console.

To access the Instance Console, click the Instances link on the Management toolbar, shown previously in figure 2.5. On the Instances screen, click the name of the instance you're interested in, and you'll be taken to the Instance Detail screen. Now choose the Console tab, and you should see a console like the one shown in figure 2.18.

Assuming you used the *cirros* image provided by DevStack to create your instance, you should be able to log in using the "cirros" user and the password "cubswin:".

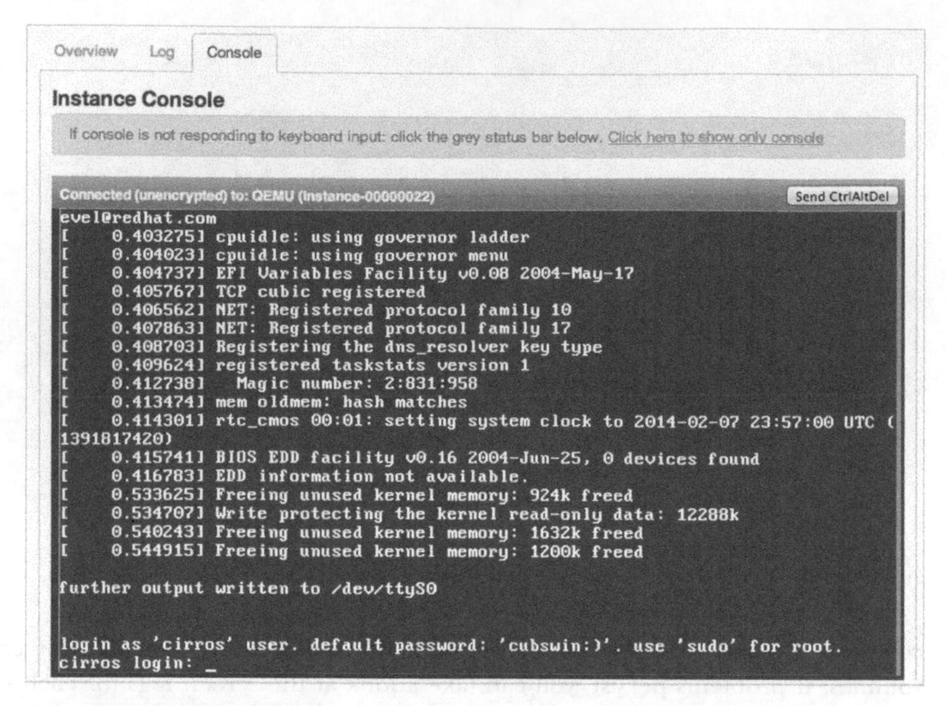

Figure 2.18 Instance Console

NAT translation

If you want to be able to communicate from your new VM to outside of your OpenStack network, you can translate requests from your VM to an outside network using the following command:

```
sudo iptables -t nat -A POSTROUTING -o eth0 -j MASQUERADE
```

After issuing this command, internet communications such as `ping 8.8.8.8` should work, assuming your OpenStack node already has connectivity.

Now that you can access your instance from the console, what about connecting to your host over the network? The Instances screen says the IP address is 10.0.0.4, but you can't SSH or even ping the server, so what gives? If you've worked with other virtualization platforms, the network was likely flat. In this context, *flat* refers to the network topology where your VM connects directly to a network when a virtual interface is added. You can configure OpenStack to behave in this way, but as you'll learn throughout this book, OpenStack networking can do much more.

For now, it's sufficient to understand that the instance address 10.0.0.4 refers to the internal OpenStack IP address of the instance. This means that if you create another instance on this internal network, the two instances can communicate using the internal address. External network access is covered next.

2.4.1 Assigning a floating IP to an instance

The final demonstration in this chapter shows how you can assign a *floating IP* to an instance. In simplistic terms, you can think of a floating IP as an external (to OpenStack) network representation of an instance. As explained in the previous section, the instance address is for communication inside an OpenStack network. If you want to communicate with the instance from networks outside OpenStack, you typically assign a floating IP. The floating IP will be the external network representation of your instance.

To assign a floating IP, go to the Instances screen, click on the More button associated with your instance, and select Associate Floating IP. The Manage Floating IP Associations pop-up window will appear, as shown in figure 2.19.

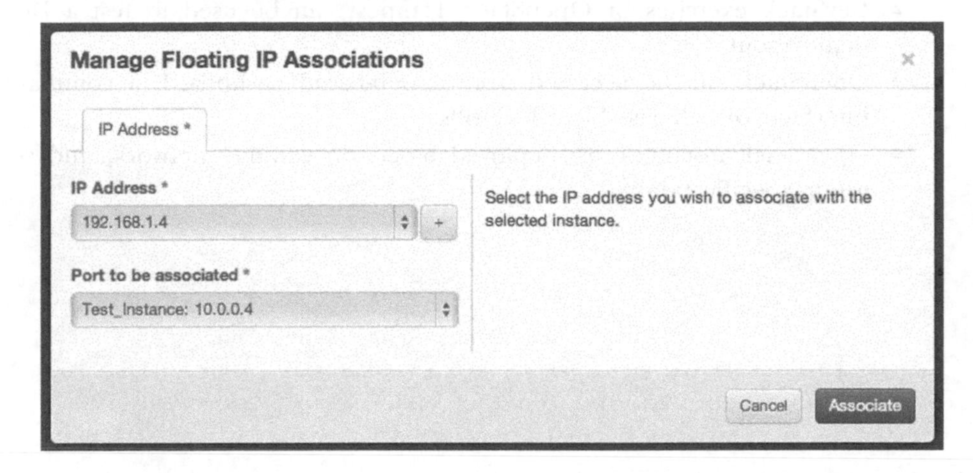

Figure 2.19 Manage Floating IP Associations

You'll want to select an address from the IP Address drop-down menu. If you see *No IP addresses available* in the drop-down, click on the + button and *Allocate* a new IP from the *Public Pool*. Once you've selected an address from the IP Address drop-down menu, click Associate, and your instance will be assigned a floating IP.

In the example shown in figure 2.19, I assigned the floating IP 192.168.1.4 to the instance with the existing IP 10.0.0.1. If you access the instance using the console, you won't see any changes, because the operating environment of the instance is unaware of the floating IP provided by OpenStack Networking. This can be confusing, but just keep in mind that there's a one-to-one relationship between the existing and floating IPs.

2.4.2 Permitting network traffic to your floating IP

There's one final step you must take to make your instance with its new floating IP accessible to the local network on your OpenStack server (so you can SSH into it). You must configure the default security group (or whatever group you applied to your instance) to allow network traffic to access your instance.

To do this, go to the Access & Security screen, previously shown in figure 2.6. Click Manage Rules. Then click Add Rule. From the Rule dropdown, select SSH and click Add. You should now be able to SSH into your instance using your assigned floating IP.

2.5 Summary

- OpenStack is a distributed cloud framework, but all components can be installed on a single server.
- DevStack is a collection of scripts that can be used to deploy a development instance of OpenStack on one or more servers.
- Component deployment through DevStack is controlled by a central configuration file.
- DevStack exercises or OpenStack Tempest can be used to test a DevStack deployment.
- OpenStack can be accessed from a web-based dashboard, a command-line interface, or web-based RESTful APIs.
- OpenStack instances are deployed based on volume, network, and security group specifications.

Learning basic OpenStack operations

This chapter covers

- Managing the OpenStack CLI to manipulate your deployment
- Exploring the OpenStack tenant model by building a new tenant
- Setting up basic tenant networking with intra-tenant configuration
- Using OpenStack networking for internal and external network configuration
- Modifying tenant quotas to control resource allocation

This chapter builds on the deployment from chapter 2 by demonstrating basic operations you'll encounter as an OpenStack administrator or user. Chapter 2 was more focused on end-user interaction with OpenStack, so those examples were based on the Dashboard, which is easy to use and can be used to perform many user and administrative functions. This chapter focuses on operational exercises, so examples are based on the OpenStack command-line interface (CLI).

If you have systems administration experience, you'll certainly appreciate the ability to script a repetitive function, such as creating a thousand users. The OpenStack APIs can also be used for these tasks, and they'll be briefly introduced. As you'll discover, if you can perform an operation with the CLI, you can easily perform the same operation with an API directly. For the examples in this chapter, we'll stick with using the CLI, but this chapter is constructed so that you can walk through the examples using either an API or the Dashboard once you understand the concepts demonstrated through the CLI.

The CLI also has the added benefit of using separate applications for each OpenStack component. While at first this might seem like a bad thing, it will help you better understand which component is responsible for what.

The basic OpenStack operations covered in this chapter can be applied to a DevStack deployment, like the one in chapter 2, or to a very large multiserver production deployment. In chapter 2 you used the `Demo` tenant (project) and the `demo` user. These and other objects were created by DevStack, but tenants, users, networks, and other objects won't be created for you automatically in manual deployment. In this chapter, we'll walk through the process of creating the necessary objects to take a test-drive in a tenant you create. By the end of the chapter, you'll know how to separate resource assignments using the OpenStack tenant model.

The chapter starts by introducing the OpenStack CLI. Then we'll progress through the process of creating a tenant, user, and networks. Finally, you'll learn about quota management from the tenant perspective. As you walk through the examples, take note of the CLI applications used in each step. You'll not only learn basic OpenStack operations, but you should get a better understanding of which OpenStack components provide what functions. In chapter 4 we'll cover OpenStack component relations in more detail.

3.1 Using the OpenStack CLI

Let's take a brief look at how you can interact with OpenStack on the command line. Before you can run CLI commands, you must first set the appropriate environment variables in your shell. Environment variables tell the CLI how and where to identify you. You can provide input for these variables directly to the CLI, but for the sake of clarity, all examples will be shown with the appropriate environment variables in place.

To set these variables, run the commands shown in the next listing in your shell. Each time you log in to a session, you'll have to set your environment variables.

Listing 3.1 Set environmental variables

```
source /opt/stack/python-novaclient/tools/nova.bash_completion
source openrc demo demo
```

Sets variables for shell completion so that pressing tab after entering "something /bo" completes "something /boot" (annotates line 1)

Run this command from ~/devstack directory. When you run OpenStack CLI commands, your identity will be (user) <demo> in (tenant) <demo>. (annotates line 2)

> **Setting environment variables manually**
>
> If you're an experienced user or you're not using DevStack, you can manually set your environmental variables by running the following commands and substituting your values for the ones in these examples:
>
> ```
> export OS_USERNAME=admin
> export OS_PASSWORD=devstack
> export OS_TENANT_NAME=admin
> export OS_AUTH_URL=http://10.0.2.32:5000/v2.0
> ```
>
> *10.0.2.32*
>
> These example commands will set the current shell user as the OpenStack `admin` user of the `admin` tenant.

To make sure your variables have been properly set, you should test if you can run an OpenStack CLI command. In your shell, run the `nova image-list` command, as shown in listing 3.2. This CLI command reads the environment variables you just set and uses them as identification. If you're properly identified and have rights to do so, the CLI will query OpenStack Compute (Nova) for your currently available `image-list`.

Listing 3.2 Setting variables and executing a first CLI command

```
devstack@devstack:~/devstack$ source \
  /opt/stack/python-novaclient/tools/nova.bash_completion
devstack@devstack:~/devstack$ source openrc demo demo
devstack@devstack:~/devstack$ nova image-list
+---+---------------------------------+--------+--------+
| ID| Name                            | Status | Server |
+---+---------------------------------+--------+--------+
| 4.| Ubuntu 12.04                    | ACTIVE |        |
| f.| cirros-0.3.1-x86_64-uec         | ACTIVE |        |
| a.| cirros-0.3.1-x86_64-uec-kernel  | ACTIVE |        |
| a.| cirros-0.3.1-x86_64-uec-ramdisk | ACTIVE |        |
+---+---------------------------------+--------+--------+
```

This simple command will list your Nova images.

You should now be able to run OpenStack CLI commands as the demo user in the demo tenant. This is the same user you used in chapter 2, so any changes you make using the CLI will be reflected in the Dashboard.

Using the command in listing 3.3, you can create a new OpenStack instance, just like you did in the Dashboard. As mentioned in chapter 2, an OpenStack instance is a VM for the purposes of this book.

Listing 3.3 Launching an instance from the CLI

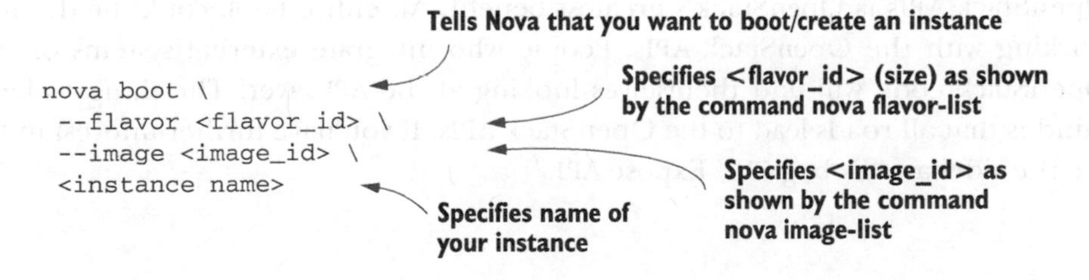

Tells Nova that you want to boot/create an instance
nova boot \
--flavor <flavor_id> \
--image <image_id> \
<instance name>
Specifies <flavor_id> (size) as shown by the command nova flavor-list
Specifies <image_id> as shown by the command nova image-list
Specifies name of your instance

When you run this command, you'll get results something like the following:

```
nova boot \
--flavor 3 \
--image 48ab76e9-c3f2-4963-8e9b-6b22a0e9c0cf \
Test_Instance_3
+---+------------------------------------+------+--------+

+--------------------------------------+---------------+
| Property                             | Value         |
+--------------------------------------+---------------+
| OS-DCF:diskConfig                    | MANUAL        |
| OS-EXT-AZ:availability_zone          | nova          |
| OS-EXT-STS:power_state               | 0             |
| OS-EXT-STS:task_state                | scheduling    |
| OS-EXT-STS:vm_state                  | building      |
| OS-SRV-USG:launched_at               | -             |
| OS-SRV-USG:terminated_at             | -             |
| accessIPv4                           |               |
| accessIPv6                           |               |
...
...
+--------------------------------------+---------------+
```

You can do everything with the OpenStack CLI that you can using the Dashboard, and more. In the preceding example, you performed the Nova boot command, which provisioned a new VM. To get help with more-advanced Nova commands, use the following command: `nova help COMMAND` (replacing `COMMAND` with the command you're interested in). There are similar command-line utilities for Keystone, Glance, Neutron, and so on.

You now have a basic idea of how the OpenStack CLI works. In later chapters, you'll mostly be working with the CLI, so learning how things work in DevStack should be helpful if things don't work as expected later.

Before we move on to the tenant examples, let's take a look at the mechanics of the OpenStack APIs.

3.2 Using the OpenStack APIs

At this point you might be wondering, "How does the OpenStack CLI work?" The answer to this is that the CLI applications call APIs specific to OpenStack components. The component-specific APIs interface with a number of sources, including other APIs and relational databases. This also holds true for the Dashboard, which you used in chapter 2. All OpenStack interactions eventually lead back to the OpenStack API layer.

It could certainly be argued that the inherent vendor neutrality provided by the OpenStack APIs is OpenStack's greatest benefit. An entire book could be devoted to working with the OpenStack APIs. People who integrate external systems or debug OpenStack code will find themselves looking at the API layer. The thing to keep in mind is that all roads lead to the OpenStack APIs. If you have further interest in them, see the sidebar, "Debug CLI/Expose API."

To get started using the OpenStack APIs directly, you can follow the example in listing 3.4. This command will query the OpenStack APIs for information, which will be returned in JavaScript Object Notation (JSON) format. Python is used to parse the JSON so it can be read on your screen.

Listing 3.4 Executing a first API command

```
curl -s -X POST http://10.0.2.32:5000/v2.0/tokens \
 -d '{"auth": {"passwordCredentials": \
 {"username":"demo", "password":"devstack"}, \
 "tenantName":"demo"}}' -H "Content-type: application/json" | \
 python -m json.tool
```

Substitute your IP for 10.0.2.32.

Debug CLI/Expose API

Every CLI command will output its API command if the debug flag is set. To enable debugging for a specific CLI command, pass the `--debug` argument before any other variables as shown here:

```
devstack@devstack:~$ nova --debug image-list

REQ: curl -i 'http://10.0.2.32:5000/v2.0/tokens'
-X POST -H "Content-Type: application/json"
-H "Accept: application/json"
-H "User-Agent: python-novaclient"
-d '{"auth": {"tenantName": "admin", "passwordCredentials":
{"username": "admin", "password": "devstack"}}}'

...
```

Now that you understand the mechanics of the OpenStack CLI and APIs, you're ready to put these skills to use. In the next section we'll walk through creating a new tenant (project) using the CLI. This is an operational function you'll perform for each new department, user, or project you want to separate from a more general tenant.

3.3 Tenant model operations

OpenStack is natively multi-tenant-aware. As mentioned in chapter 2, you can think of your OpenStack deployment as a hotel. A person can't be a resident of a hotel unless they have a room, so you can think of *tenants* as hotel rooms. Instead of beds and a TV, Hotel OpenStack provides computational resources. Just as a hotel room is configurable (single or double beds, a suite or a room, and so on), so are tenants. The number of resources (vCPU, RAM, storage, and the like), images (tenant-specific software images), and the configuration of the network are all based on tenant-specific configurations. Users are independent of tenants, but users may hold roles for specific tenants. A single user might hold the role of administrator in multiple tenants. Every time a new user is added to OpenStack, they must be assigned a tenant. Every time a

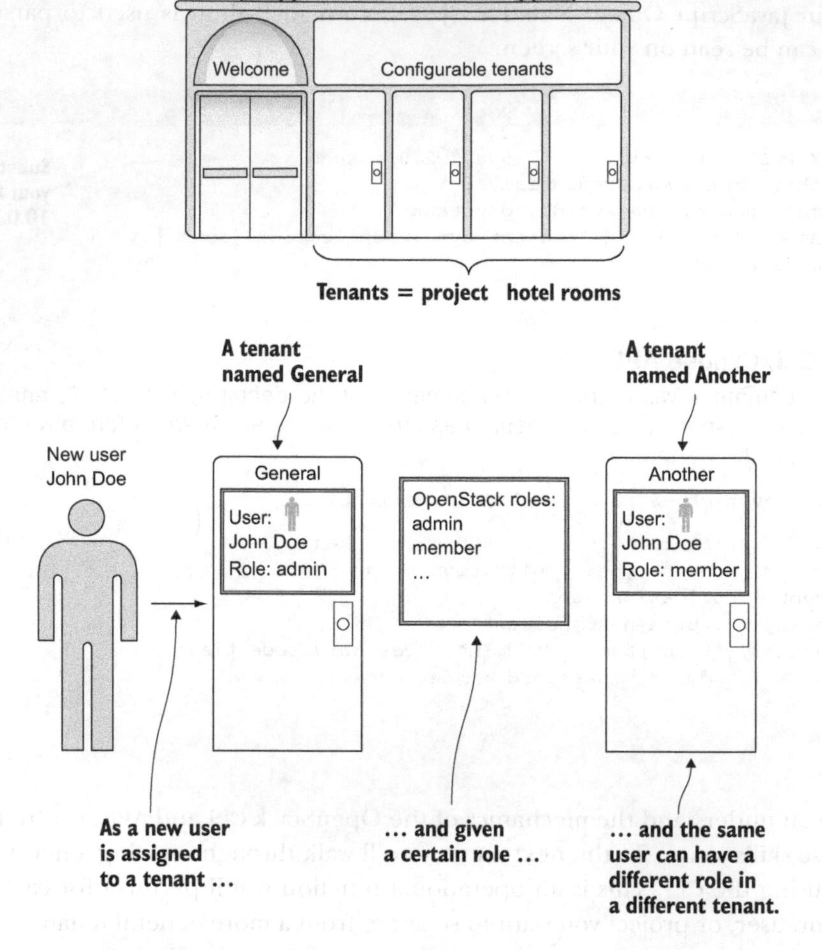

Figure 3.1 The relation of tenants, users, and roles in OpenStack Identity (Keystone). Tenants can thought of as projects or departments. Like hotel rooms, they're available in different configurations.

new instance (VM) is created, it must be created in a tenant. Management of all OpenStack resources is based on the management of tenant resources.

Because your access to OpenStack resources is based on tenant configurations, you must understand how to create new tenants, users, roles, and quotas. In chapter 2 you used DevStack, which created a few sample tenants and users for you. In the next few subsections, you'll walk through creating a new tenant and all the related objects that go with it, from scratch. As an OpenStack administrator, this will be a common task. A department or even a project might be a new tenant. Tenants will be the fundamental way that you divide and manage configurations and resources in OpenStack.

3.3.1 The tenant model

Before you start the process of creating tenant and user identity objects, you need to get an idea of how these items are related. Using our Hotel OpenStack analogy, figure 3.1 shows the interplay among tenants, members, and roles in OpenStack. You can see that a *role* is a designation independent of a tenant until a user is assigned a role in a tenant. You can see that *user* is an `admin` in the `General` tenant and a `Member` in the `Another` tenant. Notice that both users and roles have one-to-many relationships with tenants. As you move forward in this chapter's examples, you'll create several of the components shown in this figure.

As you can see in figure 3.1, tenants are the way OpenStack organizes role assignments. In OpenStack all resource configuration (users with roles, instances, networks, and so on) is organized based on tenant separation. In OpenStack jargon, the term *tenant* can be used synonymously with *project*, so think of using a tenant for a particular project or organizational division. It's worth noting that roles are defined outside of tenants, but users are created with an initial tenant assignment. It would be reasonable for a user to be created in a departmental tenant (for example, John Doe is created in the `General` tenant) and be assigned a role in another tenant (John Doe is a `Member` of `Another` tenant). This means that every tenant can have a specific user with the role `Member`, but that specific user would only have one home tenant.

I'll use the OpenStack CLI to demonstrate examples in this section. I could just as easily use the Dashboard, but demonstrations using the CLI can often be more clearly explained because the CLI forces you to direct your request to the specific CLI application that deals with that function. When using the Dashboard, it's hard to tell which component is controlling what. Once you understand the process through the CLI, using the Dashboard will be trivial.

> **USING DEVSTACK FROM CHAPTER 2** The examples covered in this chapter are executed on the OpenStack instance deployed by DevStack, as described in chapter 2. If you already have your OpenStack instance from chapter 2 set up, you're ready to run the examples in this chapter.

In listing 3.1 you set your environment variables to represent the `Demo` user in the `Demo` tenant. Because you'll be creating a new tenant, you'll need to set your environment variables to represent the `Admin` user in the `Admin` tenant, as shown in the following listing.

Listing 3.5 Prepare your shell session as `admin`

Set variables for shell "completion" (so that pressing Tab after entering "something /bo" completes "something /boot").

```
source /opt/stack/python-novaclient/tools/nova.bash_completion
source openrc admin admin
```

Run this command from ~/devstack directory. Command will set environmental variables so that when you run OpenStack CLI commands, your identity will be <user: admin> in <tenant:admin>.

ADMIN OR DEMO In the previous subsection, you set environmental variables to refer to the demo user. Under Linux, running as the *root* user when it's not necessary is considered bad practice because it's too easy to make possibly disruptive changes by accident. You can consider unnecessarily running commands as the admin user to be a similarly bad practice in OpenStack.

3.3.2 Creating tenants, users, and roles

In this section, you'll create a new tenant and user. You'll then assign a role to your new user in your new tenant.

CREATING A TENANT

Use the command shown in the following listing to create your new tenant.

Listing 3.6 Creating a new tenant

```
keystone tenant-create --name General
```

When you run the command, you'll see output like the following:

```
+-------------+----------------------------------+
|   Property  |              Value               |
+-------------+----------------------------------+
| description |                                  |
|   enabled   |               True               |
|      id     | 9932bc0607014caeab4c3d2b94d5a40c |
|     name    |             General              |
+-------------+----------------------------------+
```

You've now created a new tenant that will be referenced when creating other OpenStack objects. Take note of the tenant ID generated in this process; you'll need this ID for the next steps. Figure 3.2 illustrates the tenant you just created. The admin and Member roles were created as part of the DevStack deployment of OpenStack. The sidebar "Listing tenants and roles" explains how to list all tenants and roles for a particular OpenStack deployment.

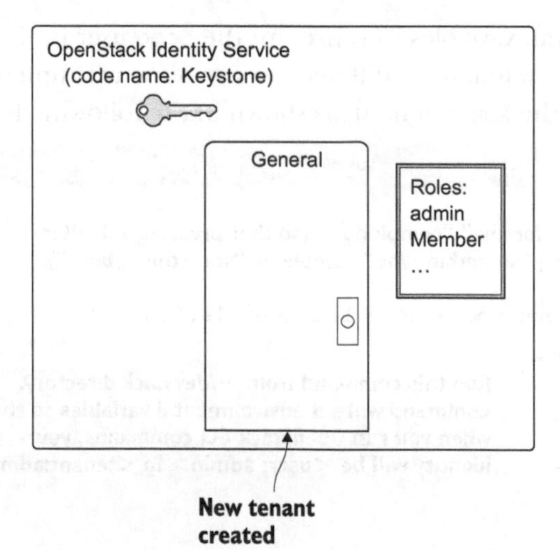

Figure 3.2 The created tenant

Listing tenants and roles

You can list all tenants on the system as follows:

```
devstack@devstack:~/devstack$ keystone tenant-list
+----------------------------------+--------------------+---------+
|                id                |        name        | enabled |
+----------------------------------+--------------------+---------+
| 9932bc0607014caeab4c3d2b94d5a40c |      General       |  True   |
| b1c52f4025d244f883dd47f61791d5cf |       admin        |  True   |
| 166c9cab0722409d8dbc2085acea70d4 |      alt_demo      |  True   |
| 324d7477c2514b60ac0ae417ce3cefc0 |        demo        |  True   |
| fafc5f46aaca4018acf8d05370f2af57 | invisible_to_admin |  True   |
| 81548fee3bb84e7db93ad4c917291473 |      service       |  True   |
+----------------------------------+--------------------+---------+
```

You can similarly list all roles on the system with the following command:

```
devstack@devstack:~/devstack$ keystone role-list
+----------------------------------+---------------+
|                id                |     name      |
+----------------------------------+---------------+
| 4b303a1c20d64deaa6cb9c4dfacc33a9 |    Member     |
| 291d6a3008c642ba8439e42c95de22d0 | ResellerAdmin |
| 9fe2ff9ee4384b1894a90878d3e92bab |   _member_    |
| 714aaa9d30794920afe25af4791511a1 |     admin     |
| b2b1621ddc7741bd8ab90221907285e0 |  anotherrole  |
| b4183a4790e14ffdaa4995a24e08b7a2 |    service    |
+----------------------------------+---------------+
```

CREATING A USER

Now that the tenant has been created, you can create a new user, as shown in the next listing.

Listing 3.7 Creating a new user

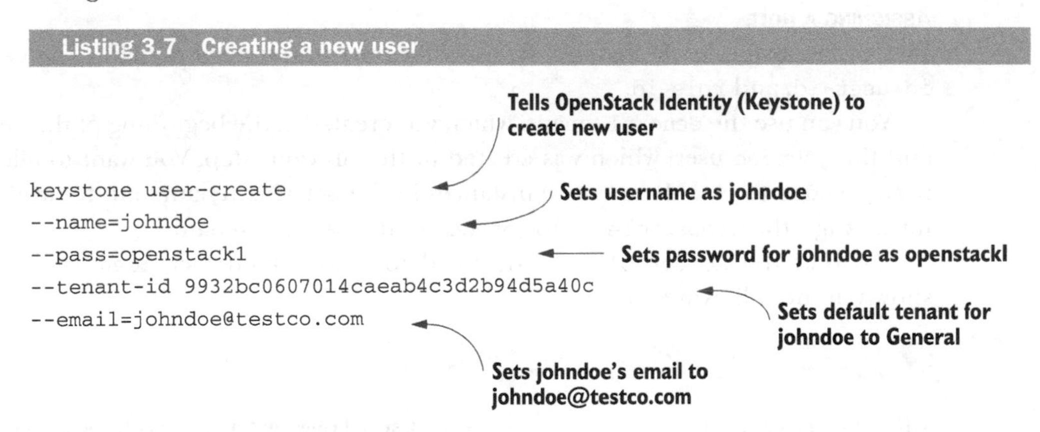
Tells OpenStack Identity (Keystone) to create new user
keystone user-create
Sets username as johndoe
--name=johndoe
--pass=openstack1
Sets password for johndoe as openstackl
--tenant-id 9932bc0607014caeab4c3d2b94d5a40c
--email=johndoe@testco.com
Sets default tenant for johndoe to General
Sets johndoe's email to johndoe@testco.com

When you run the command, you'll get output like the following:

```
+----------+----------------------------------+
| Property |              Value               |
+----------+----------------------------------+
|  email   |        johndoe@testco.com        |
| enabled  |               True               |
|    id    | 21b27d5f7ba04817894d290b660f3f44 |
|   name   |             johndoe              |
| tenantId | 9932bc0607014caeab4c3d2b94d5a40c |
+----------+----------------------------------+
```

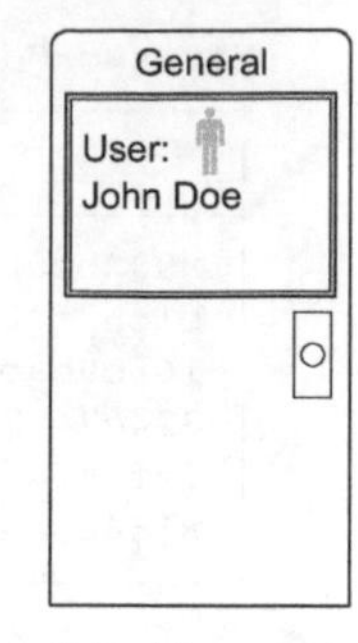

Figure 3.3 The created user

You've now created a new user. Take note of the user ID generated in this process because it will be needed in the next step. Figure 3.3 now includes the user you just created in the `General` tenant.

> **Listing users in a tenant**
>
> You can list all users in a tenant with the following command:
>
> ```
> devstack@devstack:~/devstack$ keystone user-list \
>  --tenant-id 9932bc0607014caeab4c3d2b94d5a40c
> +-------------+---------+---------+--------------------+
> |     id      |  name   | enabled |       email        |
> +-------------+---------+---------+--------------------+
> | 21b2...3f44 | johndoe |  True   | johndoe@testco.com |
> +-------------+---------+---------+--------------------+
> ```
>
> Specify tenant-id for tenant; in this case, General

You now need to add a role to your new user in your new tenant. We'll do that next.

ASSIGNING A ROLE

In order to assign a role to a user in a specific tenant, you need to define the `tenant-id`, `user-id`, and `role-id`.

You can use the `General` tenant, which was created at the beginning of this section, and the `johndoe` user, which was created in the previous step. You want to allow the user `johndoe` to be able to create instances in the `General` tenant, and to do this you must assign the `Memberrole-id` to `johndoe` in the `General` tenant.

To find the `Memberrole-id`, you need to query OpenStack `Identity` roles, as shown in the following listing.

Listing 3.8 Listing OpenStack roles

```
keystone role-list
```

Lists all roles on this OpenStack deployment

The list of roles will look something like this:

id	name
4b303a1c20d64deaa6cb9c4dfacc33a9	Member
291d6a3008c642ba8439e42c95de22d0	ResellerAdmin
9fe2ff9ee4384b1894a90878d3e92bab	_member_
714aaa9d30794920afe25af4791511a1	admin
b2b1621ddc7741bd8ab90221907285e0	anotherrole
b4183a4790e14ffdaa4995a24e08b7a2	service

You now have all of the information you need to assign your newly created user as a `Member` of your newly created tenant. Run the command shown in the following listing, substituting the appropriate IDs for your system.

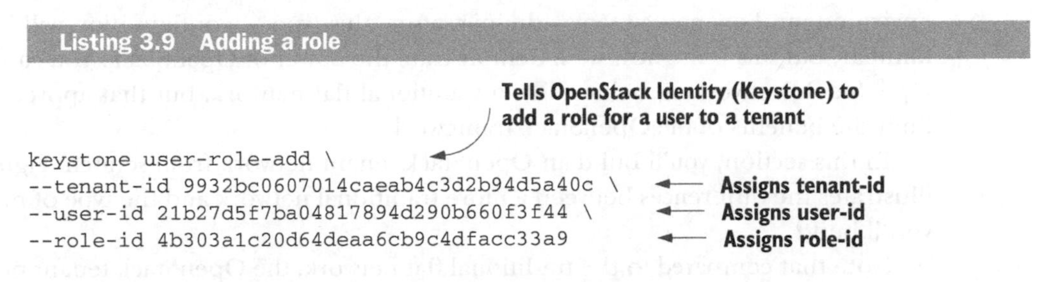

Listing 3.9 Adding a role

```
keystone user-role-add \
--tenant-id 9932bc0607014caeab4c3d2b94d5a40c \
--user-id 21b27d5f7ba04817894d290b660f3f44 \
--role-id 4b303a1c20d64deaa6cb9c4dfacc33a9
```


This command will generate no output if it's successful.

You've now assigned a role to a user in a tenant. Figure 3.4 illustrates the role you just assigned to the `johndoe` user in the `General` tenant.

At this point you can access the OpenStack Dashboard and log in using the `johndoe` user and `openstack1` password. When you do, you'll be taken to the `General` tenant/project management screen. If you try to create a new instance, you'll notice that no networks exist. You'll create a new tenant network next.

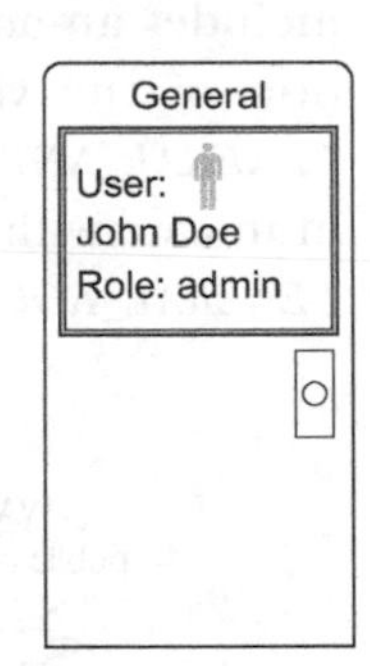

Figure 3.4 The assigned role

3.3.3 *Tenant networks*

OpenStack Networking (Neutron) is both loved and hated. To make sure you have more of the former experience than the latter, you should get your feet wet as soon as possible. In this section, we'll run through some simple tenant network configurations.

First, though, you need to understand the basic differences between how traditional "flat" networks are configured for virtual and physical machines and how OpenStack Networking will be demonstrated. The term *flat* refers to the absence of a virtual routing tier as part of the virtual server platform; in traditional configurations, the VM has direct access to a network, as if you plugged a physical device into a physical network switch. Figure 3.5 illustrates a flat network connected to a physical router.

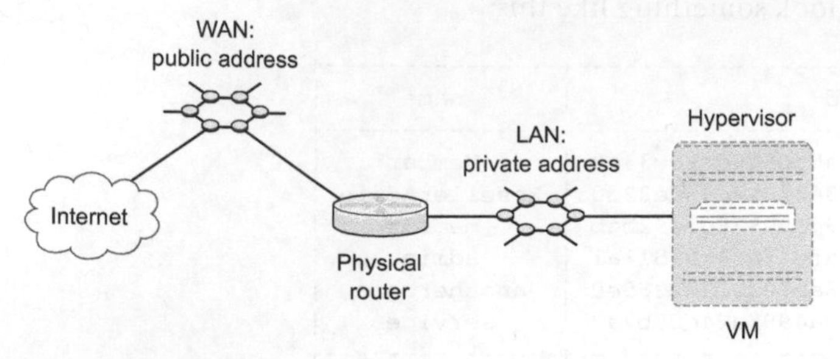

Figure 3.5 Traditional routed network

In this type of deployment, all network services (Dynamic Host Configuration Protocol (DHCP), load balancing, routing, and so on) beyond simply switching (Open System Interconnection (OSI) Model, Layer 2), must be provided outside of the virtual environment. For most systems administrators, this type of configuration will be very familiar, but this is not how we'll demonstrate the power of OpenStack. You can make OpenStack Networking behave like a traditional flat network, but that approach will limit the benefits of the OpenStack framework.

In this section, you'll build an OpenStack tenant network from scratch. Figure 3.6 illustrates the differences between a more traditional network and the type of network you'll build.

Note that compared to the traditional flat network, the OpenStack tenant network includes an additional router that resides within the virtual environment. The addition of the virtual router in the tenant separates the *internal network*, shown as *GENERAL_NETWORK*, from the *external network*, shown as *PUBLIC_NETWORK*. VMs communicate with each other using the internal network, and the *virtual router*, shown as *GENERAL_ROUTER*, uses the external network for communication outside the tenant.

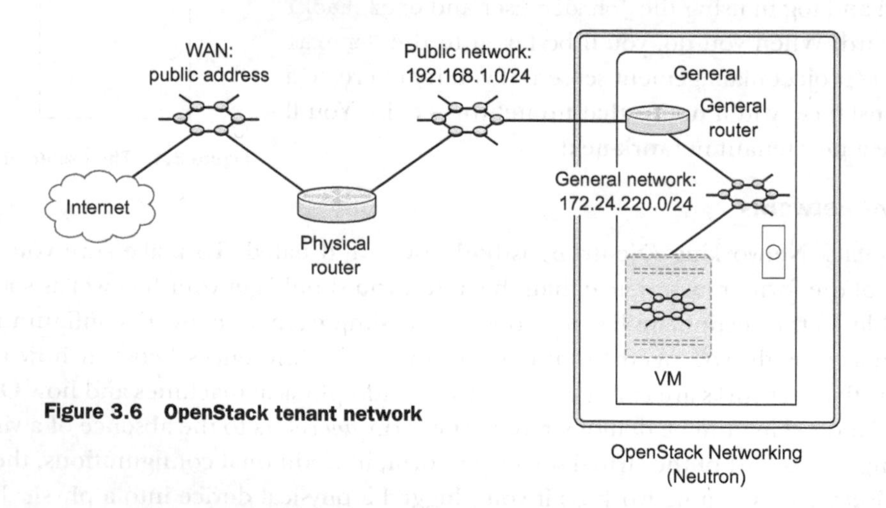

Figure 3.6 OpenStack tenant network

SET YOUR ENVIRONMENT VARIABLES The configurations in the following subsections require that the OpenStack CLI environmental variables are set. To set the environment variables, execute the commands shown in listing 3.5.

NETWORK (NEUTRON) CONSOLE

Neutron commands can be entered through the Neutron console (which is like a command line for a network router or switch) or directly through the CLI. The console is very handy if you know what you're doing, and it's a natural choice for those familiar with the Neutron command set. But for the sake of clarity, I'll demonstrate each action as a separate command using CLI commands. There are many things you can do with the Neutron CLI and console that you can't do in the Dashboard.

The distinction between the Neutron console and Neutron CLI will be made clear in the following subsections. While the examples in this chapter are executed using the CLI, you'll still need to know how to access the Neutron console. As you can see from the following listing, it's a simple matter of using the `neutron` command.

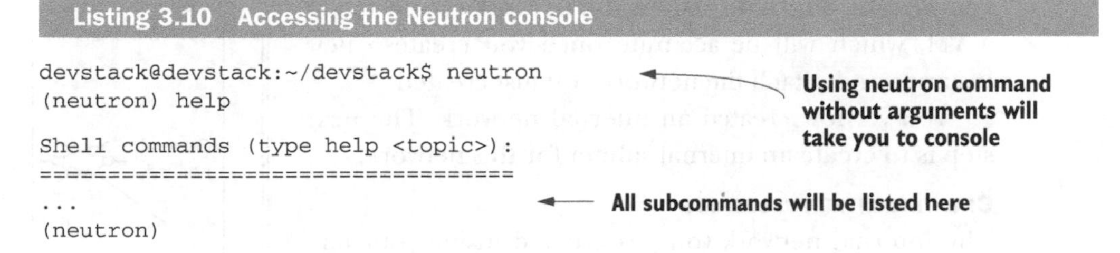

Listing 3.10 Accessing the Neutron console

```
devstack@devstack:~/devstack$ neutron
(neutron) help

Shell commands (type help <topic>):
===================================
...
(neutron)
```

Using neutron command without arguments will take you to console
All subcommands will be listed here

You can now access the Neutron interactive console. Any CLI configurations can be made in the interactive console or directly on the command line. That's how you'll create a new network.

CREATING INTERNAL NETWORKS

The first step in creating a tenant-based network is to configure the internal network that will be used directly by instances in your tenant. The internal network works on ISO Layer 2 (L2), so for the network types this is the virtual equivalent of providing a network switch to be used exclusively for a particular tenant. The next listing shows the code used to create a new network for your tenant.

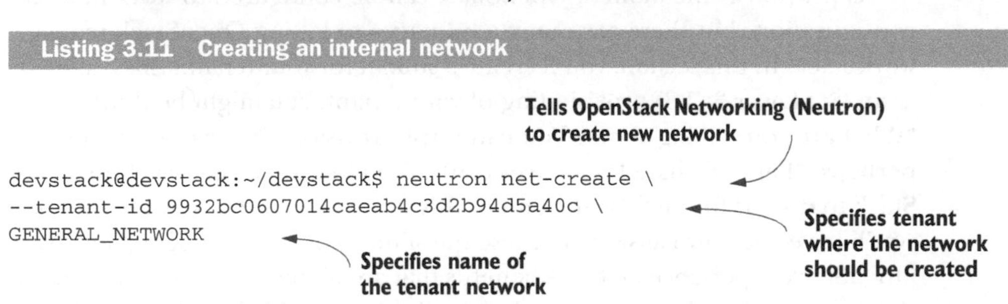

Listing 3.11 Creating an internal network

```
devstack@devstack:~/devstack$ neutron net-create \
--tenant-id 9932bc0607014caeab4c3d2b94d5a40c \
GENERAL_NETWORK
```

Tells OpenStack Networking (Neutron) to create new network
Specifies tenant where the network should be created
Specifies name of the tenant network

You'll see output like the following when you create the network:

```
Created a new network:
+---------------------------+--------------------------------------+
| Field                     | Value                                |
+---------------------------+--------------------------------------+
| admin_state_up            | True                                 |
| id                        | 35a387fd-892f-47ad-a226-e8d0f2f0636b |
| name                      | GENERAL_NETWORK                      |
| provider:network_type     | local                                |
| provider:physical_network |                                      |
| provider:segmentation_id  |                                      |
| shared                    | False                                |
| status                    | ACTIVE                               |
| subnets                   |                                      |
| tenant_id                 | 9932bc0607014caeab4c3d2b94d5a40c     |
+---------------------------+--------------------------------------+
```

Figure 3.7 illustrates the `GENERAL_NETWORK` created for your tenant. The figure shows the network connected to a VM, which will be accurate once you create a new instance and attach the network you just created.

You've now created an internal network. The next step is to create an internal subnet for this network.

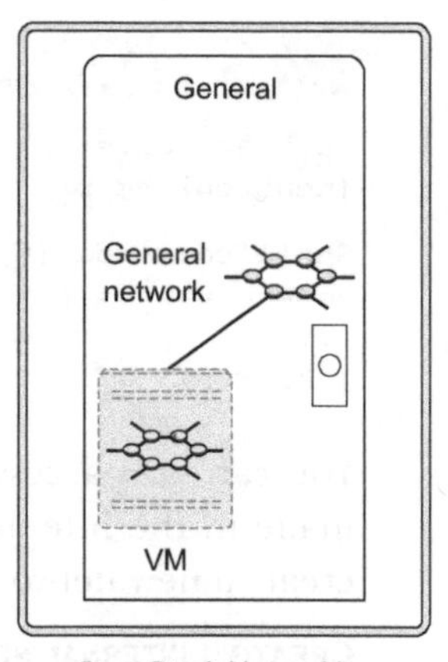

Figure 3.7 The newly created internal network

CREATING INTERNAL SUBNETS

The internal network you just created inside your tenant is completely isolated from other tenants. This is a strange concept to those who work with physical servers, or even to those who generally expose their virtual machines directly to physical networks. Most people are used to connecting their servers to the network, and network services are generally provided on a data center or enterprise level. We don't typically think about networking and computation being controlled under the same framework.

As previously mentioned, OpenStack can be configured to work in a flat network configuration, but there are many advantages to letting OpenStack manage the network stack. In this section, you'll create a subnet for your tenant; this can be thought of as an ISO Layer 3 (L3) provisioning of your tenant. You might be thinking to yourself, "What are you talking about? You can't just provision L3 services on the network!" or perhaps, "I already have L3 services centralized in my data center. I don't want OpenStack to do this for me!" By the end of this section, or perhaps by the end of the book, you'll have your own answers to these questions. For the time being, just trust that the OpenStack experience includes benefits that are either enriched or not otherwise possible without the advanced network virtualization provided by OpenStack Networking.

What does it mean to create a new subnet for a specific network? Basically, you describe the network you want to work with, and then describe the address ranges you plan on using on that network. In this case, you'll assign the new subnet to the GENERAL_NETWORK in the General tenant. You must also provide an address range for the subnet. In this context, the term *subnet* refers to both an OpenStack subnet, which is defined as part of the OpenStack network, and the IP subnet, which is defined as part of the OpenStack subnet creation process. You can use your own address range as long as it doesn't exist in the tenant or a shared tenant. One of the interesting things about OpenStack is that you could use the same address range for every internal subnet in every tenant.

The following listing shows the command used to create a subnet.

Listing 3.12 Creating an internal subnet for a network

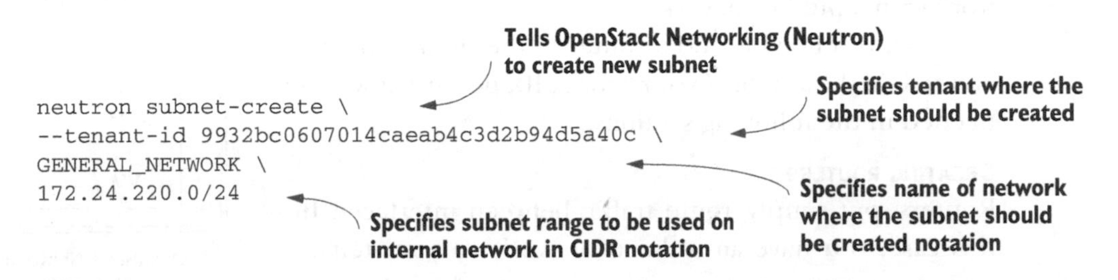
Tells OpenStack Networking (Neutron) to create new subnet
Specifies tenant where the subnet should be created
Specifies name of network where the subnet should be created notation
Specifies subnet range to be used on internal network in CIDR notation

```
neutron subnet-create \
--tenant-id 9932bc0607014caeab4c3d2b94d5a40c \
GENERAL_NETWORK \
172.24.220.0/24
```

When you run this command, you'll see output like the following:

```
Created a new subnet:
+-------------------+---------------------------------------------------+
| Field             | Value                                             |
+-------------------+---------------------------------------------------+
| allocation_pools  | {"start":"172.24.220.2","end":"172.24.220.254"}   |
| cidr              | 172.24.220.0/24                                   |
| dns_nameservers   |                                                   |
| enable_dhcp       | True                                              |
| gateway_ip        | 172.24.220.1                                      |
| host_routes       |                                                   |
| id                | 40d39310-44a3-45a8-90ce-b04b19eb5bb7              |
| ip_version        | 4                                                 |
| name              |                                                   |
| network_id        | 35a387fd-892f-47ad-a226-e8d0f2f0636b              |
| tenant_id         | 9932bc0607014caeab4c3d2b94d5a40c                  |
+-------------------+---------------------------------------------------+
```

Classless Inter-Domain Routing (CIDR)

CIDR is a compact way to represent subnets.

For internal subnets, most people use a private class C address range, which was actually a class C of the original public classful ranges. In the case of a class C range, 8 bits are used for the subnet mask, so there are 2^8 = 256 addresses, but CIDR is expressed in the form <First address>/<Size of host bit field>, where 32 bits – 8 bits = 24 bits.

> ***(continued)***
> This might seem confusing, but luckily there are many online subnet calculators you can use if you aren't up on your binary math.

You now have a new subnet assigned to your GENERAL_NETWORK. Figure 3.8 illustrates the assignment of the subnet to the GENERAL_NETWORK. This subnet is still isolated, but you're one step closer to connecting your private network with a public network.

Next, you need to add a router to the subnet you just created. Make a note of your subnet ID, because it will be needed in the following sections.

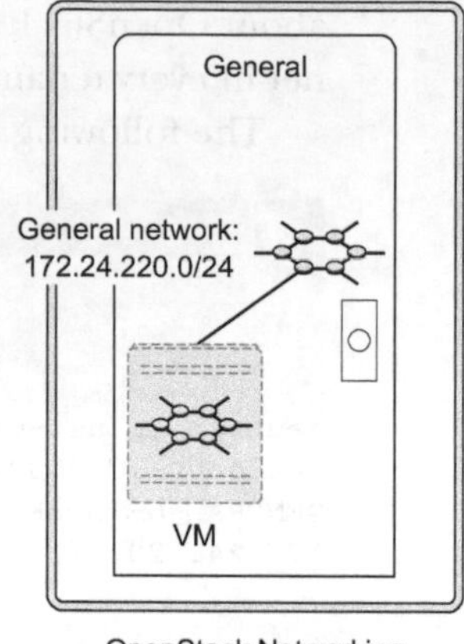
General
General network:
172.24.220.0/24
VM
OpenStack Networking
(Neutron)

Figure 3.8 The newly created internal subnet

CREATING ROUTERS

Routers, put simply, route traffic between interfaces. In this case, you have an isolated network on your tenant and you want to be able to communicate with other tenant networks or networks outside of OpenStack. The following listing shows how to create a new tenant router.

Listing 3.13 Creating a router

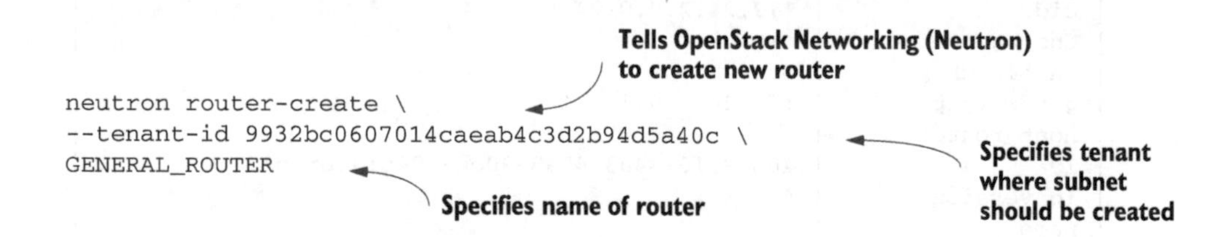
Tells OpenStack Networking (Neutron) to create new router
Specifies tenant where subnet should be created
Specifies name of router

```
neutron router-create \
--tenant-id 9932bc0607014caeab4c3d2b94d5a40c \
GENERAL_ROUTER
```

When you create the router, you'll see output like the following:

```
Created a new router:
+-----------------------+--------------------------------------+
| Field                 | Value                                |
+-----------------------+--------------------------------------+
| admin_state_up        | True                                 |
| external_gateway_info |                                      |
| id                    | df3b3d29-104f-46ca-8b8d-50658aea3f24 |
| name                  | GENERAL_ROUTER                       |
| status                | ACTIVE                               |
| tenant_id             | 9932bc0607014caeab4c3d2b94d5a40c     |
+-----------------------+--------------------------------------+
```

Figure 3.9 illustrates the router you created in your tenant.

You have a new router, but your tenant router and subnet aren't yet connected. The following listing shows how to connect your subnet to your router.

Listing 3.14 Adding a router to the internal subnet

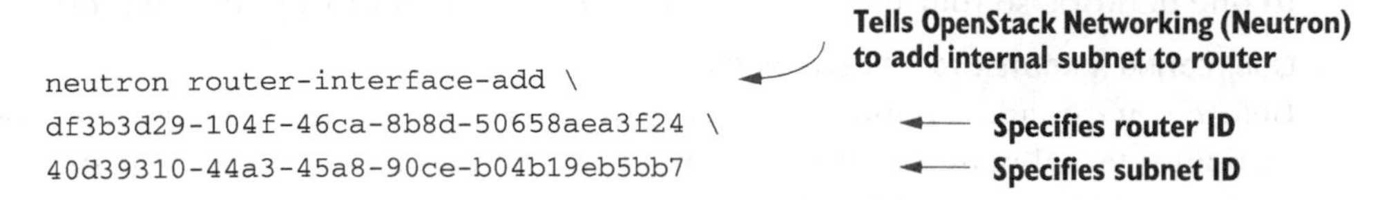

```
neutron router-interface-add \
df3b3d29-104f-46ca-8b8d-50658aea3f24 \
40d39310-44a3-45a8-90ce-b04b19eb5bb7
```


When the router is created, you'll see output like the following (IDs are autogenerated, so yours will be unique):

```
Added interface 0a1a97e3-ad63-45bf-a55f-c7cd6c8cf4b4 to
router df3b3d29-104f-46ca-8b8d-50658aea3f24
```

Figure 3.10 illustrates the new `GENERAL_ROUTER` connected to your internal network, `GENERAL_NETWORK`.

The process of adding a router to a subnet will actually create a *port* on the local virtual switch. You can think of a port as a device plugged into your virtual network port. In this case, the device is the `GENERAL_ROUTER`, the network is the `GENERAL_NETWORK`, and the subnet is 172.24.220.0/24.

> **DHCP AGENTS** In past versions of OpenStack Networking, you had to manually add Dynamic Host Configuration Protocol (DHCP) agents to your network—the DHCP agent is used to provide your instances with an IP address. In current versions, the agent is automatically added for you the first time you create an instance, but in advanced configurations it's still handy to know that agents (of all kinds) can be manipulated through Neutron.

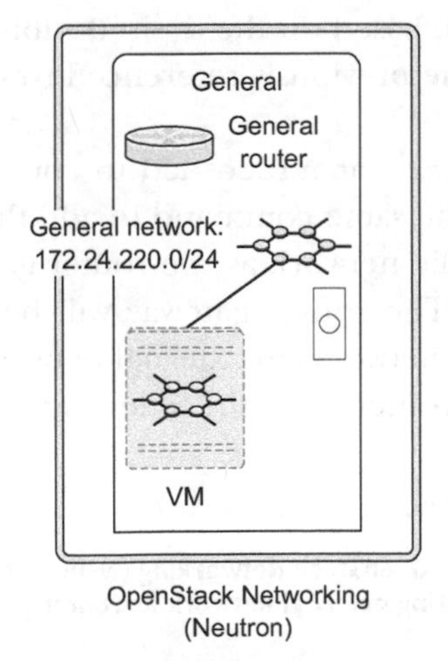

Figure 3.9 The newly created internal router

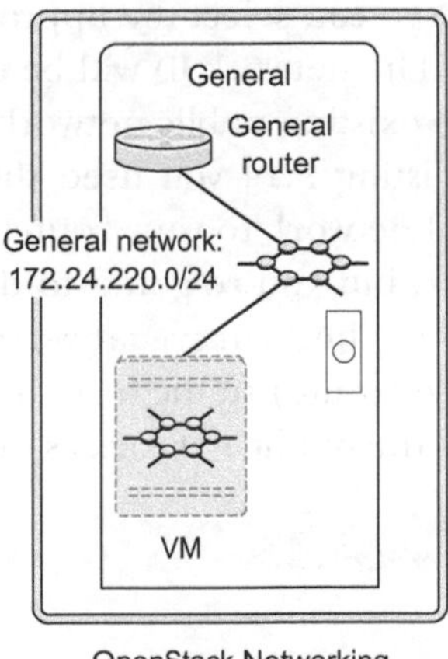

Figure 3.10 The new router connected to the internal network

The router will use the address specified during subnet creation (defaulting to the first available address), and unless you've already created an instance on this network, this will be the first port (device) on this network. If you create an instance, you should be able to communicate with the router address of 172.24.220.1, but you won't yet be able to route packets to other networks. A router isn't much good when it's only connected to one network, so your next step is to connect the router to a public network.

CONNECTING A ROUTER TO A PUBLIC NETWORK

Before you can add a public interface, you need to find it. In previous steps, you created an internal network, internal subnet, and router, so you knew their ID values. If you're working with the OpenStack deployment produced by DevStack in chapter 2, a public interface will already exist, and the following listing shows how you can list your external networks.

Listing 3.15 Listing external networks

```
neutron net-external-list
```

The preceding command will produce output like the following. If no external networks exist, jump ahead to the next subsection, "Creating an external network," and create a new external network.

```
(neutron) net-external-list
+----------------+--------+---------------------------------------+
| id             | name   | subnets                               |
+----------------+--------+---------------------------------------+
| 4eed3f..34b23d | public | e9643dc8...df4d34099109 192.168.1.0/24 |
+----------------+--------+---------------------------------------+
```

ID and subnet fields have been shortened to fit on page

You've now listed all public networks; make a note of the network ID. In the example, there will be a single public network, but in a production environment, there could be many. You can select the appropriate network ID, based on the desired subnet in the listing. The network ID will be used along with the previously referenced router ID to add the existing public network to your router.

In listing 3.14 you used the command `router-interface-add` to connect your internal network to your router. You could use the same command to add the public network, but you're going to designate this public network as the router gateway so you'll use the `router-gateway-set` command. The router gateway will be used to translate (route) traffic from internal OpenStack networks to external networks.

Add the public network as the gateway for the router using the following command.

Listing 3.16 Add existing external network as router gateway

```
neutron router-gateway-set \
df3b3d29-104f-46ca-8b8d-50658aea3f24 \

4eed3f65-2f43-4641-b80a-7c09ce34b23d
```

- Tells OpenStack Networking (Neutron) to add existing external network as router gateway
- Specifies ID of tenant router
- Specifies ID of existing external network

The preceding command will produce the following output:

```
Set gateway for router df3b3d29-104f-4
6ca-8b8d-50658aea3f24
```

Figure 3.11 illustrates the external PUBLIC_NETWORK you added to GENERAL_ROUTER as a network gateway.

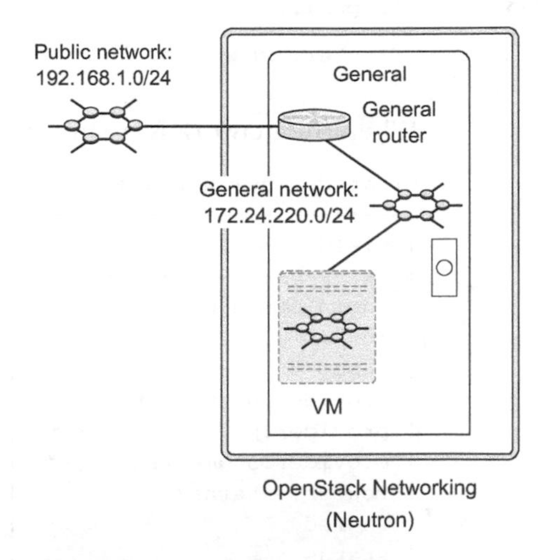

Figure 3.11 An existing network assigned as a router gateway

CREATING AN EXTERNAL NETWORK

The configurations in the following subsections require that OpenStack CLI environmental variables are set as shown in listing 3.5.

In the subsection "Creating internal networks," you created a network that was specifically for your tenant. In this subsection, you'll create a public network that can be used by multiple tenants. This public network can be attached to a private router as a network gateway, as described in the previous section.

> **Using the DevStack network from chapter 2**
>
> The examples covered in this section will be executed on an OpenStack instance deployed by DevStack, as described in chapter 2. The deployment performed in that chapter provided the necessary network configuration to allow the addition of external networks. If you already have your OpenStack instance from chapter 2, you're ready to complete the examples in this section.
>
> In upcoming chapters, you'll learn to manually make the network configurations that the previous DevStack deployment made for you.

In listing 3.15 you saw how to list networks that were designated as "external." If you're working from the DevStack deployment from chapter 2, an external network will already exist. This isn't a problem, because you can have many external networks in OpenStack Networking.

Only the admin user can create external networks, and if not specified, the new external network will be created in the admin tenant. Create a new external network as shown in the following listing.

Listing 3.17 Creating an external network

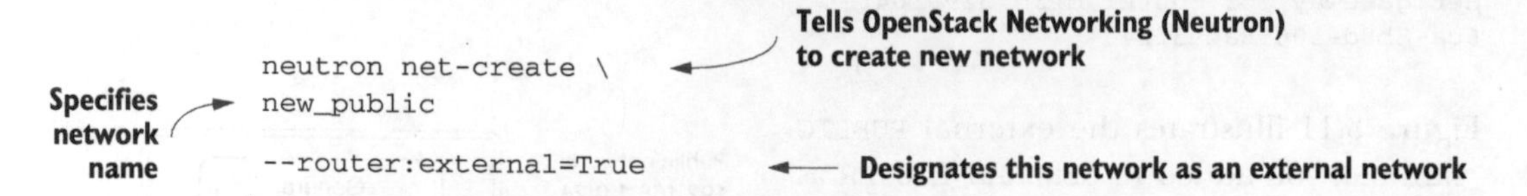

```
neutron net-create \
new_public
--router:external=True
```

Tells OpenStack Networking (Neutron) to create new network
Specifies network name
Designates this network as an external network

When the network is created, you'll see output like this:

```
Created a new network:
+---------------------------+--------------------------------------+
| Field                     | Value                                |
+---------------------------+--------------------------------------+
| admin_state_up            | True                                 |
| id                        | 8701c5f1-7852-4468-9dae-ff8a205296aa |
| name                      | new_public                           |
| provider:network_type     | local                                |
| provider:physical_network |                                      |
| provider:segmentation_id  |                                      |
| router:external           | True                                 |
| shared                    | False                                |
| status                    | ACTIVE                               |
| subnets                   |                                      |
| tenant_id                 | b1c52f4025d244f883dd47f61791d5cf     |
+---------------------------+--------------------------------------+
```

CONFIRMING THE NETWORK'S TENANT If you want to confirm that the network was created in the `admin` tenant, you can retrieve all tenant IDs as shown in the sidebar "Listing tenants and roles" in the subsection "Creating a tenant."

You now have a network that's designated as an external network, as shown in figure 3.12. This network will be in the `admin` tenant, and would not currently be visible in the `General` tenant.

Before you can use this network as a gateway for your tenant router, you must first add a subnet to the external network you just created. You'll do that next.

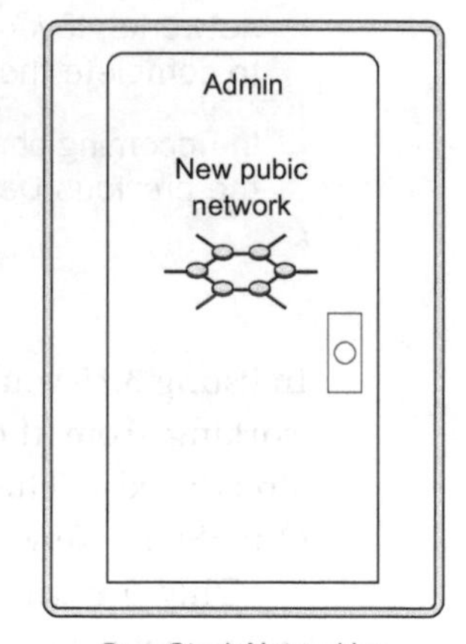

Admin
New pubic network
OpenStack Networking (Neutron)

Figure 3.12 Created external network

CREATING AN EXTERNAL SUBNET

You must now create an external subnet, shown next.

Listing 3.18 Creating an external subnet

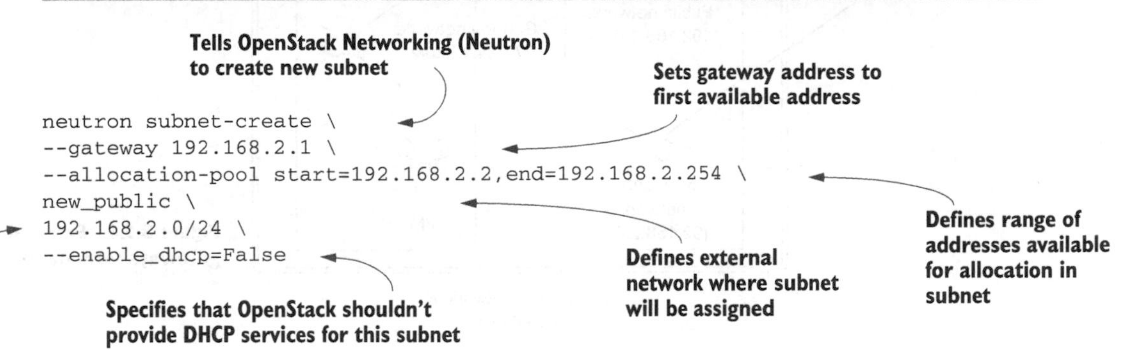
Tells OpenStack Networking (Neutron) to create new subnet
Sets gateway address to first available address
Defines range of addresses available for allocation in subnet
Defines subnet in CIDR format
Defines external network where subnet will be assigned
Specifies that OpenStack shouldn't provide DHCP services for this subnet

```
neutron subnet-create \
--gateway 192.168.2.1 \
--allocation-pool start=192.168.2.2,end=192.168.2.254 \
new_public \
192.168.2.0/24 \
--enable_dhcp=False
```

When the subnet is created, you'll see output like the following:

```
Created a new subnet:
+-------------------+------------------------------------------------------+
| Field             | Value                                                |
+-------------------+------------------------------------------------------+
| allocation_pools  | {"start": "192.168.2.2", "end": "192.168.2.254"}     |
| cidr              | 192.168.2.0/24                                       |
| dns_nameservers   |                                                      |
| enable_dhcp       | False                                                |
| gateway_ip        | 192.168.2.1                                          |
| host_routes       |                                                      |
| id                | 2cfa7201-d7f3-4e0c-983b-4c9f3fcf3caa                 |
| ip_version        | 4                                                    |
| name              |                                                      |
| network_id        | 8701c5f1-7852-4468-9dae-ff8a205296aa                 |
| tenant_id         | b1c52f4025d244f883dd47f61791d5cf                     |
+-------------------+------------------------------------------------------+
```

You now have the subnet 192.168.2.0/24 assigned to the external `new_public` network. The subnet and external network you just created, as shown in figure 3.13, can now be used by an OpenStack Networking router as a gateway.

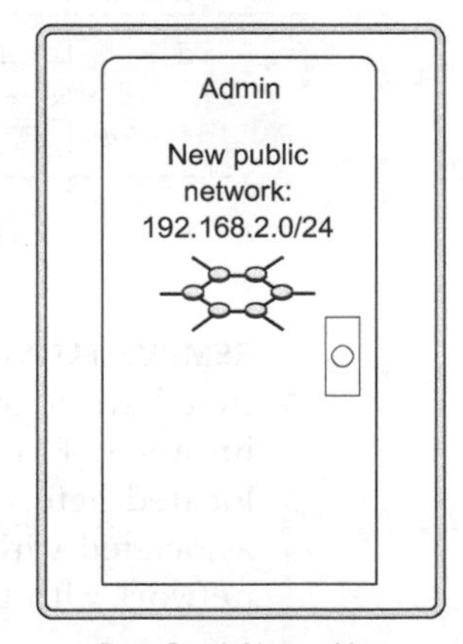
Admin
New public network: 192.168.2.0/24
OpenStack Networking (Neutron)

Figure 3.13 The newly created external subnet

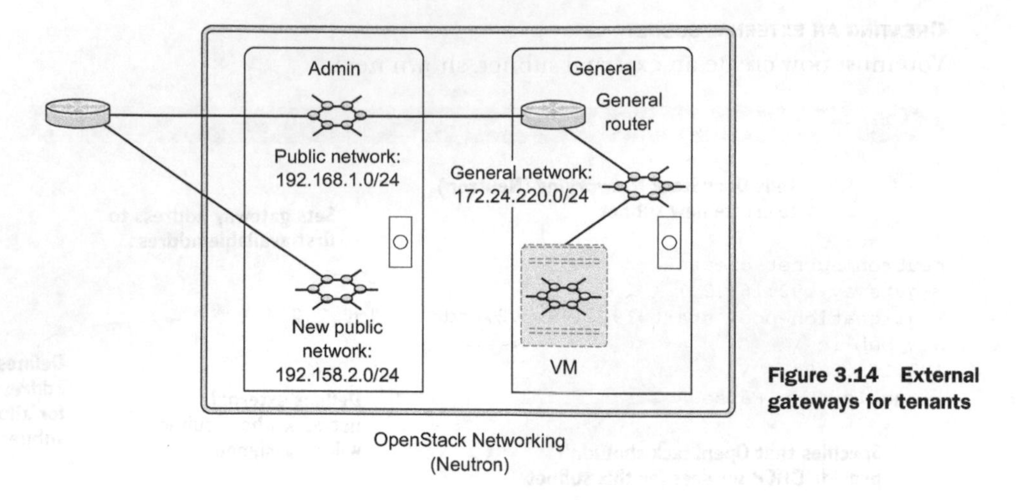

Figure 3.14 External gateways for tenants

Figure 3.14 shows your current state, assuming you followed the examples in the previous subsections.

Even if your network doesn't resemble the figure, the main idea here is to illustrate that there are two possible external networks that could be assigned as gateways for the tenant router. `new_public` and `PUBLIC_NETWORKS` are separate virtual networks and have been assigned different subnets. Currently the `PUBLIC_NETWORK` is assigned as the gateway for your `GENERAL_ROUTER`. This means that any instance network traffic that's not directly network-connected to your tenant (such as access to the internet or other tenants) will use this (gateway) network as its link to the outside world.

Listing routers

To list all the routers in the system, use the `neutron router-list` command, like this:

```
devstack@devstack:~/devstack$ neutron router-list
+--------+----------------+-----------------------------------------+
| id     | name           | external_gateway_info                   |
+--------+----------------+-----------------------------------------+
| df..24 | GENERAL_ROUTER | {"network_id": "4e..3d", "enable_snat": ..|
+--------+----------------+-----------------------------------------+
```

REMOVE FLOATING ADDRESSES Floating addresses are external IP addresses that have a one-to-one relationship with internal IP addresses assigned to instances. Floating IP addresses must be removed from instances and be deallocated before removing the router gateway. Floating addresses are directly associated with the external network, so any attempt to remove an external network with these associations still in place will result in failure.

Let's assume you want to change the current gateway from `PUBLIC_NETWORK` to `new_public`. You must first remove the old gateway, and then add the new one. The following listing shows how to clear the existing gateway.

Listing 3.19 Clearing a router gateway

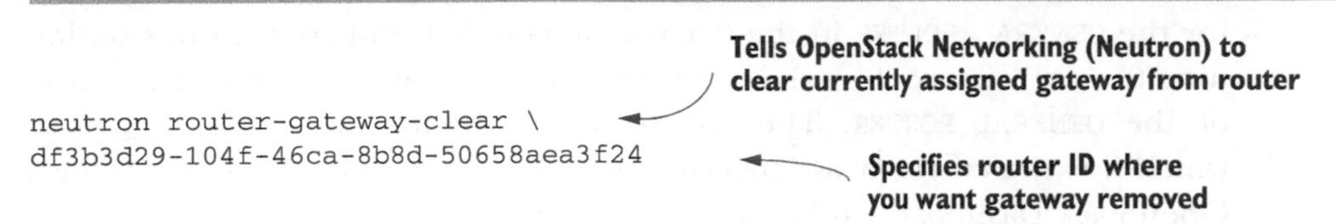
Tells OpenStack Networking (Neutron) to clear currently assigned gateway from router
Specifies router ID where you want gateway removed

```
neutron router-gateway-clear \
df3b3d29-104f-46ca-8b8d-50658aea3f24
```

When the gateway is removed, you'll get the following confirmation:

```
Removed gateway from router df3b3d29-104f-46ca-8b8d-50658aea3f24
```

Once the existing gateway has been removed, your tenant will be configured in the state shown in figure 3.15, where no external networks are connected to the tenant.

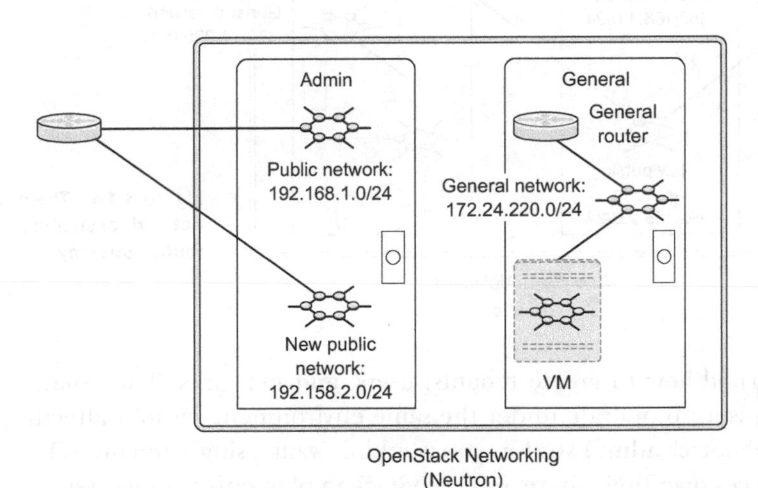
Admin
Public network:
192.168.1.0/24
New public
network:
192.158.2.0/24
General
General
router
General network:
172.24.220.0/24
VM
OpenStack Networking
(Neutron)

Figure 3.15 Removed router gateway

Your tenant network configuration is now back to where it was before you added the existing external network as a gateway in the previous section. You can now add the `new_public` external network as a gateway instead of `PUBLIC_NETWORK`. The following listing shows the commands required to complete the network configuration.

Listing 3.20 Adding a new external network as the router gateway

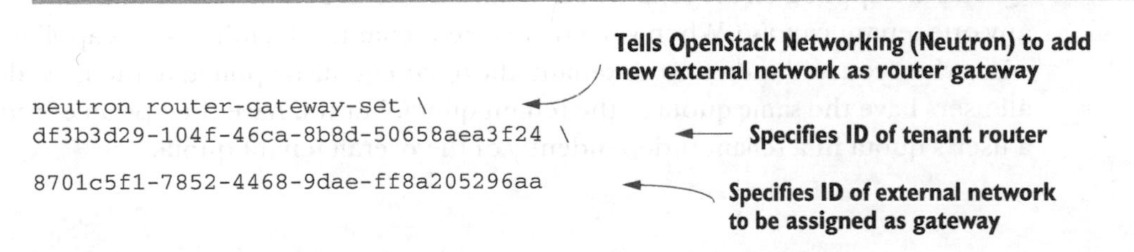
Tells OpenStack Networking (Neutron) to add new external network as router gateway
Specifies ID of tenant router
Specifies ID of external network to be assigned as gateway

```
neutron router-gateway-set \
df3b3d29-104f-46ca-8b8d-50658aea3f24 \
8701c5f1-7852-4468-9dae-ff8a205296aa
```

When the gateway is set, you'll see output like the following:

```
Set gateway for router
df3b3d29-104f-46ca-8b8d-50658aea3f24
```

Figure 3.16 illustrates the assignment of the new network, `new_public`, as the gateway for the `GENERAL_ROUTER` in the `General` tenant. You can confirm this setting by running the `neutron router-show <router-id>` command, where `<router-id>` is the ID of the `GENERAL_ROUTER`. The command will return the `external_gateway_info`, which lists the currently assigned gateway network. Optionally, you can log in to the OpenStack Dashboard and look at your tenant network. The `PUBLIC_NETWORK` will no longer be there, and it will be replaced by the `new_public` network.

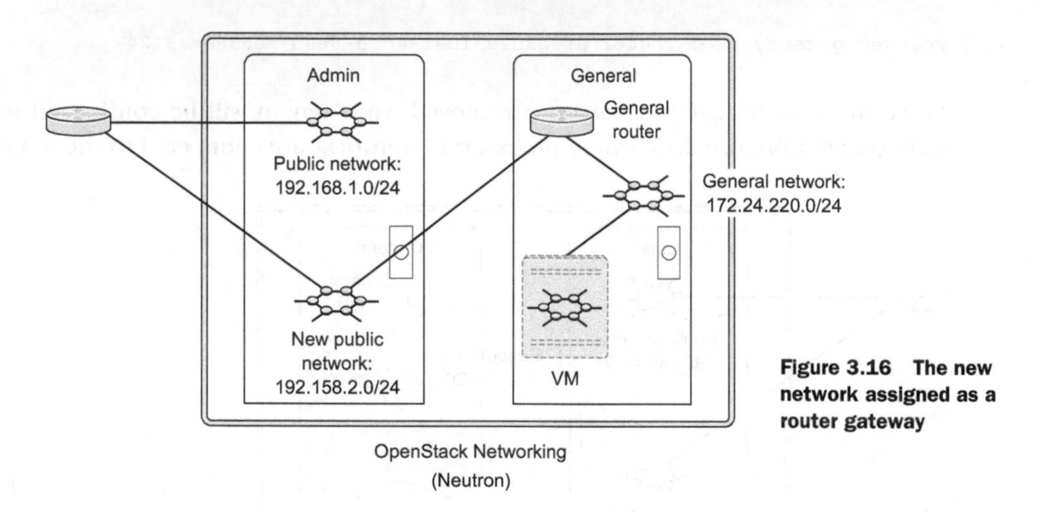

Figure 3.16 The new network assigned as a router gateway

You've now learned how to create tenants, users, and networks. The tenant model allows multiple users to operate under the same environment without affecting each other. As an OpenStack administrator, you wouldn't want a single tenant to be able to use more resources than their share, so OpenStack implements a quota system for several of the major components, including Compute, Block Storage (Cinder), Object Storage (Swift), and Networking services. In this book, we'll work with Block Storage, not Object Storage, so all references to *Storage* identify Cinder, not Swift.

We'll look at quotas next.

3.4 Quotas

Quotas are applied on the tenant and tenant-user level to limit the amount of resources any one tenant can use. When you create a new tenant, a default quota is applied. Likewise, when you add users to your tenant, the tenant quota is applied to them. By default, all users have the same quota as the tenant quota, but you have the option of reducing a user's quota in a tenant independently of the overall tenant quota.

Consider the case where you have an application administrator and database administrator sharing the same tenant for a project. You might want to assign half of the tenant resource to each user. On the other hand, if you increase a user's quota in a tenant in excess of the tenant quota, the tenant quota will increase to match the new user value.

Quota management is an important operational component for an OpenStack administrator to understand. In the rest of this subsection, you'll work through some CLI exercises to display and update quotas for tenants and tenant users via the Compute component. As with the majority of tenant configurations, you can also make these changes through the Dashboard or directly with an API.

3.4.1 Tenant quotas

In order to modify quota settings, you'll need to know the tenant ID you want to work with and a user ID that's currently in that tenant. The following example shows how you can list all tenants on the system and find the tenant ID:

```
devstack@devstack:~/devstack$ keystone tenant-list
+----------------------------------+--------------------+---------+
|                id                |        name        | enabled |
+----------------------------------+--------------------+---------+
| 9932bc0607014caeab4c3d2b94d5a40c |      General       |  True   |
| b1c52f4025d244f883dd47f61791d5cf |       admin        |  True   |
| 166c9cab0722409d8dbc2085acea70d4 |      alt_demo      |  True   |
| 324d7477c2514b60ac0ae417ce3cefc0 |        demo        |  True   |
| fafc5f46aaca4018acf8d05370f2af57 | invisible_to_admin |  True   |
| 81548fee3bb84e7db93ad4c917291473 |      service       |  True   |
+----------------------------------+--------------------+---------+
```

The following listing shows the command you can use to show the quota settings for a particular tenant.

Listing 3.21 Showing the Compute quota for a tenant

```
nova quota-show  \
 --tenant 9932bc0607014caeab4c3d2b94d5a40c
```

- `nova quota-show` — Tells OpenStack Compute (Nova) to show quota information
- `--tenant` — Specifies tenant ID for quota query

The quota information will be displayed as follows. (Consult the OpenStack operations manual for current quota items and unit values: http://docs.openstack.org/openstack-ops/content/projects_users.html.)

```
+-----------------------------+-------+
| Quota                       | Limit |
+-----------------------------+-------+
| instances                   | 10    |
| cores                       | 10    |
| ram                         | 51200 |
| floating_ips                | 10    |
```

```
| fixed_ips                    | -1    |
| metadata_items               | 128   |
| injected_files               | 5     |
| injected_file_content_bytes  | 10240 |
| injected_file_path_bytes     | 255   |
| key_pairs                    | 100   |
| security_groups              | 10    |
| security_group_rules         | 20    |
+------------------------------+-------+
```

You now know the quota for a particular tenant. To retrieve the default quota assigned to new tenants, simply omit the tenant ID from the command (`nova quota-show`).

Now suppose you're an OpenStack administrator and you've created a tenant for a department in your company. This department has just been tasked with deploying a new application, which will require resources that exceed their existing tenant quota. In this case, you'd want to increase the quota for the entire tenant. The command for doing this is shown in the following listing.

Listing 3.22 Updating the Compute quota for tenant

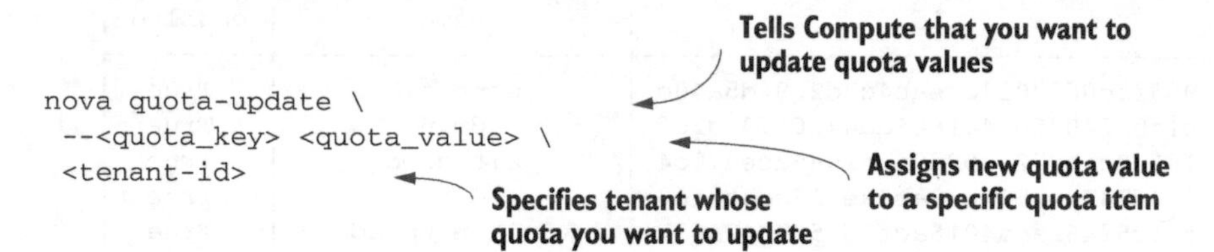

A list of quota keys can be obtained by displaying the quota values, as shown previously in listing 3.21. In the following example, the `cores<quota_key>` is updated:

```
devstack@devstack:~/devstack$ nova quota-update \
 --cores 20 \
 9932bc0607014caeab4c3d2b94d5a40c
```

Sets <quota-key> to cores and <quota-value> to 20

Specifies ID for General tenant

You've now successfully updated your tenant quota, and your users can now start assigning additional resources. If you rerun the command shown in listing 3.21, you'll see that the quota has been updated.

Next we'll look at how you can work with quotas on the tenant-user level. As you'll soon see, the ability to apply quotas on the user level for a specific tenant can be very useful in managing resource utilization.

3.4.2 Tenant-user quotas

In some cases, there might be only a single user in a tenant. In these cases, you'd only need quota management on the tenant level. But what if there are multiple users in one tenant? OpenStack provides the ability to manage quotas for individual users on a tenant level. This means that an individual user can have separate quotas for each tenant of which they are a member.

Suppose that one of your users with a role on a specific tenant is only responsible for a single instance. Despite being responsible for only one instance, this user has on several occasions added additional instances to this tenant. The additional instances count against the overall tenant quota, so although the user in question *should* only have one instance, they might actually have several. To prevent this from happening, you can adjust the user's quota for the tenant, and not the entire tenant quota. The following listing displays the existing quota for a particular user.

Listing 3.23 Showing the Compute quota for a tenant user

```
nova quota-show \
 --user <user_id> \          <- Specifies user ID in a tenant for query
 --tenant <tenant_id>        <- Specifies tenant ID for query
```

The following example shows the user ID related to `johndoe`, which you created in the subsection "Creating a user," and the tenant ID related to the `General` tenant, which you created in the subsection "Creating a tenant." Your actual IDs will differ from the examples listed here.

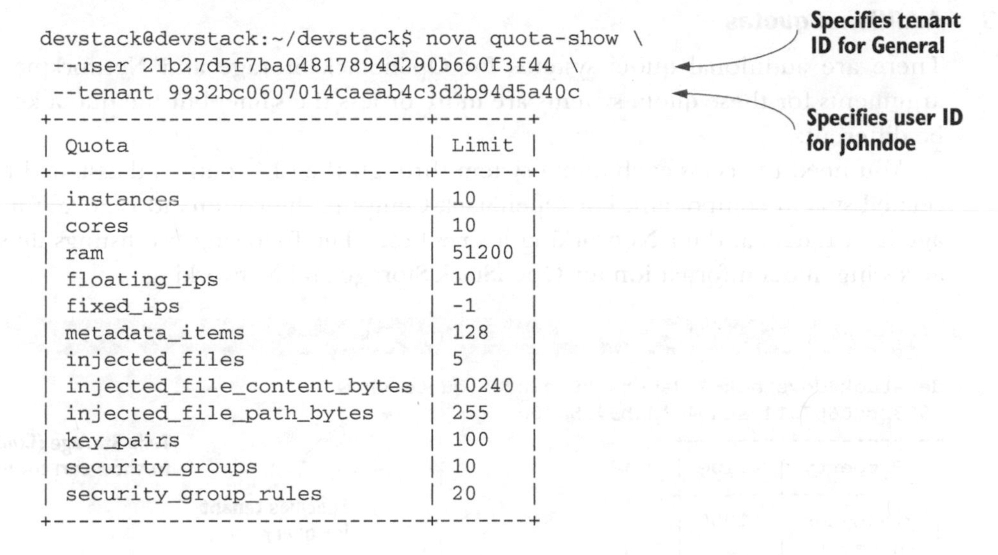


```
devstack@devstack:~/devstack$ nova quota-show \
 --user 21b27d5f7ba04817894d290b660f3f44 \
 --tenant 9932bc0607014caeab4c3d2b94d5a40c
+-----------------------------+-------+
| Quota                       | Limit |
+-----------------------------+-------+
| instances                   | 10    |
| cores                       | 10    |
| ram                         | 51200 |
| floating_ips                | 10    |
| fixed_ips                   | -1    |
| metadata_items              | 128   |
| injected_files              | 5     |
| injected_file_content_bytes | 10240 |
| injected_file_path_bytes    | 255   |
| key_pairs                   | 100   |
| security_groups             | 10    |
| security_group_rules        | 20    |
+-----------------------------+-------+
```

As you can see, the user quotas are the same size as the original tenant quota. By default, users added to a tenant can use all resources assigned to that tenant. For this tenant, you updated the *cores* value in a previous example, but that only updated the tenant quota, which, as you can see, doesn't automatically increase a user's tenant quota.

Assume that user `johndoe` is a problem user that you want to restrict to running a single instance in the `General` tenant. The next listing shows the command you can use to do this.

Listing 3.24 Updating the Compute quota for a tenant user

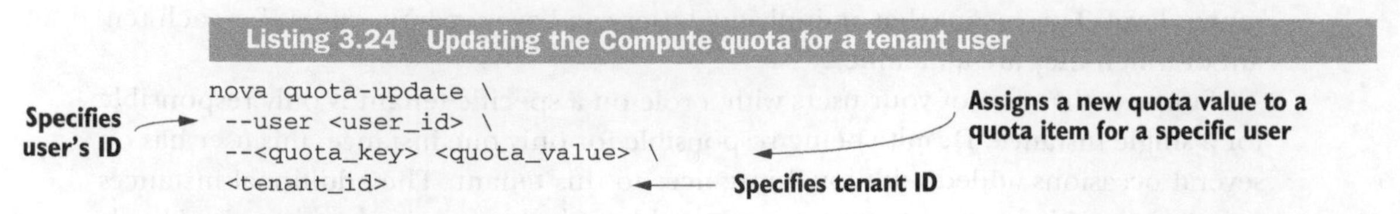


```
nova quota-update \
 --user <user_id> \
 --<quota_key> <quota_value> \
 <tenant_id>
```

In the following example, the user *johndoe* is configured an instance-quota-metric limitation of 1 instance (instance - quota = 1) in the `General` tenant:

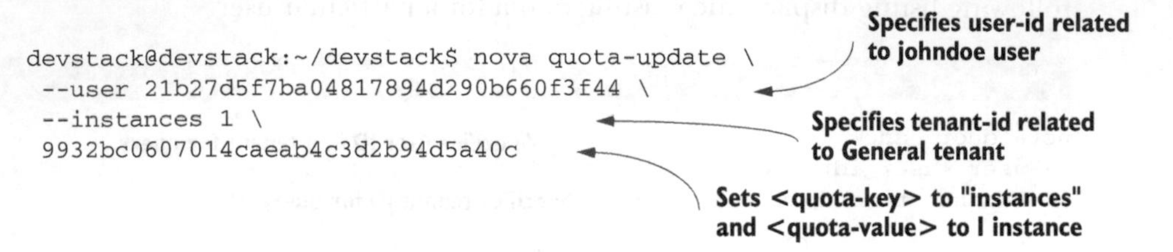


```
devstack@devstack:~/devstack$ nova quota-update \
 --user 21b27d5f7ba04817894d290b660f3f44 \
 --instances 1 \
 9932bc0607014caeab4c3d2b94d5a40c
```

You have now restricted the user `johndoe` to running a single instance. You might further want to restrict the resources this user can utilize for their individual instance, such as limiting the number of cores to 4.

3.4.3 Additional quotas

There are additional quota systems for OpenStack Storage and Networking. The arguments for these quota systems are more or less the same, but the quota keys will be different.

You need to access each quota system through the CLI command assigned to the related system component. For OpenStack Compute the command is `nova`, for Storage it's `cinder`, and for Networking it's `neutron`. The following two listings illustrate accessing quota information for OpenStack Storage and Networking.

Listing 3.25 Showing Storage tenant quota

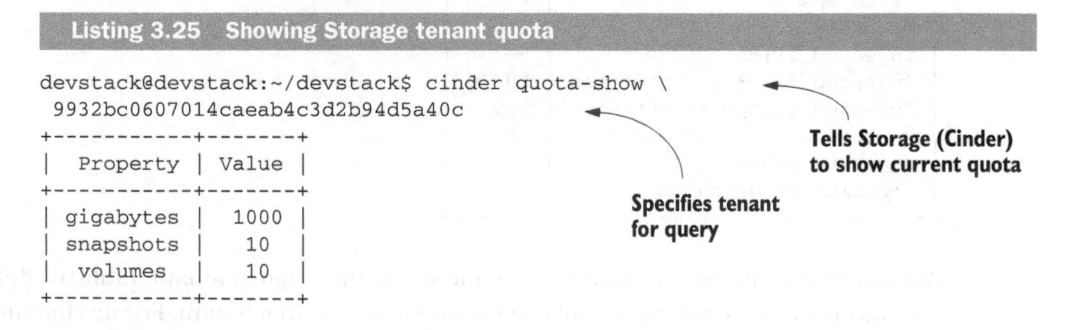


```
devstack@devstack:~/devstack$ cinder quota-show \
 9932bc0607014caeab4c3d2b94d5a40c
+-----------+-------+
|  Property | Value |
+-----------+-------+
| gigabytes |  1000 |
| snapshots |   10  |
|  volumes  |   10  |
+-----------+-------+
```

The following example demonstrates how to display the current Neutron quota for a specific tenant.

Listing 3.26 Showing Networking tenant quota

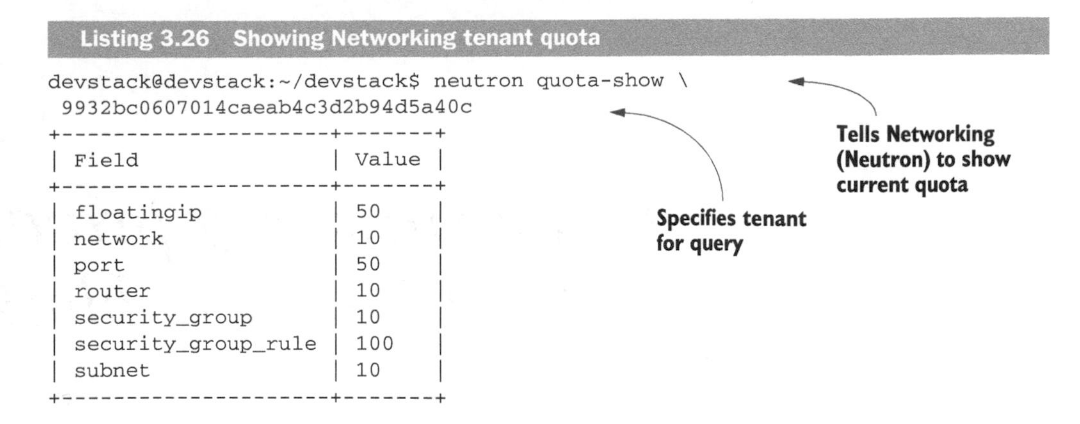
Tells Networking (Neutron) to show current quota
Specifies tenant for query

```
devstack@devstack:~/devstack$ neutron quota-show \
 9932bc0607014caeab4c3d2b94d5a40c
+---------------------+-------+
| Field               | Value |
+---------------------+-------+
| floatingip          | 50    |
| network             | 10    |
| port                | 50    |
| router              | 10    |
| security_group      | 10    |
| security_group_rule | 100   |
| subnet              | 10    |
+---------------------+-------+
```

As you can see, the CLI commands for Storage and Networking quotas are very similar to the ones you used with Compute. You can also use the Dashboard for quota configuration.

3.5 Summary

- The Dashboard is intended for end users.
- The CLI and APIs are intended for administration, scripting, and repetitive tasks.
- Anything you can do with the Dashboard, you can do with the CLI or an API.
- The CLI can be configured to output the API-level calls that are used for a specific command.
- Resources managed by OpenStack are reserved and provisioned based on tenants (projects).
- The terms *tenant* and *project* are used interchangeably in OpenStack, but projects related to resources, users, and rights shouldn't be confused with OpenStack projects like Compute, Network, and so on.
- Roles determine the rights of a user in a specific tenant.
- Users are assigned a home tenant, but they might hold many roles in other tenants.
- Tenant networks and subnets are generally isolated private networks for a specific tenant.
- Public networks and subnets are generally shared between tenants and are used for external (public) network access.
- Layer 3 services (DHCP, metadata services, and the like) can be provisioned on networks.
- Virtual routers are used to route network traffic from tenant (private) networks to public networks.
- Quotas are assigned to both tenants and to specific users in a tenant.

4 Understanding private cloud building blocks

This chapter covers

- Understanding OpenStack core project interoperability
- Exploring the relationship between vendor hardware and OpenStack
- Learning from a manual OpenStack install

In chapter 1 you were introduced to OpenStack. You learned how OpenStack fits into the cloud ecosystem, reasons for adopting the technology, and the focus of this book. In chapter 2 you went from those high-level concepts directly into a hands-on test-drive of the OpenStack framework using DevStack. In chapter 3 you worked through some examples that you might encounter working as an OpenStack operator and gained further insight into the structure of the framework.

In this chapter, we'll shift back to the high-level concepts. If the first chapter was to introduce and inform you, the second to get you excited about the technology, and the third to make you operationally comfortable, the fourth gives you a foundational understanding of what's really going on inside the OpenStack framework.

This chapter won't be as thought-provoking as chapter 1 or as fun as chapters 2 and 3. But regardless of whether you're a system administrator, developer, IT architect, or even a CTO, this is the *most important* chapter for your understanding of the OpenStack framework. If you'll be dealing with OpenStack in the trenches, this chapter will build your OpenStack foundation, which will be deepened in chapters 5 through 8. If you'll be working with OpenStack on a high level, even if you're simply responsible for a vendor-managed solution, this chapter will help you understand the collection of interacting components that make up an OpenStack deployment.

What are you waiting for? Let's get started!

4.1 How are OpenStack components related?

Since the first public release of OpenStack in 2010, the framework has grown from a few core components to nearly ten. There are now hundreds of OpenStack-related projects, each with various levels of interoperability. These projects range from OpenStack library dependencies to projects where the OpenStack framework is the dependency.

In an effort to provide structure around the diverse set of projects, the OpenStack Foundation created several project designations, including core, incubated, library, gating, supporting, and related. These project designations and their descriptions can be found in table 4.1.

Table 4.1 Project designations

Project designation	Description
Core	Official OpenStack projects (most people use these)
Incubated	Core projects in development (on track to become core)
Library	Dependencies of core projects
Gating	Integration test suites and deployment tools
Supporting	Documentation and development infrastructure
Related	Unofficial projects (self-associated projects)

Incubated projects, once fully developed and accepted, will eventually function in the same way core projects do. Library functions will be abstracted (not observable) by core project interaction. Gating and supporting projects aren't used to provide resources in a deployed system, so you don't need to worry about those. That leaves the related projects, which as the name implies, have some affiliation with OpenStack, even if the affiliation is self-nominated.

4.1.1 Understanding component communication

Often when someone refers to "OpenStack," they're referring to projects with a "core" designation. Core projects can use the OpenStack trademark and must pass all "must-pass" tests, as defined by the OpenStack Foundation. Simply put, core components are those that almost everyone will use in an OpenStack deployment. Projects like Compute, Networking, Storage, shared services, and Dashboard are examples of projects with a core designation, as shown in table 4.2.

Table 4.2 Core projects

Project	Codename	Description
Compute	Nova	Manages virtual machine (VM) resources, including CPU, memory, disk, and network interfaces
Networking	Neutron	Provides resources used by VM network interfaces, including IP addressing, routing, and software-defined networking (SDN)
Object Storage	Swift	Provides object-level storage accessible via RESTful APIs
Block Storage	Cinder	Provides block-level (traditional disk) storage to VMs
Identity Service (shared service)	Keystone	Manages role-based access control (RBAC) for OpenStack components; provides authorization services
Image Service (shared service)	Glance	Manages VM disk images; provides image delivery to VMs and snapshot (backup) services
Telemetry Service (shared service)	Ceilometer	Centralized collection for metering and monitoring OpenStack components
Orchestration Service (shared service)	Heat	Template-based cloud application orchestration for OpenStack environments
Database Service (shared service)	Trove	Provides users with relational and non-relational database services
Dashboard	Horizon	Provides a web-based GUI for working with OpenStack

In addition to various project designations, there are also several topologies in which you can deploy project components. If more of a specific resource (Storage, Compute, Networking, and so on) is required, more component-specific servers can be added. We'll discuss the project designations and their related components in section 4.1.2.

DASHBOARD AUTHENTICATION COMMUNICATION

Let's jump right in and take a look at how core components communicate. We'll walk through the process of accessing the OpenStack Dashboard, reviewing the VM creation options, and creating a virtual machine.

You must first provide the Dashboard with your login credentials and obtain an authentication token. The authentication token is saved as a cookie in your web browser and used with subsequent requests. As shown in figure 4.1, you obtain an authentication token from the Identity Service. While you can use the Dashboard (instead of the CLI or APIs) to navigate through the rest of this example, we won't show the interaction with the Dashboard. Even during the login process, the Dashboard just displays interactions between the web browser and the OpenStack APIs. We're primarily concerned with component interaction on the API level.

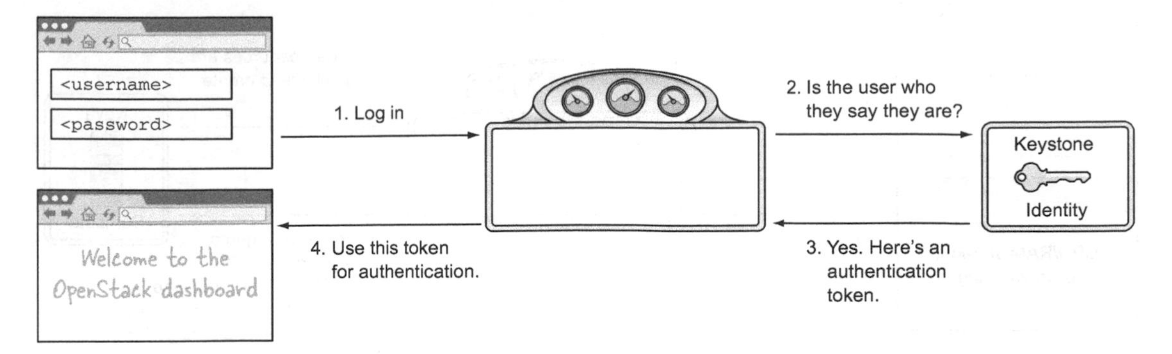

Figure 4.1 Dashboard login

Once you have your authentication token, you can take the second step and access the Compute component to create your virtual machine (VM).

Resource query and request communication

As explained in chapter 3, OpenStack works on a tenant model. If the OpenStack deployment is a hotel of resources, you can think of tenants as rooms in the hotel. Each tenant (room) is assigned a resource quota (a number of towels, beds, and so on). OpenStack users (guests) are assigned to tenants (rooms) through the use of roles. The identity information is kept by the Identity component, and the quota information is maintained by the Compute component.

In the Dashboard, when you click Launch Instance, the Compute component is queried to determine what resources and configurations are available in your current tenant. Based on the available options, you describe the VM you want and submit the configuration for creation.

The communication between components during a VM creation request is shown in figure 4.2. Because the creation of a VM isn't instantaneous, the process is executed asynchronously, so after you submit a VM for provisioning, you're returned to the Dashboard. In the Dashboard, your browser will periodically update the VM status information.

Resource provisioning communication

When VM creation requests are submitted, the Compute service component will interact with other components to provision the requested VM. The first thing that happens is that the VM object record is registered with the Compute service component. This object record contains information about the VM status and configuration—the VM object isn't the VM instance, only a record describing the instance.

When components communicate in the VM creation process, they reference common objects, like this VM object. For instance, the Compute service component might request a storage assignment from the Storage service component. The Storage service component would then provision the requested storage and provide a reference to a `Storage` object, which would then be referenced in the VM object record.

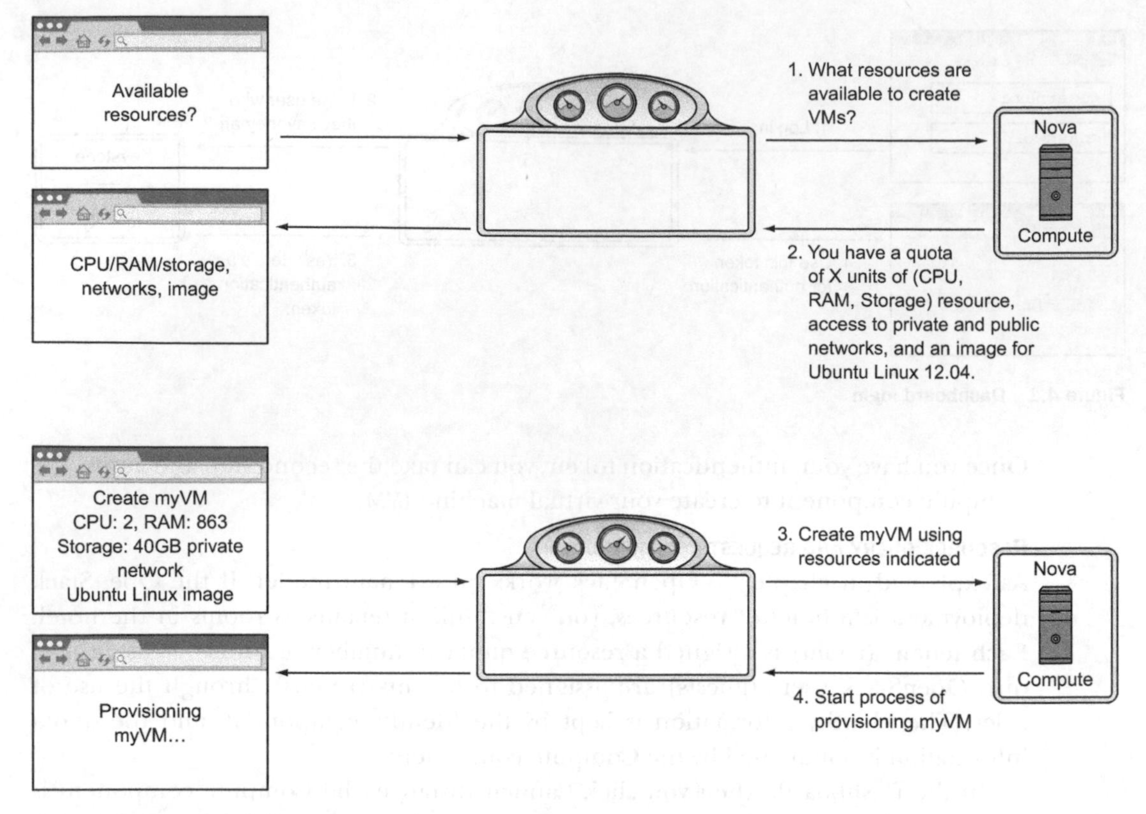

Figure 4.2 Resource query and request

As shown in figure 4.3, the Compute service component communicates with other core components to provision and assign resources to the VM object. Compute will first request infrastructure components like Storage and Networking. When the virtual infrastructure has been assigned to the VM and referenced in the VM object, the Image Service will prepare the virtual storage volume with the requested image or snapshot. At this point the VM creation process is complete and the Compute component can spawn the VM.

As demonstrated in the previous figures, core components work in concert to provide OpenStack services. OpenStack interactions, even those in the Dashboard, eventually find their way to the OpenStack APIs.

As you'll see next, related projects often use API calls exclusively to interact with OpenStack.

RELATED PROJECT COMMUNICATION

Let's take a look at how Ubuntu Juju, a related project, interacts with OpenStack. Juju is a cloud automation package that uses OpenStack for virtual infrastructure. Juju automates the deployment and configuration of applications on virtual infrastructure using application-specific charms.

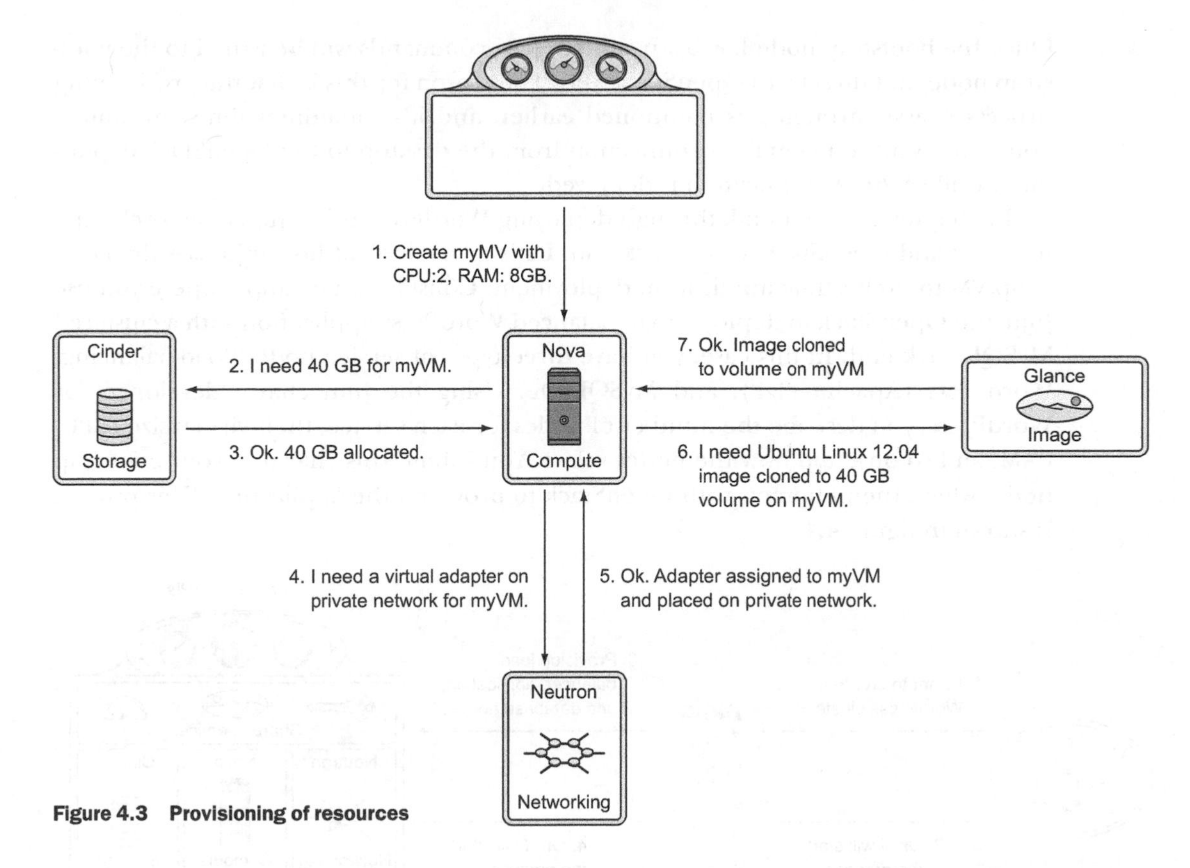

Figure 4.3 Provisioning of resources

For the lack of a better description, Juju *charms* are collections of installation scripts that define how services and applications integrate into virtual infrastructure. Because infrastructure, including networks and storage, can be provisioned programmatically using OpenStack, Juju can deploy entire application suites from a charm. Simply put, Juju turns newly provisioned VM instances into applications. We discuss this process in detail in later chapters, but essentially you tell an application charm how large you want your instances to be and how many instances you want, and it does the work to deploy your applications.

The first step in using Juju in your OpenStack deployment is to deploy what Juju calls its *bootstrap*, using the Juju CLI. The bootstrap is a VM that Juju uses to control its automation processes. The deployment of the bootstrap, from a component perspective, is similar to what you've seen in recent figures (see figures 4.1, 4.2, and 4.3). The difference here is that in place of the web browser making the request, it's the Juju application.

JUJU NODES FROM THE OPENSTACK PERSPECTIVE Juju nodes run the Ubuntu Linux operating system and include Juju-specific automation tools. From the OpenStack perspective, a Juju node is no different than any other VM provided by OpenStack. As a related project, Juju makes use of resources provided by OpenStack, but that's where the integration ends.

Once the bootstrap node has been started, Juju commands will be issued to the bootstrap node, not directly to OpenStack APIs. The reason for this is that the provisioning process is asynchronous, as mentioned earlier, and it's sometimes time-consuming. You don't want to maintain a connection from the desktop to the OpenStack deployment while a 20-VM application is deployed.

In chapter 12 you'll walk through deploying WordPress using Juju as an orchestration tool and OpenStack as the back end. Let's take a look at how Juju uses the bootstrap VM to orchestrate application deployment. Consider an example where you use Juju and OpenStack to deploy a load-balanced WordPress application with a clustered MySQL back end. In this case, you have three types of service nodes: load-balancing, WordPress (Apache/PHP), and MySQL DB. Using the Juju charm developed for WordPress, you describe the number of nodes for each service, their virtual size (CPU, RAM, and so on), and how the nodes relate. You submit this charm to your bootstrap node, which then interacts with OpenStack to provision the application. This process is shown in figure 4.4.

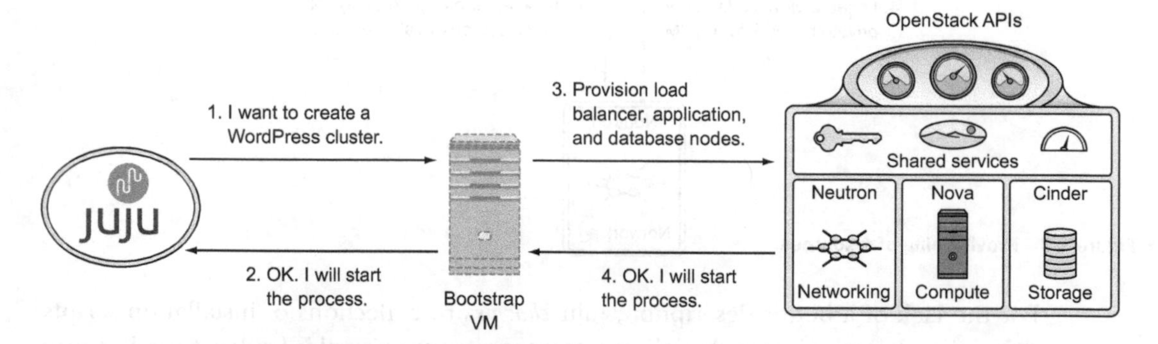

Figure 4.4 OpenStack interacting with a related project

Let's assume that OpenStack, through the direction of the bootstrap node, successfully provisions all the required virtual infrastructure. At this point you have a collection of VMs, but no applications. The bootstrap node polls OpenStack, watching for its requested VMs to come online. Once the VMs are online, it will start a process outside the OpenStack framework to complete the install. As shown in figure 4.5, the bootstrap node will communicate directly with the newly provisioned VMs. From this point forward, OpenStack simply provides the virtual infrastructure and is unaware of the application roles assigned to each VM.

We've now discussed how the components of OpenStack communicate on the logical level. In the figures, we've illustrated component communication, as if everything was communicating inside a single big node (physical instance). In practice, however, OpenStack components will be distributed across many physical commodity servers in a multi-node topology.

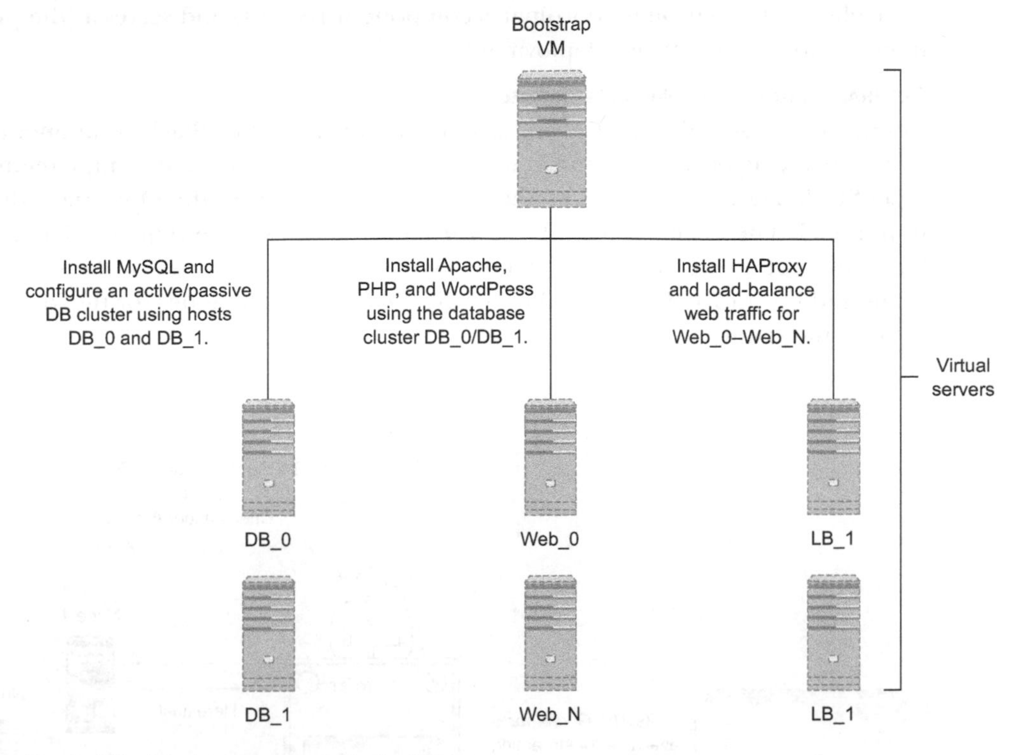

Figure 4.5 Juju bootstrap controls the VMs

4.1.2 Distributed computing model

Let's take a look at the OpenStack component distribution model. In distributed computing, there are several component distribution methods.

In a *mesh* distribution, control and data are distributed on the node level, and no central authority exists. This method is fully distributed, but maintaining concurrency across nodes is more difficult than in a central-control model. Mesh distributions are most often used when workloads are self-contained and require little coordination beyond collecting results.

On the other end of the spectrum, a *hub-and-spoke* distribution passes all control and data through a central node, like spokes around a hub. Hub-and-spoke topologies are generally limited in scale, due to the aggregation of both the control and data plane to a central node. Hub-and-spoke is most often used for workloads with a high degree of node-to-node communication and coordination.

The OpenStack distribution model shares characteristics of both mesh and hub-and-spoke distributions. Like mesh, once OpenStack provisions the virtual infrastructure, the infrastructure will continue to function without the involvement of a central controller. But like hub-and-spoke, component interaction is coordinated through a central API service. The node that maintains the API services is known as the OpenStack controller. The controller coordinates component requests and serves as the primary interface for an OpenStack deployment.

GENERAL DISTRIBUTED COMPONENT MODEL

Briefly, let's suspend our thinking around the idea of OpenStack components, and focus on the hybrid mesh and hub-and-spoke distribution model implemented by OpenStack. Figure 4.6 illustrates the interaction of nodes in the OpenStack distribution model. The client contacts the controller to make service requests. The controller, while not an operational dependency of the nodes, is aware of the system-wide status and inventory. The controller selects the appropriate nodes for the job and distributes the request.

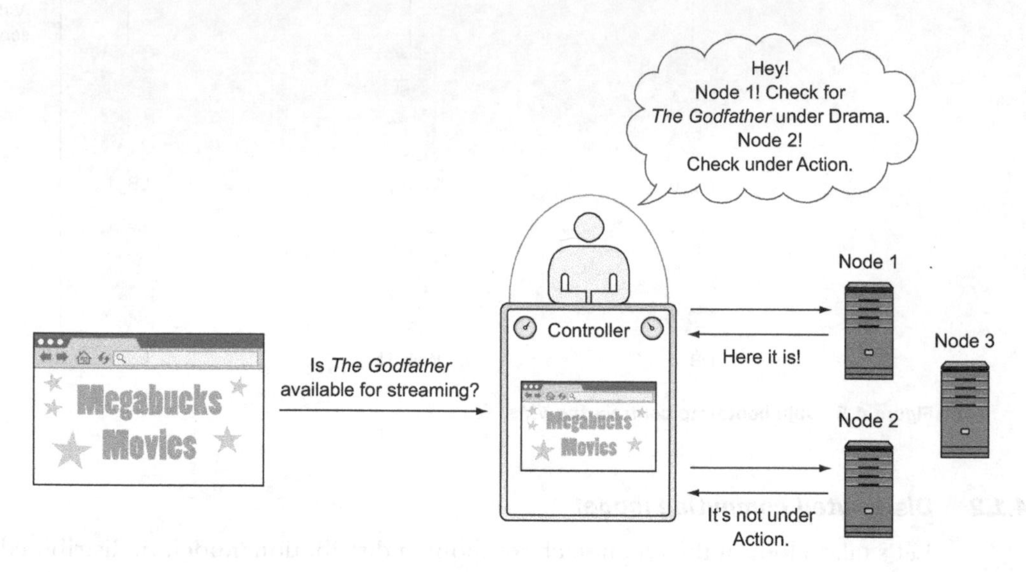

Figure 4.6 Distributed component model

OPENSTACK'S DISTRIBUTED COMPONENT MODEL

The general distributed component model presented in figure 4.6 is representative of the way OpenStack components communicate. Let's discuss one final abstract example of this model before we look at OpenStack specifics. Suppose a distributed component model, like the one shown in figure 4.6, was implemented in a content management system, like the ones used to stream movies on demand. Consider two movies streaming simultaneously to two users. The initial requests to stream a movie were made from the clients to a controller, and the controller directed two nodes to stream the two movies to the clients. Now, suppose that while the movies are streaming, the controller experiences a catastrophic failure. The movie streams wouldn't be interrupted, and neither the clients nor the nodes would be aware of this event. In this type of distributed model, new requests can't be fulfilled until a controller is available, but existing operations will continue.

Now let's think about how OpenStack components behave. This time we'll think about components in relation to the OpenStack distribution model. The control portion of the component will reside on the control node, and the provisioning components will be distributed on the resource nodes. Figure 4.7 introduces OpenStack components into the distributed model.

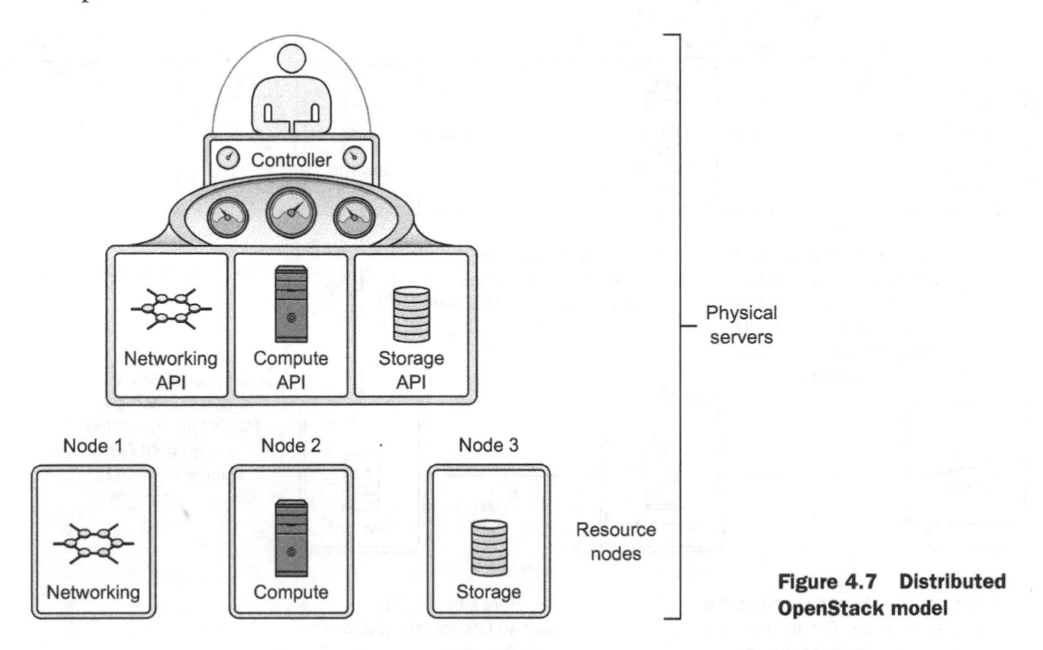

Figure 4.7 Distributed OpenStack model

DISTRIBUTED COMPONENT INTERACTION IN VM PROVISIONING

In the OpenStack distributed model, many resource nodes can exist for a single controller. OpenStack components are actually collections of services. As previously stated, some services run on controller nodes and some on resource nodes. Depending on the component, there might be several services that run on the controller and several more on resource nodes. For the Compute component alone, there are six controller services. In comparison, the Compute resource nodes generally run a single Compute component.

Let's take a look at what happens when a VM request is made. Figure 4.8 illustrates the node-level interaction of distributed OpenStack components required to create a VM. From a component perspective, nothing has changed from the previous figures. What we want to demonstrate is how OpenStack components communicate when components are distributed on multiple nodes.

VM-LEVEL COMPONENT COMMUNICATION

In a multi-node deployment, you'll have multiple nodes for each primary node type (compute, storage, network). The ratio of compute, network, and storage nodes will be dependent on your requirements for these resources. Specific node types might

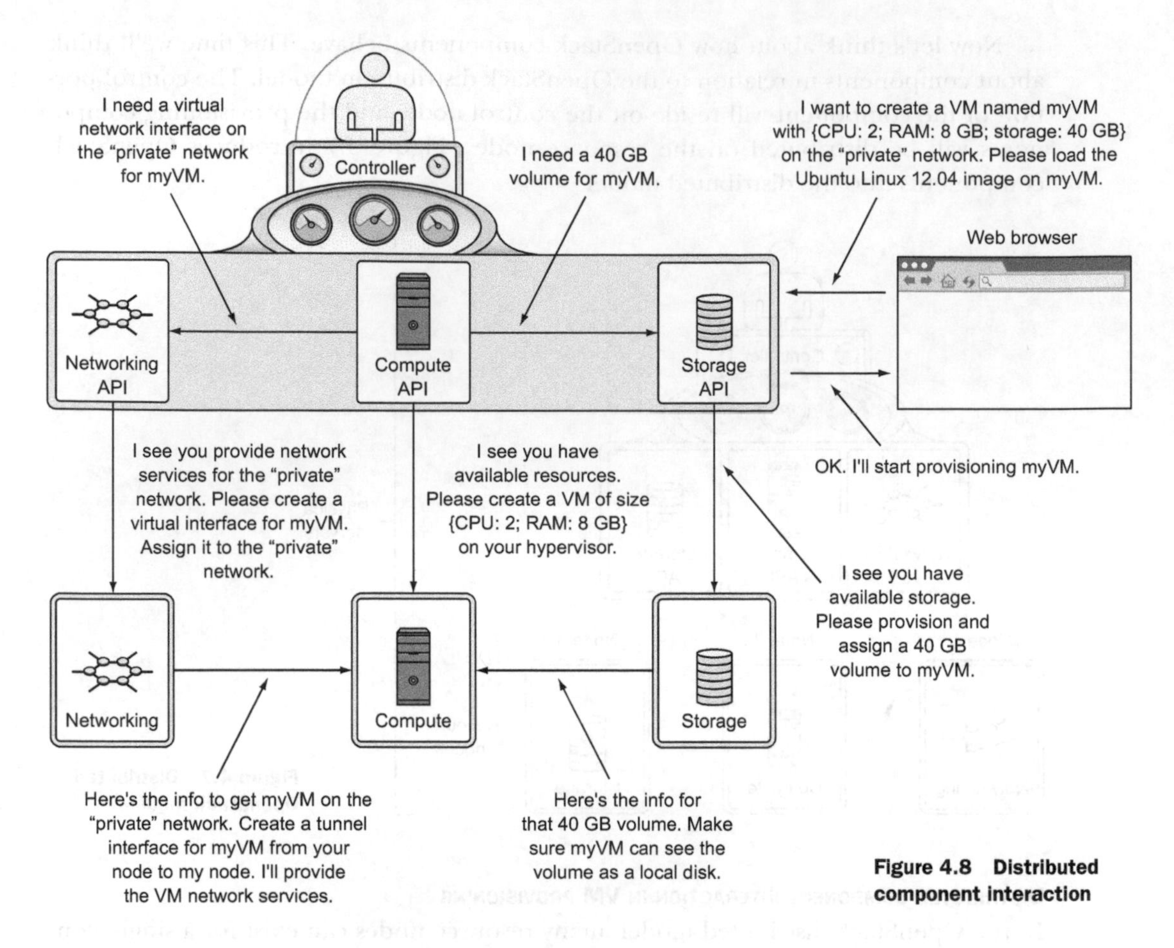

Figure 4.8 Distributed component interaction

additionally be connected to other vendor components, such as storage nodes to vendor storage systems and network nodes to vendor network devices. The way specific vendor resources are used by OpenStack is explained in section 4.2.

We've described OpenStack component relations from the component communication level and the distributed services level. Now we'll take a look at what's going on from the perspective of the VM.

Virtual machines, as the name implies, are virtualized representations of resources that would be available on a single physical machine. A VM runs an operating system (OS), just like a physical system, and any OS running on a general-purpose VM will expect virtual hardware to behave exactly like physical hardware resources. This is to say, the OS reads and writes to network and storage devices the same way it writes to CPU registers or RAM. When a physical machine runs a hypervisor, the hypervisor does the work of translating multiple virtual address spaces to a single physical address space. In a distributed OpenStack component, not only do you have virtual resources, but they're also distributed on separate physical nodes. You need to understand how the distribution of resources relates to what is seen by the VMs.

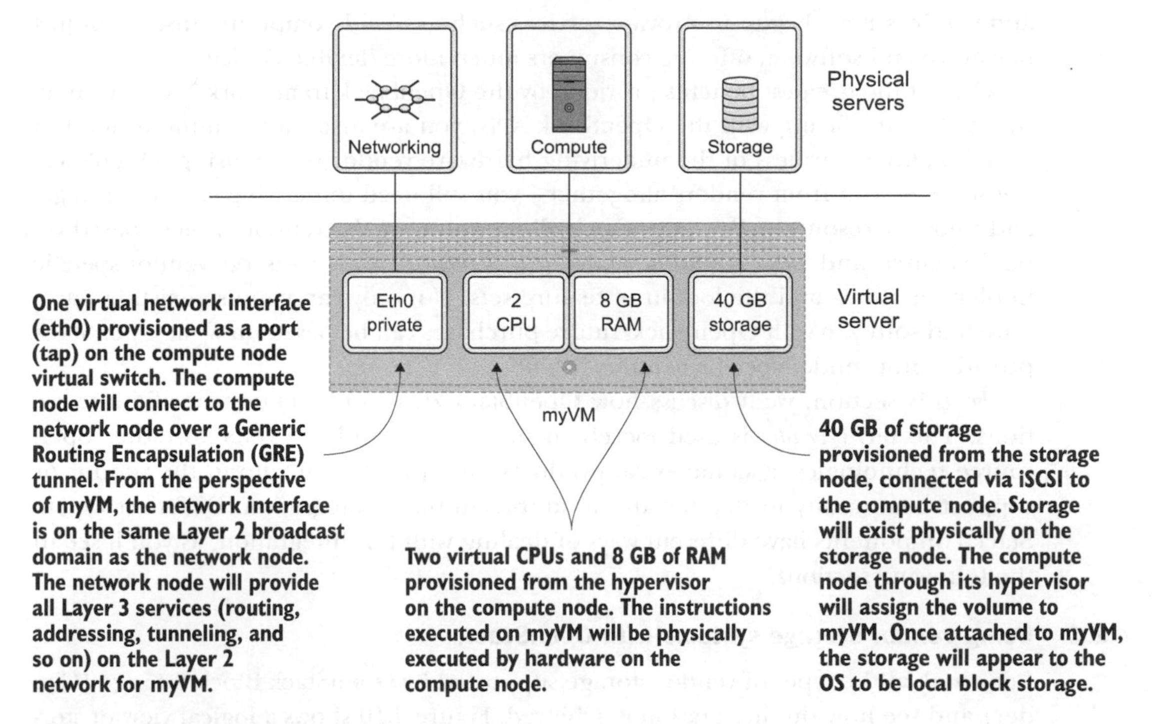

Networking
Compute
Storage
Physical servers
Eth0 private
2 CPU
8 GB RAM
40 GB storage
Virtual server
myVM
One virtual network interface (eth0) provisioned as a port (tap) on the compute node virtual switch. The compute node will connect to the network node over a Generic Routing Encapsulation (GRE) tunnel. From the perspective of myVM, the network interface is on the same Layer 2 broadcast domain as the network node. The network node will provide all Layer 3 services (routing, addressing, tunneling, and so on) on the Layer 2 network for myVM.
Two virtual CPUs and 8 GB of RAM provisioned from the hypervisor on the compute node. The instructions executed on myVM will be physically executed by hardware on the compute node.
40 GB of storage provisioned from the storage node, connected via iSCSI to the compute node. Storage will exist physically on the storage node. The compute node through its hypervisor will assign the volume to myVM. Once attached to myVM, the storage will appear to the OS to be local block storage.

Figure 4.9 Component-VM relations

Although VM resources are provided by multiple component-specific nodes, from the perspective of the VM all resources are provided by a single piece of hardware. Figure 4.9 illustrates how resources from component-specific resource nodes are combined to create a single VM.

You can think of the VM as *living* on a specific compute node, but the actual data will live on a storage node, and data communicated (Layer 3) by the VM lives (passes through) the network node.

> **DISTRIBUTED VIRTUAL ROUTING (DVR)** Until recent releases of OpenStack, L3 network functions like routing were typically performed by a small number of dedicated network nodes. The Neutron/DVR subproject has emerged to manage the distribution of routing across compute and dedicated network nodes.

The OpenStack distributed architecture and component design allows for very efficient deployment of virtual infrastructure. The OpenStack framework provides you with the ability to manage many nodes across component-node types from a single system.

4.2 How is OpenStack related to vendor technologies?

For many years, the vendors that provided compute, storage, and network hardware focused on marketing faster and more capable hardware. More recently, though, hardware has been viewed as a commodity, software has become more interoperable, and vendors have begun to provide services such as cloud computing instead of just hardware and software, offering consumers much more flexible choices.

One of the greatest benefits provided by the OpenStack framework is vendor neutrality. By interfacing with the OpenStack APIs, you are assured a minimum level of functionality regardless of the underlying hardware vendor you're using. OpenStack doesn't free you from vendors altogether—you still need underlying servers, storage, and network resources. But OpenStack allows you to make vendor choices based on performance and price without taking into account sunk costs on vendor-specific implementations and the lock-in of feature sets. Not only can you use existing hardware and software with OpenStack, future purchases can be based on what OpenStack provides, not vendor-specific features.

In this section, we'll discuss how OpenStack deals with vendor-specific integrations. The term *vendor* is used loosely in this context and can refer to either open source technologies or commercial products. In OpenStack it's up to the vendor or support community to develop the vendor-technology integration. Different OpenStack components have different ways of dealing with this integration, as you'll see in the following sections.

4.2.1 Using vendor storage systems with OpenStack

Let's look at the types of vendor storage supported by OpenStack Block Storage (Cinder) and see how this integration is achieved. Figure 4.10 shows a logical view of storage resource assignments and management.

This figure shows the CPU and RAM portions of a VM being provided by a commodity server. It also shows that the storage assigned to the VM is not on the commodity server; it's provided by a separate storage system. As you'll soon learn, there are many ways to provide that virtual block device to a VM.

STORAGE SYSTEM IN DEVSTACK In chapter 2 you walked through deploying DevStack, but you didn't do any specific configuration for storage. In that single-node DevStack deployment, the storage resources were consumed from the same computer as the compute resources. However, in multi-node production deployments, compute and storage resources are isolated on specific storage and compute nodes and/or appliances.

The use of storage vendors and technologies isn't limited to OpenStack Block Storage (Cinder) and could even be used with OpenStack Object Storage (Swift). We'll look at Cinder because the storage it manages is used as part of a VM, and this chapter focuses on the integration between OpenStack and vendor components. This is not to say that OpenStack Object Storage is any less complex, just that it's more self-contained and isn't used directly by a running VM (and thus is less relevant in this chapter).

HOW STORAGE IS USED BY VMS

In OpenStack and other environments that provide infrastructure as a service (IaaS), virtual block storage devices are provisioned and assigned to VMs. The operating

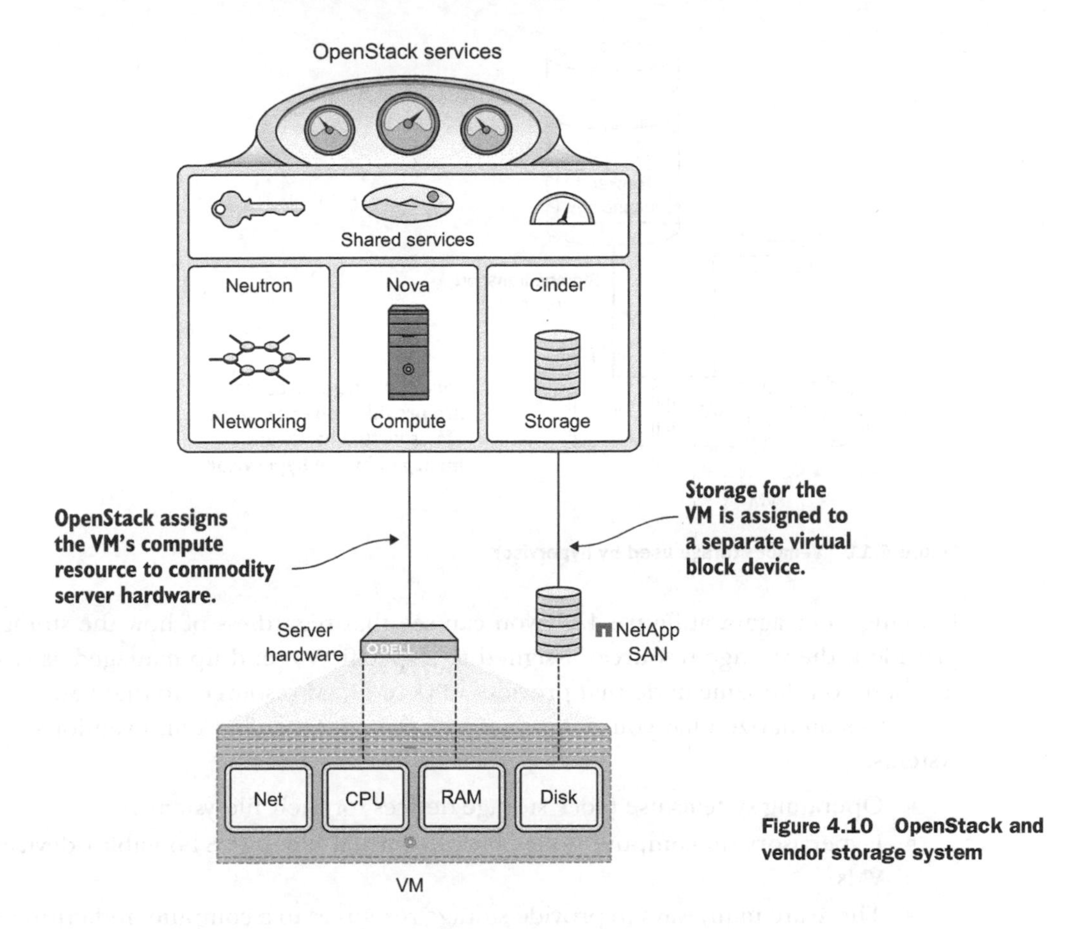

Figure 4.10 OpenStack and vendor storage system

systems running in the VMs manage the filesystems on their virtual block devices or volumes.

You might be wondering, "If the compute portion of a VM is provided by a server and the storage is provided by a separate server or storage appliance, how are they connected to provide a single VM?" The simple answer is that all resources eventually make their way to the VM as virtual hardware, which is then connected together on the hypervisor level. Take a look at figure 4.11, which shows a technical view of the logical view shown previously in figure 4.10.

In this figure, a vendor storage system is directly connected to a compute node (connected through Peripheral Component Interconnect Express (PCI-E), Ethernet, Fiber Channel (FC), Fiber Channel over Ethernet (FCoE), or vendor-specific communication link). The compute node and the storage system communicate using a *storage transport protocol* such as Internet Small Computer System Interface (iSCSI), Network File System (NFS), or a vendor-specific protocol. In short, storage can be provided to the compute node running the hypervisor using many different methods, and it's the compute node's job to present those resources to the virtual machine.

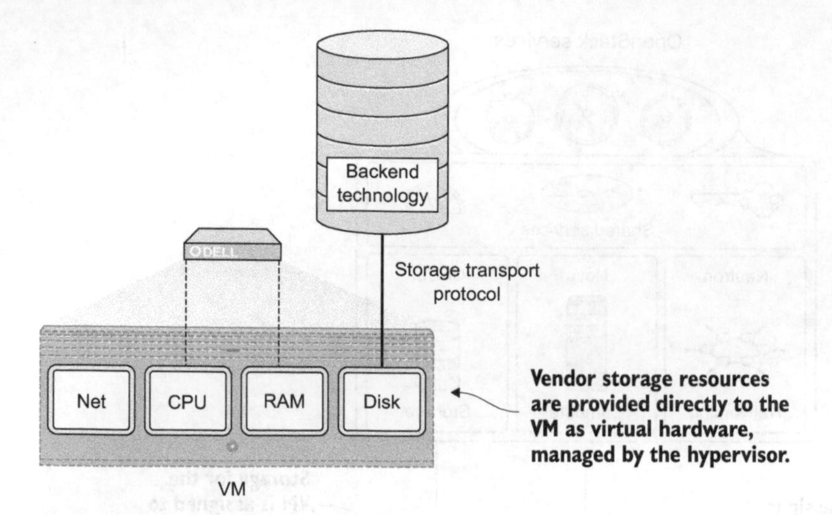

Figure 4.11 Vendor storage used by hypervisor

Looking once again at figure 4.11, you can see that regardless of how the storage is provided, the storage resources assigned to a specific VM end up managed as virtual hardware on the same node that provides CPU and RAM resources to that VM.

Let's summarize what you've learned so far about OpenStack and vendor storage systems:

- Operating systems use block storage devices for their filesystems.
- Hypervisors on compute nodes provide virtual block (OS-bootable) devices to VMs.
- There are many ways to provide storage resources to a compute node running a hypervisor.
- Vendor storage systems can be used to provide storage resources to compute nodes.
- OpenStack manages the relationship between the hypervisor, the compute node, and the storage system.

In the next section you'll learn just how OpenStack manages these resources.

HOW OPENSTACK SUPPORTS VENDOR STORAGE

You might be thinking, "OK, I understand how the storage is used, but how is it managed by OpenStack?" Cinder is a modular system allowing developers to create plug-ins (drivers) to support any storage technology and vendor. These modules might be developed by a product development team in a corporation or a community effort.

Figure 4.12 shows Cinder using a plug-in to manage a vendor storage system.

As you learned earlier in this chapter, individual OpenStack components have specific responsibilities. In the case of Cinder, its responsibility is to translate the request for storage from OpenStack Compute into an actionable request using the vendor-specific API of a storage system.

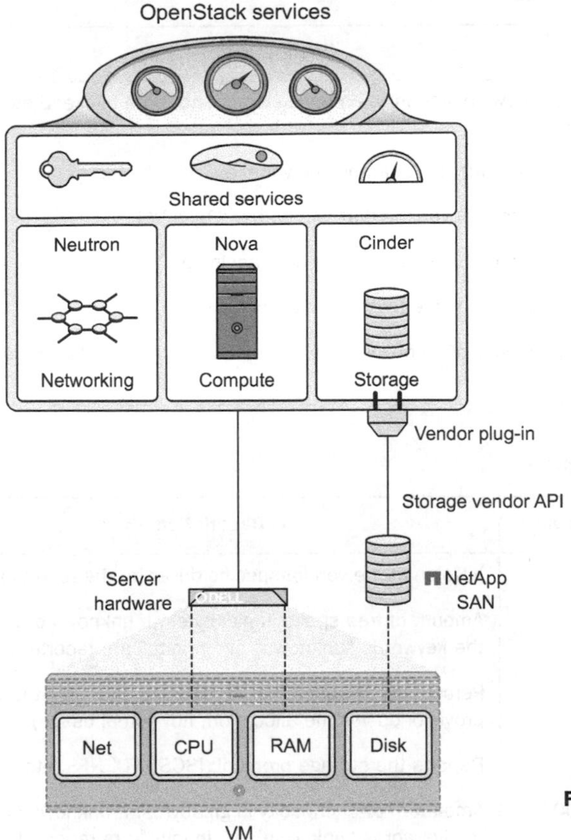

Figure 4.12 Cinder manages vendor storage.

Obviously, if you're going to translate one language or API to another, you need a minimum number of defined functions that can be related to each other. For each OpenStack release, there are a minimum number of required features and statistical reports for each plug-in. If plug-ins aren't maintained between releases, and additional functions and reports are required, they're deprecated in subsequent releases. The current lists of minimum features and reports (at the time of writing) are found in tables 4.3 and 4.4. The most current list of plug-in requirements can be found on the GitHub repository: http://docs.openstack.org/developer/cinder/devref/drivers.html. However, as of the time of this writing, the list of Cinder plug-in *minimum features* hasn't changed since the Icehouse release.

Table 4.3 Minimum features

Feature name	Description
Volume create/delete	Creates/deletes a volume for a VM on a backend storage system
Volume attach/detach	Attaches/detaches a volume to/from a VM on a backend storage system
Snapshot create/delete	Takes a running snapshot of a volume on a backend storage system

Table 4.3 Minimum features *(continued)*

Feature name	Description
Volume from snapshot	Creates a new volume from a previous snapshot on a backend storage system
Get volume stats	Reports the statistics on a specific volume
Image to volume	Copies image to a volume that can be used by a VM
Volume to image	Copies a volume used by a VM to a binary image
Clone volume	Clones one VM volume to another VM volume
Extend volume	Extends the size of a VM volume without destroying the data on the existing volume

Table 4.4 Minimum reporting statistics

Statistic name	Example	Description
driver_version	1.0a	Version of the vendor-specific driver for the reporting plug-in.
free_capacity_gb	1000	Amount of free space in gigabytes. If unknown or infinite, the keywords "unknown" or "infinite" are reported.
reserved_percentage	10	Percentage of space that is reserved but not yet used (thin provisioned volume allocation, not actual usage).
storage_protocol	iSCSI	Reports the storage protocol: iSCSI, FC, NFS, etc.
total_capacity_gb	102400	Amount of total capacity in gigabytes. If unknown or infinite, the keywords "unknown" or "infinite" are reported.
vendor_name	Dell	Name of the vendor that provides the backend storage system.
volume_backend_name	Equ_vol00	Name of the volume on the vendor backend. This is needed for statistical reporting and troubleshooting.

EXAMPLES OF VENDOR STORAGE IN OPENSTACK As previously stated, support for vendor storage is provided by plug-ins in Cinder. Plug-ins have already been developed by and for many vendors, including Coraid, Dell, EMC, GlusterFS, HDS, HP, Huawei, IBM, NetApp, Nexenta, Ceph, Scality, SolidFire, VMware, Microsoft, Zadara, and Oracle. In addition to commercial vendors, Cinder also supports storage provided by Linux Logical Volume Manager (LVM) and NFS mounts. An up-to-date Cinder support matrix can be found here: https://wiki.openstack.org/wiki/CinderSupportMatrix.

UNKNOWN OR INFINITE FREE SPACE In table 4.4, under *free_capacity_gb*, you'll notice that the values *unknown* and *infinite* can be used as free space values. Situations where these values are necessary might exist, but from a general operations perspective you should be aware that these are valid values for a storage driver.

4.2.2 Using vendor network systems with OpenStack

In OpenStack, it's common for compute resources to be provided by server hardware, storage resources by vendor storage systems, and networks by one or more vendors simultaneously. Obviously, if a VM is *running* on a specific server, that server is providing all of the computational resources (CPU, RAM, I/O, and so on) for that VM. Because a server can support more than one VM, this relationship is one-to-many from the perspective of the server and one-to-one from the perspective of the VM. That is to say that from a computation standpoint, the only resources consumed will come from the server hosting the VM.

As discussed in the previous section, although storage resources are technically removed from the compute node, from the perspective of the VM this is also a one-to-one relationship. In general, you'll have a single node running on a single volume that appears to the VM to be from a single container of virtual hardware.

Figure 4.13 shows the logical view of network resource assignments and management first introduced in chapter 1.

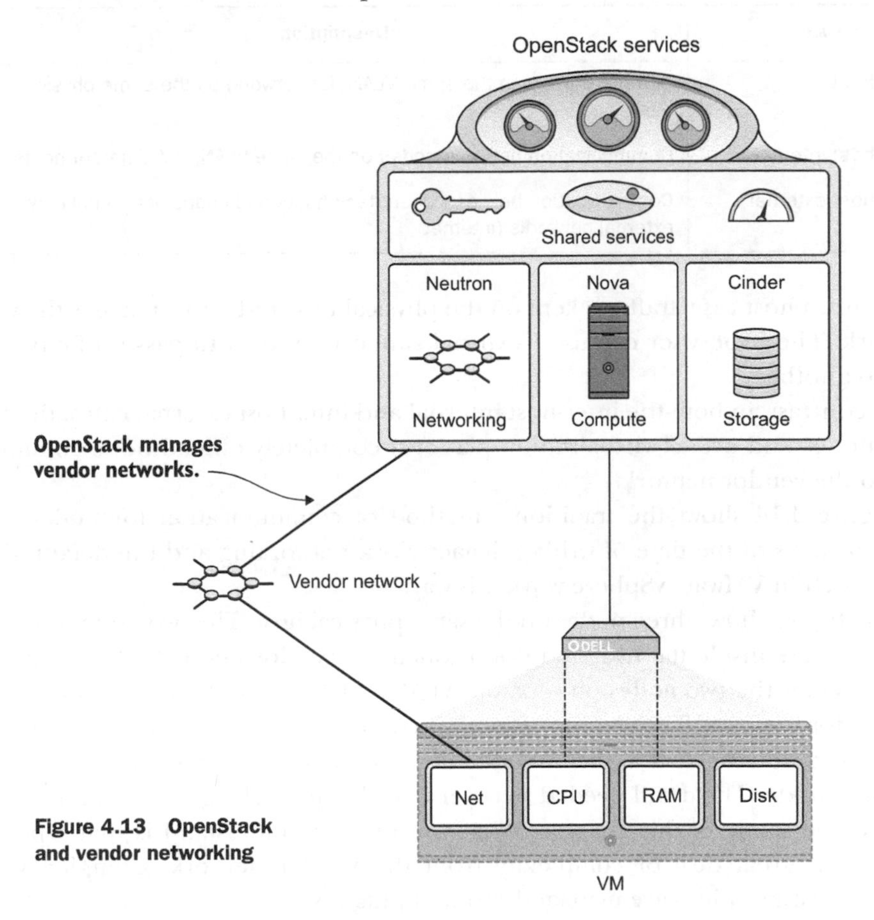

Figure 4.13 OpenStack and vendor networking

This figure represents a simplistic view of networking that suggests network resources are to be consumed in the same one-to-one way as compute and storage. Unfortunately, networking is not that simple. What the figure doesn't show are the layers of management that go into connecting two endpoints on a network. This section describes OpenStack Networking (Neutron) and how it manages vendor networks.

We'll look first at how VMs use networking.

How networking is used by VMs

Obviously, a network isn't very useful with a single VM, so you can expect that there will be, at a minimum, two VMs/nodes communicating. The way in which two nodes communicate depends on their relation to one another in the overall network. Table 4.5 summarizes several communication cases experienced in traditional virtual environments. These are described as traditional cases because software-defined networking (SDN), regardless of vendor, has blurred the lines of this paradigm.

Table 4.5 Node communication cases

Case	Description
Intra-host	Communication on the same VLAN (L2 network) on the same physical host
Inter-host-internal	Communication between nodes on the same VLAN, but different hosts
Inter-host-external	Communication between OpenStack hosts and endpoints on unknown external networks (internet)

In the intra-host case, traffic is kept on the physical host and never reaches the vendor network. The hypervisor can use its virtual switch (network) to pass traffic from one host to another.

In contrast, in both the inter-host-internal and inter-host-external cases, the hypervisor nodes and overall virtualization platform completely offload node communication to the vendor network.

Figure 4.14 shows the traditional method of communication for nodes on the same host. As of the time of writing, legacy Nova networking and the default distributed switch in VMware vSphere work this way.

The figure shows three nodes on the same physical host. The two nodes on VLAN_1 communicate inside the host and don't touch the vendor network. But communication between the two nodes on separate VLANs, VLAN_1 and VLAN_2, is offloaded to the vendor network. The vendor network is completely in charge of making sure this communication makes it to the intended destination, even when the endpoints are on the same node. The detail needed to cover how the networking works in these cases is beyond the scope of this chapter. What you need to understand is that OpenStack abstracts a great deal of complexity from the vendor network. Complex vendor-specific configurations are managed through plug-ins.

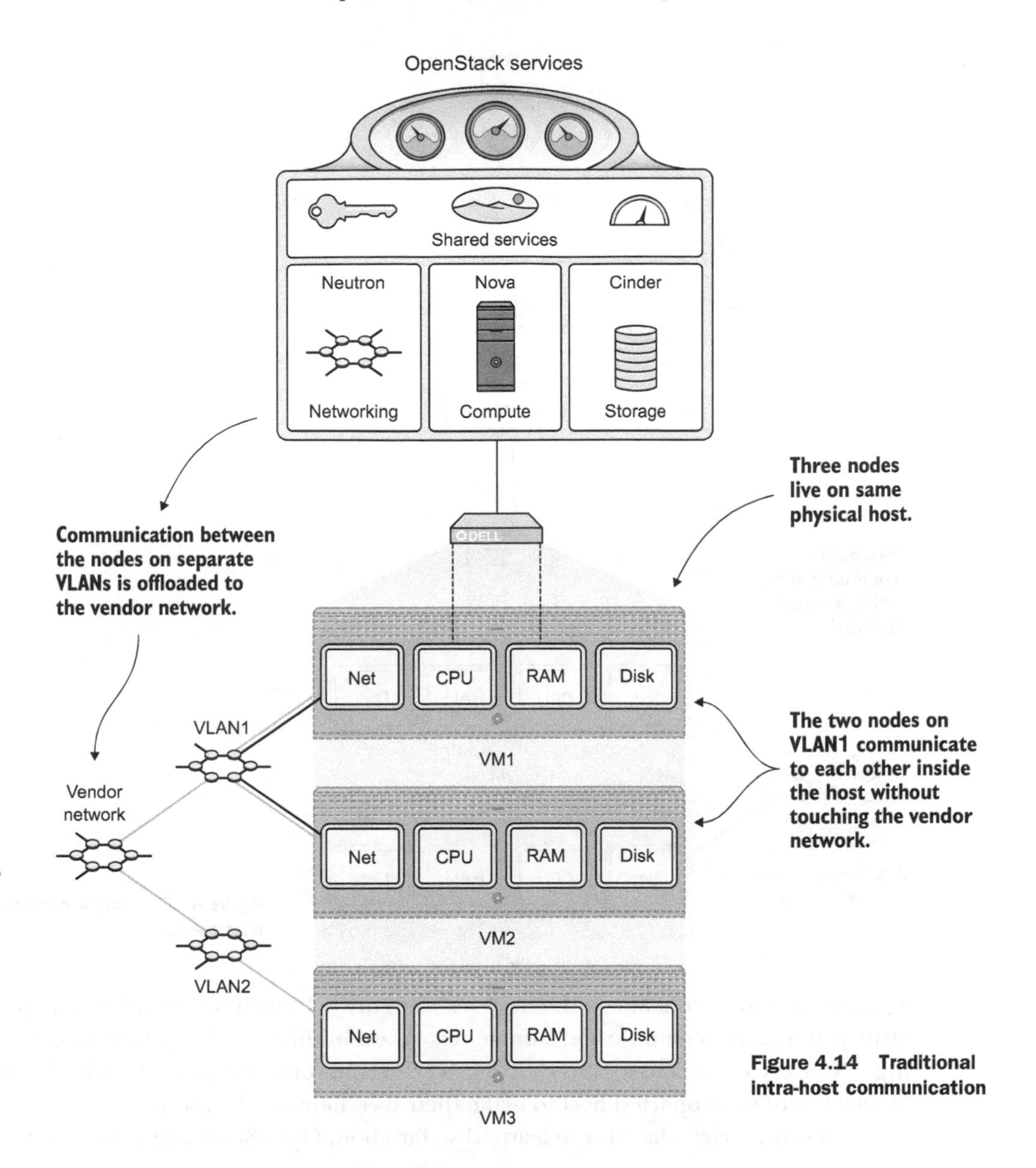
OpenStack services
Shared services
Neutron
Nova
Cinder
Networking
Compute
Storage
Three nodes live on same physical host.
Communication between the nodes on separate VLANs is offloaded to the vendor network.
DELL
Net
CPU
RAM
Disk
VM1
VLAN1
Vendor network
The two nodes on VLAN1 communicate to each other inside the host without touching the vendor network.
VM2
VLAN2
VM3

Figure 4.14 Traditional intra-host communication

By now it should be clear that vendor networking is more complicated than simply *provisioning* resources, which is what vendor storage systems do. Take a look at figure 4.15, which shows two hosts communicating using a vendor network. Of course, you could configure OpenStack to behave like a traditional virtualization framework and simply offload all the communication to the vendor network, but this is undesirable for a cloud platform. The details of why it's undesirable are beyond the scope of this chapter, but suffice it to say that this approach will not scale and will be a limiting factor in how you manage and provision resources.

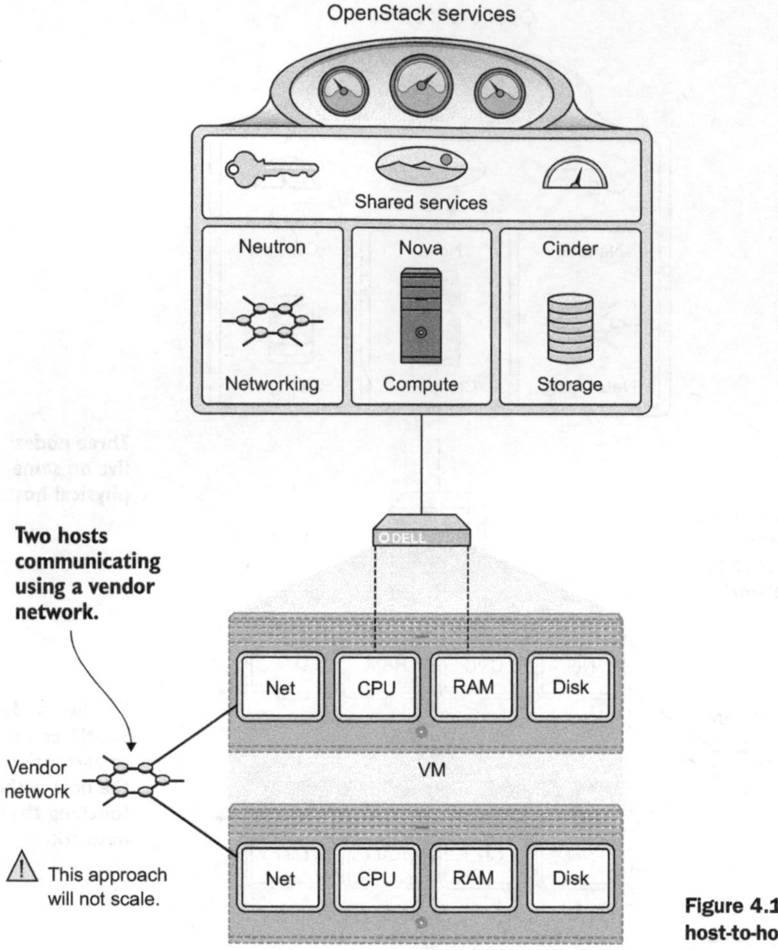

Figure 4.15 Vendor networking host-to-host

Suppose you want to manage the network in figure 4.16 with the same level of granularity you manage compute and storage resources. In this model, OpenStack Networking (Neutron) interfaces directly with vendor network components, which allows Neutron and its supported host to make their own network decisions.

Let's summarize what you've learned so far about OpenStack and vendor networking systems:

- Traditional hypervisors and virtualization frameworks unintelligently offload many functions to vendor networking.
- Traditional hypervisors and virtualization frameworks have little or no knowledge of how networking was performed, even for their own VMs.
- Managing vendor networking is more complicated than controlling a one-to-one relationship, like with vendor storage.
- Neutron is the codename for OpenStack Networking.

- Neutron integrates with vendor networking components to make networking decisions for OpenStack.

We'll take a look at how Neutron interfaces with vendor network components in the next section.

How OpenStack supports vendor networking

Just as Cinder uses vendor-specific plug-ins to communicate with vendor storage systems, Neutron uses plug-ins to manage vendor networking. As previously stated, plug-ins translate between OpenStack APIs and vendor-specific APIs. The relationship between Neutron and the vendor network is shown in figure 4.16.

You might wonder just what is being managed on the vendor network. The answer to this question is *it depends*. There are many networking vendors who produce many types of networking devices. These devices must interoperate at least on the level of network communication. After all, what good is a network if you can't communicate between networks and devices?

Software-defined networking (SDN) supports the idea of a separation of network management and communication functions. Because OpenStack Networking is a type

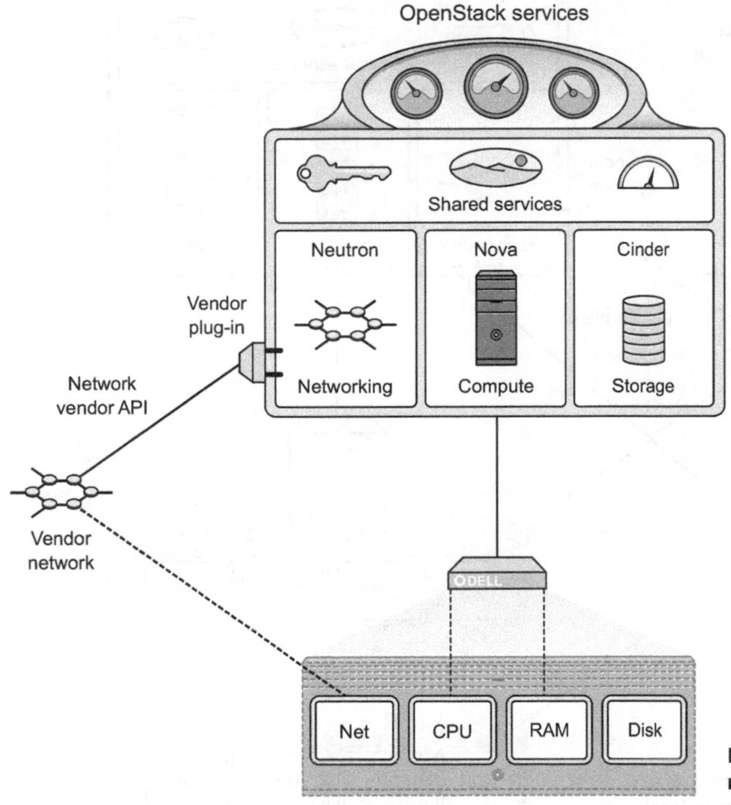

Figure 4.16 Neutron manages vendor networking.

of SDN, this so-called separation of the *control plane* and *data plane* is at the heart of OpenStack Networking when dealing with vendor hardware and software.

OPENSTACK NETWORKING ALSO PROVIDES L3 SERVICES In the context of vendor networking, OpenStack functions as a network controller. But it's worth noting that OpenStack networking does provide L3 services in the form of virtual routing, DHCP, and other services.

Figure 4.17 shows Neutron managing network devices in the control plane through the use of vendor-specific plug-ins. As you can see, the data plane never touches Neutron. In fact, Neutron might have no low-level insight into how the communication between the two nodes is happening. But Neutron knows that both nodes are on the specific network hardware that it manages, so Neutron can configure the endpoints to communicate, regardless of how the communication navigates the data plane.

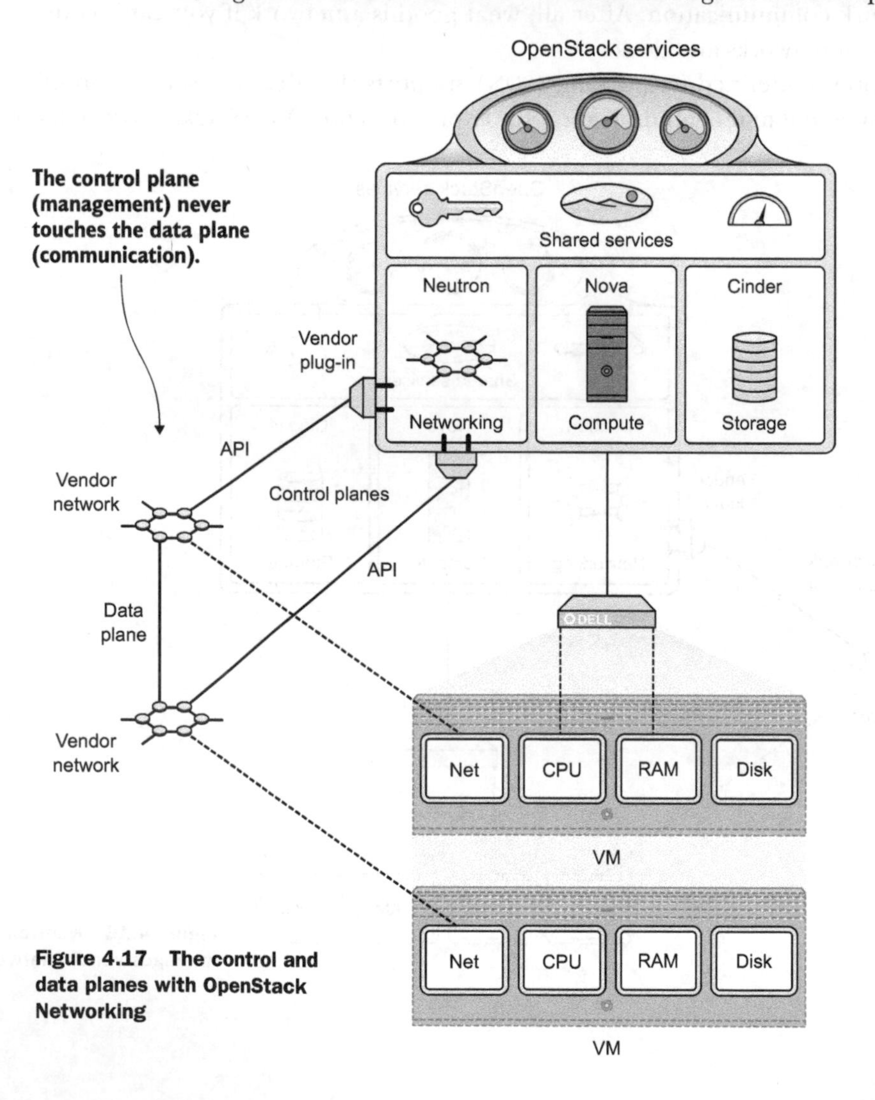

Figure 4.17 The control and data planes with OpenStack Networking

Understanding SDN and OpenStack Networking

This is a very complicated topic, and you'll probably want to go back and reread this section a few times. You aren't expected to fully understand SDN, but it's important that you understand the basic role of Neutron in relation to vendor networking. Chances are that your local network expert (unless this is you), doesn't know any more about SDN, and by relation OpenStack Networking, than you do. OpenStack/Neutron works on the control plane to manage communication between VMs that it manages, but it doesn't control the data plane related to communication between endpoints.

This is a new way of thinking about networking that really turns the traditional network world on its head. I've just introduced the topic in this section to give you some insight into how OpenStack can manage vendor networks without you having to hand over your enterprise or data center to OpenStack control. As previously stated, OpenStack can be configured to work very traditionally in terms of network integration, but the framework is well positioned to take advantage of the SDN model and technologies. The Open Networking Foundation (www.opennetworking.org) was founded to promote SDN and is a good starting point for gaining a deeper understanding of SDN.

In the next section, you'll learn about the types of vendor networking used in OpenStack.

EXAMPLES OF VENDOR NETWORKING IN OPENSTACK

In early versions of OpenStack, networking was provided in a traditional way, with networking being managed by OpenStack Compute (Nova). As demand for network control outside the scope of OpenStack Compute grew, OpenStack Networking (originally Quantum, and later Neutron) was developed as a separate project.

As previously described, Neutron manages vendor networking using vendor-specific plug-ins. As the community added more and more support for vendor networking, the need for further modularity through a standard plug-in module was identified. The benefits of modular plug-ins include reduced redundant code, easier vendor integration, and standardization of core network functions.

With the release of OpenStack Havana in late 2013, the Neutron Modular Layer 2 (ML2) plug-in was introduced. The ML2 plug-in is divided into *type* and *mechanism* drivers. Figure 4.18 shows the hierarchy of the ML2 plug-in with the type and mechanism drivers.

The *type* drivers, as the name suggests, are related to the type of network the plug-in manages. You can think of the type driver as how Neutron manages the endpoints. For example, Neutron could specify a tunnel be created between endpoints without knowing anything about the network between the endpoints. This gets us back to the discussion about the separation of control and data planes.

The *mechanism* drivers are responsible for managing the virtual and physical network devices that are attached to endpoints. These drivers create, update, and delete network and port resources based on the requirements of the type driver.

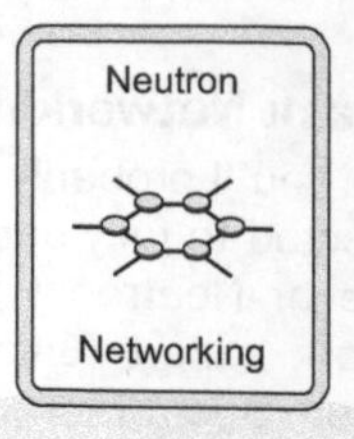

ML2 plug-in			API extension				
Type driver			Mechanism driver				
GRE	VXLAN	VLAN	Arista	Cisco	Linux bridge	OVS	L2 pop

Figure 4.18 Network management with the Neutron ML2 plug-in

The goal of the ML2 plug-in is to replace many of the monolithic plug-ins that exist today.

> **EXAMPLES OF VENDOR NETWORKING IN OPENSTACK** Neutron plug-ins have been developed for many vendors, including Arista, Cisco, Nicira/VMware, NEC, Brocade, IBM, and Juniper. In addition, ML2 drivers have been developed for Big Switch/Floodlight, Arista, Mellanox, Cisco, Brocade, Nicira/VMware, and NEC.

The next section will touch on what you've learned in the first part of this book and what will be covered in the second part.

4.3 Why walk through a manual deployment?

In chapter 1 you were introduced to OpenStack. In that introduction you learned how OpenStack fits into the cloud ecosystem, why you might want to adopt the technology, and what the focus of this book will be. In chapter 2, motivated by the fantastic possibilities described in the first chapter, you took a limited test-drive of the OpenStack framework, working through some exercises that didn't require an in-depth knowledge of the framework. Chapter 3 presented more examples, but this time from an operational perspective, giving you further insight into the structure of the framework. Finally, in this chapter you learned how OpenStack works through its framework of components and interoperates with vendor hardware and software.

You've covered a great deal in four chapters. If you completed all of the exercises and have a working DevStack deployment, congratulate yourself! You might (unfortunately) already be considered an OpenStack expert in many organizations. But although the first part of this book may be sufficient to make you look like an expert, there's much more to learn before you take the leap to a multi-node production deployment.

Part 2 of this book covers deploying OpenStack manually, going through each command and configuration, and explaining both the steps involved and what they mean. If your view is more high-level, or you plan on relying on a vendor for your OpenStack support, you can skip to part 3, where we'll cover topics related to design, implementation, and even the economics of OpenStack production deployments. This being said, even if you expect a fully managed OpenStack solution to be provided by a vendor, there's certainly value in knowing what's going on under the covers. I recommend at least reviewing part 2, even if you don't plan on personally deploying a production OpenStack environment.

4.4 Summary

- OpenStack is a framework that consists of many projects.
- OpenStack project designations range from core (integral parts of OpenStack) to related (projects that have some relation).
- OpenStack works using a collection of distributed core components.
- Core components communicate with each other using their respective APIs.
- OpenStack can manage vendor-provided hardware and software.
- OpenStack manages vendor-provided hardware and software through component plug-ins.

Part 2

Walking through a manual deployment

In the second part of the book, you'll step through a manual deployment of several core OpenStack components. Although it's important that you understand the underlying component interactions that make up OpenStack, this part of the book is not intended as a blueprint for OpenStack deployment. The OpenStack foundation does a good job of providing detailed documentation (http://docs.openstack.org/) for each software release. The goal of this part of the book is to build your confidence in the underlying system, through low-level exposure of the components and configurations. The intent here is to help you understand the underlying OpenStack architecture well enough to make informed decisions when designing a production deployment.

5 Walking through a Controller deployment

This chapter covers

- Installing controller prerequisites
- Deploying shared services
- Configuring controller-side Block Storage, Networking, Compute, and Dashboard services

In the first two chapters, you were introduced to OpenStack and took a test-drive of the framework using the Horizon web interface. Chapter 3 introduced you to some basic operational tasks using the command-line interface (CLI). In chapter 4 you learned how OpenStack components are related and distributed in a multi-node environment. That first part of the book was designed to help you gain an understanding of what OpenStack can do, to get you comfortable with the operation of the framework, and to give you a foundational understanding of how the components of the framework interact. In this second part of the book, you'll take a deep dive into the components themselves.

By the end of this part of the book, you'll have gained a familiarity with the configuration, use, and placement of individual OpenStack core components.

THIS BOOK IS NOT ... This book does not focus on best practices for OpenStack operation or architecture. These topics, while important, are very much dependent on both the release version of OpenStack and the requirements of the user. The purpose of this book is to help you build a foundational understanding of the OpenStack framework that will transcend individual requirements and persist through many future releases of OpenStack.

The first part of the book was based on a single-node deployment of OpenStack using DevStack to install and configure the OpenStack components and dependencies. This second part of the book is based on a multi-node manual deployment of OpenStack, so DevStack won't be used. In this part of the book, you'll install component software using a package management system provided by the Linux distribution and configure the components manually. Through this process, you'll gain an understanding of the dependencies, configuration, relationships, and use of individual OpenStack components.

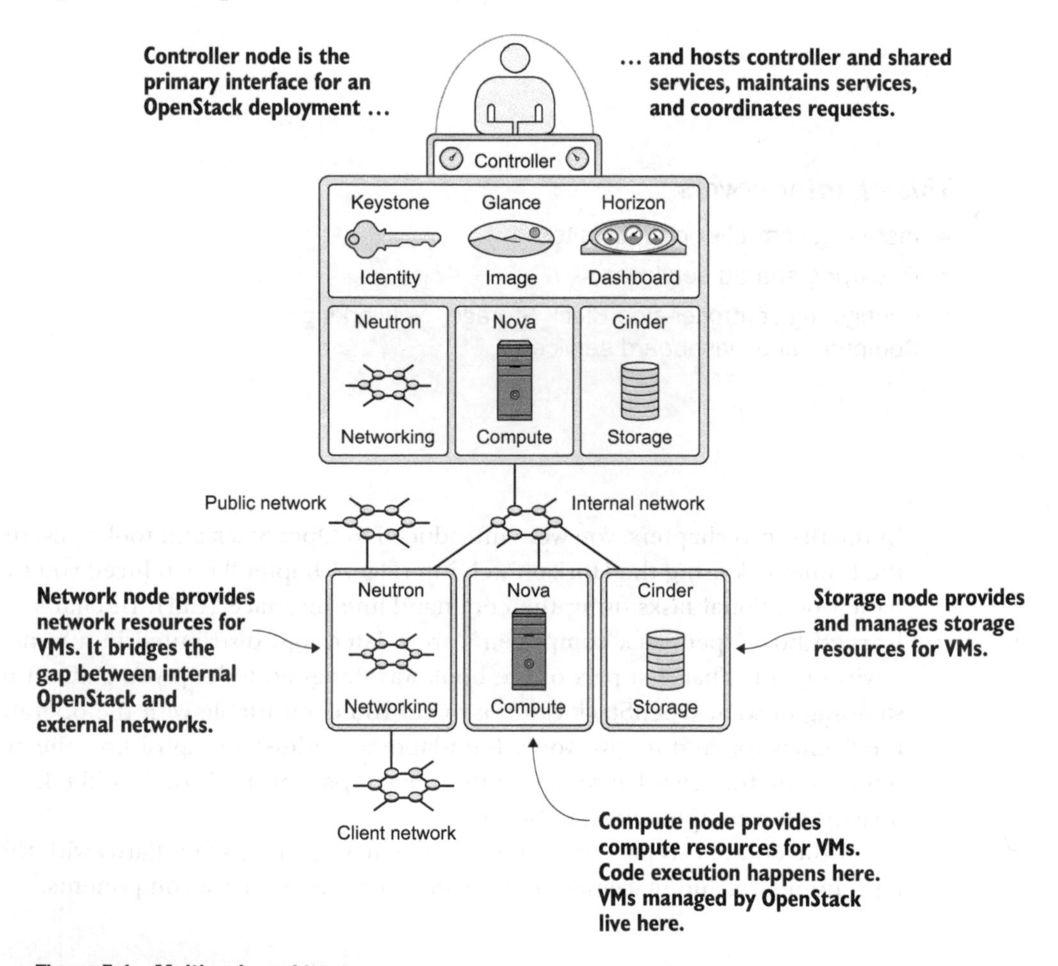

Figure 5.1 Multi-node architecture

Figure 5.1 shows the architecture that you'll re-create in this part of the book. In the figure you can see four nodes:

- *Controller*—The node that hosts controller and other shared services. This node maintains the server-side API services. The controller coordinates component requests and serves as the primary interface for an OpenStack deployment.
- *Network*—The node that provides network resources for virtual machines. This node bridges the gap between internal OpenStack networks and external networks.
- *Storage*—The node that provides and manages storage resources for virtual machines.
- *Compute*—The node that provides compute resources for virtual machines. Code execution will occur on these nodes. You can think of virtual machines managed by OpenStack as living on these nodes.

As explained in chapter 4, in the OpenStack distribution model, resource nodes get instructions from the controller (see figure 5.1). As you can see in figure 5.2, you'll be building out different portions of a multi-node deployment in each chapter. In this chapter, you'll build the controller node (shown at the top of the figure). In subsequent chapters, you'll build out the other nodes (network, storage, and compute) to complete a manual multi-node deployment of OpenStack.

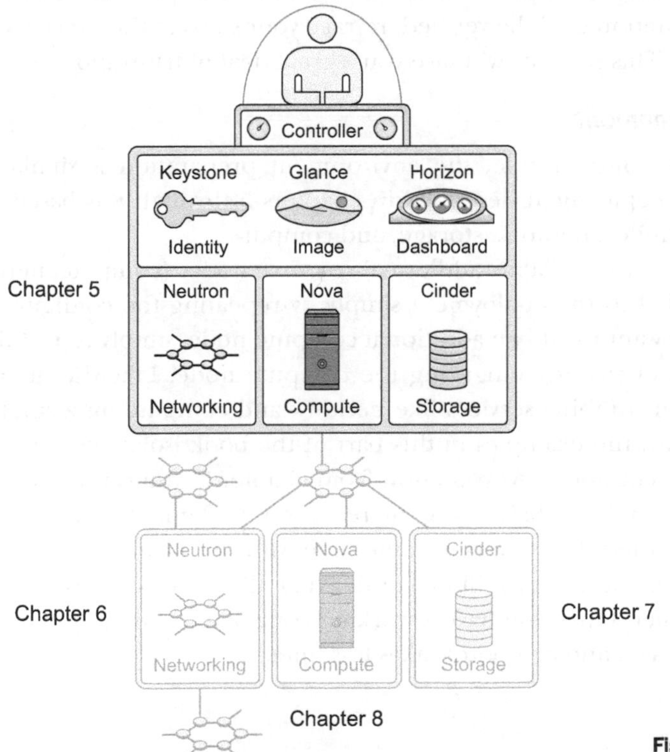

Figure 5.2 Deployment roadmap

5.1 Deploying controller prerequisites

Before continuing in this chapter, you must have access to a freshly installed Ubuntu 14.04 physical or virtual node. An installation tutorial for Ubuntu 14.04 is provided in the appendix.

> **WHAT OS DISTRIBUTION SHOULD I USE?** The examples in the second part of the book (chapters 5 through 8) are intended to be used with the Ubuntu 14.04 Long Term Support (LTS) version of Ubuntu Server. This version of Ubuntu includes the Icehouse release of OpenStack and guarantees support until April 2019.

In the chapter 2 deployment, DevStack installed and configured OpenStack dependencies for you. In this chapter, you'll manually install these dependencies. Luckily, you can use a package management system to install the software (no compiling required), but you must still manually configure the components.

> **PROCEED WITH CARE** Working in a multi-node environment greatly increases deployment and troubleshooting complexity. A small, seemingly unrelated, mistake in the configuration of one component or dependency can cause issues that are very hard to track down. Read each section carefully, making sure you understand what you're installing and configuring.

Several of the following examples include a verification step, which shouldn't be skipped. If a configuration can't be verified, retrace your steps to the previous verified point and start over. This practice will save you a great deal of frustration.

5.1.1 Preparing the environment

Aside from network configuration, the environment preparation is similar for all nodes. The manual deployment described in chapters 5 through 8 is based on four physical nodes: controller, network, storage, and compute.

If additional nodes are available, additional resource nodes (compute, network, or storage) can be added to the deployment simply by repeating the configuration for that resource. If you want to add an additional compute node, simply repeat the steps in chapter 8, which discusses configuring the compute node. Likewise, if you have fewer nodes, you can combine services like network and compute on a single node. For the sake of clarity, the examples in this part of the book isolate core OpenStack services on independent nodes. As you know from chapter 2, you can put OpenStack on a single node, but multi-node is where the real fun (and benefit) begins.

It's time to get started. Give yourself some time with this chapter. The controller install can take a while because you'll be configuring all the backend services as you bring up the controller. Once you have a working controller, setting up the resource nodes (network, storage, and compute) takes less time.

5.1.2 Configuring the network interface

You'll want to configure the network interface on the controller node so one interface is used for client-facing traffic and another is used for internal OpenStack management. Technically, you could use a single interface on the controller, but as you'll soon learn, OpenStack allows you to specify several networks (public, internal, and admin) for OpenStack operation.

REVIEWING THE NETWORK

The first step in configuring the network interfaces is to determine what physical interfaces exist on your server. Then you can configure interfaces to be used in your OpenStack environment. You can list the interfaces with the `ifconfig -a` command, as shown in the following listing.

Listing 5.1 Listing the interfaces

```
$ ifconfig -a
em1       Link encap:Ethernet  HWaddr b8:2a:72:d3:09:46
          inet addr:10.33.2.50  Bcast:10.33.2.255
          inet6 addr: fe80::ba2a:72ff:fed3:946/64 Scope:Link
          UP BROADCAST RUNNING MULTICAST  MTU:1500  Metric:1
          RX packets:950 errors:0 dropped:0 overruns:0 frame:0
          TX packets:117 errors:0 dropped:0 overruns:0 carrier:0
          collisions:0 txqueuelen:1000
          RX bytes:396512 (396.5 KB)  TX bytes:17351 (17.3 KB)
          Interrupt:35

em2       Link encap:Ethernet  HWaddr b8:2a:72:d3:09:47
          BROADCAST MULTICAST  MTU:1500  Metric:1
          RX packets:0 errors:0 dropped:0 overruns:0 frame:0
          TX packets:0 errors:0 dropped:0 overruns:0 carrier:0
          collisions:0 txqueuelen:1000
          RX bytes:0 (0.0 B)  TX bytes:0 (0.0 B)
          Interrupt:38
```

You might see many interfaces, but for now we'll focus on interfaces `em1` and `em2`, which will be used for the public and internal networks. On the example controller, `em1` will be used as the public interface and `em2` will be used as the internal interface. An address to be used as the OpenStack public address has already been assigned on `em1`, and interface `em2` will be used as the internal interface. Specific network addresses, VLANs, and interface functions are explained in the next section.

Next, you need to configure the physical interfaces on your controller node.

CONFIGURING THE NETWORK

Under Ubuntu, the interface configuration is maintained in the /etc/network/interfaces file. If you're using another Linux distribution, you'll need to check the network interface configuration(s) for your specific distribution and version.

You'll build a working configuration for the controller node based on the italicized addresses in table 5.1.

Table 5.1 Network address table

Node	Function	Interface	IP address/subnet mask
Controller	*Public interface/node address*	*em1*	*10.33.2.50/24*
Controller	*OpenStack internal*	*em2*	*192.168.0.50/24*
Network	Node address	em1	10.33.2.51/24
Network	OpenStack internal	em2	192.168.0.51/24
Network	VM interface/network	p2p1	None: Assigned to OpenStack Networking
Storage	Node address	em1	10.33.2.52/24
Storage	OpenStack internal	em2	192.168.0.52/25
Compute	Node address	em1	10.33.2.53/24
Compute	OpenStack internal	em2	192.168.0.53/24

The terms in the Function column are explained here:

- *Public interface*—Accessed by tenant users, Horizon, and public API calls.
- *Node address*—Primary address of the node. This address doesn't need to be public, but for the sake of simplicity, we'll put the public interface for the controller and the node interfaces for the resource nodes on the same network.
- *OpenStack Internal*—Interface used to carry OpenStack component traffic between component nodes including AMPQ, internal API, and so on.

NETWORK INTERFACE NAME Network interface names vary based on the order and location of the hardware in the server. For instance, Lan-On-Motherboard interfaces are reported in the format em<port number> (ethernet-on-motherboard <1,2 ..>), whereas PCI add-in interfaces are reported in the format p<slot number>p<port number>_<virtual function instance>.

In order to modify the network configuration, or any privileged configuration, you must use sudo privileges (`sudo vi /etc/network/interfaces`). As you may know, the `sudo` command allows normal users (if they're in the `sudo` group) to execute commands with elevated privileges.

The following listing shows the network interface configuration example. Modify your interface configuration, based on the values found in table 5.1, or you can use your own address scheme.

Listing 5.2 Modifying interface configuration in /etc/network/interfaces

```
# The loopback network interface
auto lo
iface lo inet loopback

# The Public/Node network interface
```

```
auto em1
iface em1 inet static
        address 10.33.2.50
        netmask 255.255.255.0
        network 10.33.2.0
        broadcast 10.33.2.255
        gateway 10.33.2.1
        dns-nameservers 8.8.8.8
        dns-search testco.com

# The OpenStack Internal Interface
auto em2
iface em2 inet static
        address 192.168.0.50
        netmask 255.255.255.0
```

You should now refresh your network settings to update any changes in your network configuration. First, though, if you changed the address of the primary interface, you should reboot the server at this point, because you'll lose connectivity to the system after the refresh. If you didn't change the settings of your primary interface, you shouldn't experience an interruption.

The following listing shows the command used to refresh the network settings, along with the output.

Listing 5.3 Refreshing networking settings

```
$ sudo ifdown em2 && sudo ifup em2
```

Your network configuration should now be active. The interface will automatically be brought online based on your configuration. This process can be repeated for each interface that requires a configuration refresh.

In order to confirm that the configuration was applied, you should once again check your interfaces, as shown in the following listing.

Listing 5.4 Checking the network for updates

```
$ifconfig -a
em1       Link encap:Ethernet  HWaddr b8:2a:72:d3:09:46
          inet addr:10.33.2.50  Bcast:10.33.2.255  Mask:255.255.255.0
          inet6 addr: fe80::ba2a:72ff:fed3:946/64 Scope:Link
          UP BROADCAST RUNNING MULTICAST  MTU:1500  Metric:1
          RX packets:3014 errors:0 dropped:0 overruns:0 frame:0
          TX packets:656 errors:0 dropped:0 overruns:0 carrier:0
          collisions:0 txqueuelen:1000
          RX bytes:2829516 (2.8 MB)  TX bytes:94684 (94.6 KB)
          Interrupt:35

em2       Link encap:Ethernet  HWaddr b8:2a:72:d3:09:47
          inet addr:192.168.0.50  Bcast:192.168.0.255  Mask:255.255.255.0
          inet6 addr: fe80::ba2a:72ff:fed3:947/64 Scope:Link
          UP BROADCAST RUNNING MULTICAST  MTU:1500  Metric:1
          RX packets:1 errors:0 dropped:0 overruns:0 frame:0
```

```
TX packets:6 errors:0 dropped:0 overruns:0 carrier:0
collisions:0 txqueuelen:1000
RX bytes:64 (64.0 B)  TX bytes:532 (532.0 B)
Interrupt:38
```

At this point you should be able to remotely access the controller server, and the controller server should have internet access. The remainder of the install can be performed either remotely using SSH or directly from the console.

5.1.3 Updating packages

Ubuntu 14.04 LTS includes the Icehouse (2014.1) release of OpenStack, which includes the following components:

- *Nova*—The OpenStack Compute project, which works as an IaaS cloud fabric controller
- *Glance*—Provides services for VM images, discovery, retrieval, and registration
- *Swift*—Provides highly scalable, distributed, object store services
- *Horizon*—The OpenStack Dashboard project, which provides a web-based admin and user GUI
- *Keystone*—Provides identity, token, catalog, and policy services for the OpenStack suite
- *Neutron*—Provides network management services for OpenStack components
- *Cinder*—Provides block storage as a service to OpenStack Compute
- *Ceilometer*—Provides a central point of record for resource utilization metrics
- *Heat*—Provides application-level orchestration of OpenStack resources

I WANT TO USE A DIFFERENT OS OR OPENSTACK VERSION You might be inclined to use a different Linux distribution or release that provides a more current OpenStack version. However, it's highly recommended that you stick with the versions specified. Once you understand OpenStack on fundamental and operational levels, you can always migrate to new releases.

The Ubuntu Linux distribution uses the APT system for package management. The APT package index is a database of all available packages defined in the /etc/apt/sources.list file. You should make sure the local database is synchronized with the latest packages available in the repository for your specific Linux distribution. Prior to installing OpenStack, you should also upgrade any repository items, including the Linux kernel, that might be out of date.

The following listing demonstrates how to update and upgrade packages on your server.

Listing 5.5 Updating and upgrading packages

```
sudo apt-get -y update
sudo apt-get -y upgrade
```

Once you've updated and upgraded the packages, you should reboot the server to refresh any packages or configurations that might have changed.

Listing 5.6 Rebooting the server

```
sudo reboot
```

Now it's time to install the OpenStack software dependencies.

5.1.4 Installing software dependencies

In the context of OpenStack, a *dependency* is software that's not part of the OpenStack project but is required by OpenStack components. This includes software used to run OpenStack code (Python and modules), the queueing system (RabbitMQ), and the database platform (MySQL), among other things.

In this section, you'll walk through the deployment of OpenStack software dependencies. You'll start with the installation of RabbitMQ.

INSTALLING RABBITMQ

RabbitMQ is an Advanced Message Queuing Protocol (AMQP) –compliant queuing system that allows for guaranteed delivery and ordering of messages in large distributed systems. OpenStack uses the RabbitMQ messaging service as its default queuing system, allowing OpenStack component functions that require quick and ordered messages to communicate.

You can use APT or an equivalent package management system for your Linux distribution to install RabbitMQ. The following listing demonstrates the installation using APT.

Listing 5.7 Installing RabbitMQ

```
sudo apt-get -y install rabbitmq-server
```

When you run the preceding command, you'll see output like the following:

```
...
The following extra packages will be installed:
  erlang-asn1 erlang-base erlang-corba ...
  libltdl7 libodbc1 libsctp1 lksctp-tools ...
...
Setting up rabbitmq-server (3.2.4-1) ...
Adding group `rabbitmq' (GID 118) ...
Done.
Adding system user `rabbitmq' (UID 111) ...
Adding new user `rabbitmq' (UID 111) with group `rabbitmq' ...
Not creating home directory `/var/lib/rabbitmq'.
 * Starting message broker rabbitmq-server
```

If you see the error `[* FAILED - check /var/log/rabbitmq/startup...]`, make sure your hostname in /etc/hostname matches the host entry found in /etc/hosts and restart if necessary.

RabbitMQ automatically creates a user named guest, with administrative privileges. You'll want to change the password for the guest account; the following example changes the password to *openstack1*.

Listing 5.8 Configuring the RabbitMQ guest password

```
$ sudo rabbitmqctl change_password guest openstack1
Changing password for user "guest" ...
...done.
```

You must now verify that RabbitMQ is running properly.

Listing 5.9 Verifying the RabbitMQ status

```
$sudo rabbitmqctl status
Status of node rabbit@controller ...
[{pid,2452},
 {running_applications,[{rabbit,"RabbitMQ","3.2.4"},
                        {mnesia,"MNESIA  CXC 138 12","4.11"},
                        {os_mon,"CPO  CXC 138 46","2.2.14"},
                        {xmerl,"XML parser","1.3.5"},
                        {sasl,"SASL  CXC 138 11","2.3.4"},
                        {stdlib,"ERTS  CXC 138 10","1.19.4"},
                        {kernel,"ERTS  CXC 138 10","2.16.4"}]},
...
...done.
```

You now have a fully functional deployment of RabbitMQ ready for OpenStack use.

INSTALLING MYSQL

OpenStack uses a traditional relational database to store configurations and status information. By default, OpenStack is configured to use an embedded SQLite database for all components, but because of the performance and general accessibility of MySQL, I'll show you how to configure components to use MySQL in place of SQLite. This will allow you to use a MySQL server for your backend configuration and status data store. All OpenStack components deployed in chapters 5 through 8 that use a database will use the central database deployed in this step.

Using APT or an equivalent package management system for your Linux distribution, install MySQL as demonstrated in the following listing.

Listing 5.10 Installing MySQL binaries

```
$ sudo apt-get -y install python-mysqldb mysql-server
Reading package lists... Done
Building dependency tree
Reading state information... Done
Suggested packages:
  python-mysqldb-dbg
  ...
The following NEW packages will be installed:
  ...
```

```
  mysql-server python-mysqldb
...
Setting up libaio1:amd64 (0.3.109-3) ...
Setting up libmysqlclient18:amd64 (5.5.29-0ubuntu1) ...
Setting up libnet-daemon-perl (0.48-1) ...
Setting up libplrpc-perl (0.2020-2) ...
Setting up libdbi-perl (1.622-1) ...
Setting up libdbd-mysql-perl (4.021-1) ...
Setting up mysql-client-core-5.5 (5.5.38-0ubuntu1) ...
Setting up libterm-readkey-perl (2.30-4build4) ...
Setting up mysql-client-5.5 (5.5.38-0ubuntu1) ...
Setting up mysql-server-core-5.5 (5.5.38-0ubuntu1) ...
Setting up mysql-server-5.5 (5.5.38-0ubuntu1) ...
Setting up libhtml-template-perl (2.91-1) ...
Setting up python-mysqldb (1.2.3-1ubuntu1) ...
Setting up mysql-server (5.5.38-0ubuntu1) ...
Setting up mysql-server (5.5.38-0ubuntu0.14.04.1) ...
Setting up python-mysqldb (1.2.3-1build1) ...
```

When prompted, enter `openstack1` as the root MySQL password. You are, of course, free to select any password you want as long as you consistently use the same password in all the examples—the examples will assume you use this password.

In order for external services (those on other nodes using the internal network) to contact the local MySQL instance, you must change the address that MySQL binds to on startup. Using your favorite text editor, open /etc/mysql/my.cnf and change the bind-address to `0.0.0.0` as follows.

Listing 5.11 Modifying /etc/mysql/my.cnf

```
# Instead of skip-networking the default is now to listen
#only on localhost which is more compatible and is not
#less secure.

#bind-address           = 127.0.0.1
#Bind to Internal Address of Controller
bind-address            = 0.0.0.0
```

MYSQL PERFORMANCE Explaining the performance tuning of MySQL is beyond the scope of this book, but you need to be aware of the impact MySQL can have on OpenStack performance. Because state and configuration information is maintained in MySQL, a poorly performing MySQL server can greatly impact many aspects of OpenStack performance. In multi-user and production environments, it's recommended that you take the time to understand and configure settings in /etc/mysql/my.cnf related to performance.

You'll now want to restart MySQL and check its operation.

Listing 5.12 Restart and verify that MySQL is running and accessible

```
sudo service mysql restart
sudo service mysql status
mysqladmin -u root -h localhost -p status
```

When you run these commands, it will look something like this:

```
$ sudo service mysql restart
[mysql stop/waiting
mysql start/running, process 17396

$ service mysql status
mysql start/running, process 17396

$ mysqladmin -u root -h localhost -p status
Enter password: <enter openstack1 as set in previous step>
Uptime: 193  Threads: 1  Questions: 571  Slow queries: 0
Opens: 421  Flush tables: 1  Open tables: 41
Queries per second avg: 2.958
```

At this point you should have a running MySQL instance. If the instance fails to start and there were no errors during the install process, you should check the /etc/mysql/my.cnf file for any inadvertent typos resulting from the `[bind-address = 0.0.0.0]` modification.

ACCESSING THE MYSQL CONSOLE

The MySQL console is typically accessed from the MySQL client application. The `mysql` command takes several arguments, including `-u <username>`, `-h <hostname>`, and `-p <password>`.

You can either leave the password blank and then be prompted, or enter it as part of the command. There must be no spaces between the `-p` argument and the password. For example, if your password is `openstack1`, the command to access the MySQL console would be `mysql -u root -popenstack1`.

The following listing shows the login with a password prompt.

Listing 5.13 Logging in to the MySQL server as root

```
$ mysql -u root -p
Enter password: <enter mysql root password>
...
<verbose text removed>
...
mysql>
```

Now that you've confirmed that MySQL is running and you're able to access the console, you're ready to move forward with the component install. Refer back to listing 5.13 throughout part 2 of the book whenever you need to create databases and grant user rights.

5.2 Deploying shared services

OpenStack shared services are those services that span Compute, Storage, and Network services and are shared by OpenStack components. These are the official OpenStack shared services:

- *Identity Service (Keystone)*—Provides identity, token, catalog, and policy services for the OpenStack suite.
- *Image Service (Glance)*—Provides services for VM image discovery, retrieval, and registration.
- *Telemetry Service (Ceilometer)*—Provides a central service for monitoring and measurement information in the OpenStack suite.
- *Orchestration Service (Heat)*—Enables applications to be deployed using scripts from VM resources managed by OpenStack.
- *Database Service (Trove)*—Provides cloud-based relational and non-relational database services using OpenStack resources.

In chapters 5 through 8, we'll limit our walk-through to the first two of these shared services (Identity and Image Services), which are required for basic VM provisioning. Through the deployment of these two services, you should gain enough understanding to deploy the other optional services, should you choose. Several of these optional services are covered in detail in the third part of the book.

5.2.1 Deploying the Identity Service (Keystone)

OpenStack Identity Service, as the name implies, is the system of record for all identities (users, roles, tenants, and so on) across the OpenStack framework. It provides a common shared identity service for authentication, authorization, and resource inventory for all OpenStack components. This service can be configured to integrate with existing backend services such as Microsoft Active Directory (AD) and Lightweight Directory Access Protocol (LDAP), or it can operate independently. It supports multiple forms of authentication, including username and password, token-based credentials, and AWS-style (REST) logins.

Users with administrative roles will use the Identity Service (Keystone) to manage user identities across all OpenStack components, performing the following tasks:

- Creating users, tenants, and roles
- Assigning resource rights based on role-based access control (RBAC) policies
- Configuring authentication and authorization

Users with non-administrative roles will primarily interact with Keystone for authentication and authorization.

Keystone maintains the following objects:

- *Users*—As you might expect, these are the users of the system, such as the `admin` or `guest` user.
- *Tenants* —These are the projects (tenants) that are used to group resources, rights, and users together.
- *Roles*—These define what a user can do in a particular tenant.

- *Services*—This is the list of service components registered with a Keystone instance, such as the Compute, Network, Image, and Storage services. You can think of this as the listing of services provided by an OpenStack deployment.
- *Endpoints*—These are the URL locations of service-specific APIs registered with a particular Keystone server. You can think of this as the contact information for services provided by an OpenStack deployment.

In the following sections, you'll install the Keystone packages from the repository and then configure the service.

INSTALLING IDENTITY SERVICE (KEYSTONE)

The first step is to install the Keystone package with related dependencies with the following command.

Listing 5.14 Installing the Keystone package

```
sudo apt-get -y install keystone
```

When you run the preceding command, you'll get output like the following:

```
Reading package lists... Done
Building dependency tree
Reading state information... Done
The following extra packages will be installed:
  dbconfig-common python-keystone python-keystoneclient
  python-passlib python-prettytable
Suggested packages:
  python-memcached
The following NEW packages will be installed:
  dbconfig-common keystone python-keystone
  python-keystoneclient python-passlib python-prettytable
0 upgraded, 6 newly installed, 0 to remove and 0 not
upgraded. Need to get 751 kB of archives.
After this operation, 3,682 kB of additional disk space will
be used.
Preconfiguring packages ...
. . .
keystone start/running, process 6692

openstack@openstack1:~$ id keystone
uid=114(keystone) gid=124(keystone) groups=124(keystone)
```

The install process will retrieve the Keystone binaries, set up an account for the Keystone service named `keystone`, and place the default configuration files in the /etc/keystone directory.

CONFIGURING THE KEYSTONE DATA STORE

By default, Keystone settings are stored locally in a SQLite database. In this case, you'll be deploying a multi-node system, so SQLite isn't appropriate because you can't remotely access the SQLite database and performance is limited. In place of SQLite, you'll use MySQL.

The first step is to log in to the database server as described in subsection "Accessing the MySQL console." For the remainder of this part of the book, MySQL accounts will be identified with "_dbu" appended to the service name, such as *keystone_dbu*, for the keystone service. You want to create the keystone database and the MySQL user `keystone_dbu`, and then grant that user access to the *keystone* database.

In MySQL, the user creation and rights authorization functions can be completed in the same step. The command `keystone_dbu.* TO $keystone'@'\%'` specifies that the MySQL user `keystone_dbu` has access to all objects under the keystone database from any remote address.

The following listing shows the commands for creating the database, creating the user, and granting access.

Listing 5.15 Creating a database and granting access

```
mysql> CREATE DATABASE keystone;
Query OK, 1 row affected (0.00 sec)

mysql> GRANT ALL ON keystone_dbu.* TO 'keystone'@'localhost' \
    -> IDENTIFIED BY 'openstack1';
Query OK, 0 rows affected (0.00 sec)
```

You can verify that the database was created by issuing the next command.

Listing 5.16 Verifying the database and user

```
show grants for 'keystone_dbu'@'localhost';
```

In the following command output, you can see that the MySQL user `keystone_dbu` now has access to the keystone database:

```
+--------------------------------------------------------+
| Grants for keystone@%
+--------------------------------------------------------+
| GRANT USAGE ON *.* TO 'keystone_dbu'@'localhost' *removed password*
| GRANT ALL PRIVILEGES ON `keystone`.* TO 'keystone'@'localhost'
+--------------------------------------------------------+
2 rows in set (0.00 sec)
```

You now want to exit the MySQL shell. To exit the shell at any time, type `quit` at the `mysql>` prompt and press Enter.

> **SQL ACCESS AND RIGHTS CHANGE** In the previous database creation example, you might have noticed that the rights assigned not only allowed access to the database from the localhost, but also from any host. This is to allow components on remote servers to access the database directly. In a production environment, you'd want to either distribute database resources to local hosts or allow specific hosts database-level access, not all hosts.

By default, Keystone's data store is SQLite, so now you need to configure it to use MySQL. Keystone is configured via a primary configuration file, /etc/keystone/keystone.conf. To change the data store to MySQL, change the `connection` line under the `[sql]` section in /etc/keystone/keystone.conf as follows.

Listing 5.17 Modifying /etc/keystone/keystone.conf

```
[sql]
#connection = sqlite:////var/lib/keystone/keystone.db
connection = mysql://keystone_dbu:openstack1@localhost:3306/keystone
mysql_sql_mode=TRADITIONAL
```

The format of the MySQL connection string is `[db_username]:[db_username_password]@[db_hostname]:[db_port]/[db_name]`.

The changes won't become active until Keystone has been restarted. The process for restarting Keystone is shown in the following listing.

Listing 5.18 Restarting Keystone

```
$ sudo service keystone restart
keystone stop/waiting
keystone start/running, process 7868
```

At this point you've configured a MySQL user and database for the Keystone service. You've also configured Keystone to use MySQL and have restarted the service. But before you can use Keystone, you'll need to initialize the database with the Keystone schema, as shown in the next section.

Initializing the Keystone database

You've created the Keystone database and configured the service to use it, but the database is empty, so you need to initialize it. The initialization process builds a database schema at the location configured in the /etc/keystone/keystone.conf file.

You can initialize the Keystone database with the following command. There will be no output if it's successful.

Listing 5.19 Initializing the data store

```
sudo keystone-manage db_sync
```

You now have a Keystone service running with a MySQL backend data store. Next, you need to create OpenStack objects (users, roles, tenants, and so on) using Keystone.

Initializing Keystone variables

The next step is to populate Keystone with users, tenants, and roles. In order to gain access to the Keystone service, you must first set a few temporary environmental variables that Keystone will use for authentication.

Using your favorite text editor, create a file named keystone.auth in your home directory with the data shown in the following listing.

Listing 5.20 Creating keystone.auth

```
#This file contains environmental variables used to access Keystone

# Host address
HOST_IP=192.168.0.50 #The Management Address

# Keystone definitions
KEYSTONE_REGION=RegionOne
ADMIN_PASSWORD=admin_pass
SERVICE_PASSWORD=service_pass
export SERVICE_TOKEN="ADMIN"
export SERVICE_ENDPOINT="http://192.168.0.50:35357/v2.0"
SERVICE_TENANT_NAME=service
```

Once the file is created, you'll use the `source` command to run the keystone.auth script and then use the `set` command to verify that your environmental variables have been set. The following listing shows how to run the script and verify the variables.

Listing 5.21 Setting and confirming keystone.auth variables

```
$ source ~/keystone.auth
$ set | grep SERVICE
SERVICE_ENDPOINT=http://192.168.0.50:35357/v2.0
SERVICE_PASSWORD=service_pass
SERVICE_TENANT_NAME=service
SERVICE_TOKEN=ADMIN
```

You can now do a quick check to see if Keystone is functional. Run the command `keystone discover` to display known Keystone servers and API versions.

Listing 5.22 Checking Keystone operation

```
$ keystone discover
Keystone found at http://localhost:35357
    - supports version v3.0 (stable) here http://localhost:35357/v3/
No handlers could be found for logger "keystoneclient.generic.client"
    - supports version v2.0 (stable) here http://localhost:35357/v2.0/
        - and s3tokens: OpenStack S3 API
        - and OS-EP-FILTER: OpenStack Keystone Endpoint Filter API
        - and OS-FEDERATION: OpenStack Federation APIs
        - and OS-KSADM: OpenStack Keystone Admin
        - and OS-SIMPLE-CERT: OpenStack Simple Certificate API
        - and OS-EC2: OpenStack EC2 API
```

At this point you should see the Keystone service that was just installed. If you see no errors, you're ready to start preparing Keystone for other OpenStack services.

CREATING KEYSTONE SERVICES AND ENDPOINTS

Services for which Keystone will manage authorization and authentication must be specified. Because the process for creating Keystone services and endpoints will be the same across all OpenStack components, you can now create them for components that you'll install in later chapters.

In OpenStack, many services can be distributed across many nodes, but not all OpenStack deployments run all possible OpenStack services. To identify a service that's available in a particular deployment, you must register the service in Keystone, identifying the type of service and where it can be found in the deployment.

In practice, you identify a service by creating (registering) a new service with Keystone. The service location is specified by creating an endpoint for the new service in Keystone. Keystone will maintain a list of all active services and their endpoints.

Your first step is to create the service and endpoint for Keystone itself using the following command.

Listing 5.23 Creating the Keystone service

```
keystone service-create --name=keystone \
  --type=identity --description="Identity Service"
```

When you run the preceding command, you'll get output like the following:

```
+-------------+----------------------------------+
|   Property  |              Value               |
+-------------+----------------------------------+
| description |         Identity Service         |
|      id     | 541cffe246434a2e8d97653303df4ffd |
|     name    |             keystone             |
|     type    |             identity             |
+-------------+----------------------------------+
```

At this point the service has been created in Keystone and has been assigned a service ID that will be referenced in the creation of the endpoint.

In section 5.15.5 you learned that OpenStack allows you to specify several types of networks for component communication. You specify a network by assigning a URL (address) when you register an endpoint. The differences between these endpoints are the API functions exposed on each interface. Which API functions are exposed on which assignment varies based on the particular service. These are the possible endpoint assignments:

- `publicurl`—Intended for end-user communication, such as CLI and Dashboard communication.
- `internalurl`—Intended for component-to-component communication, such as a resource service (nova-compute) communicating with its corresponding controller service (nova-server).
- `adminurl`—Intended for communicating with services using the `admin` user, such as bootstrapping the initial configuration of Keystone using the `admin` account.

In addition to specifying `publicurl`, `internalurl`, or `adminurl`, you must also provide a region when creating endpoints. Regions are considered discrete OpenStack environments with unique API endpoints and services, but they all share a single Keystone

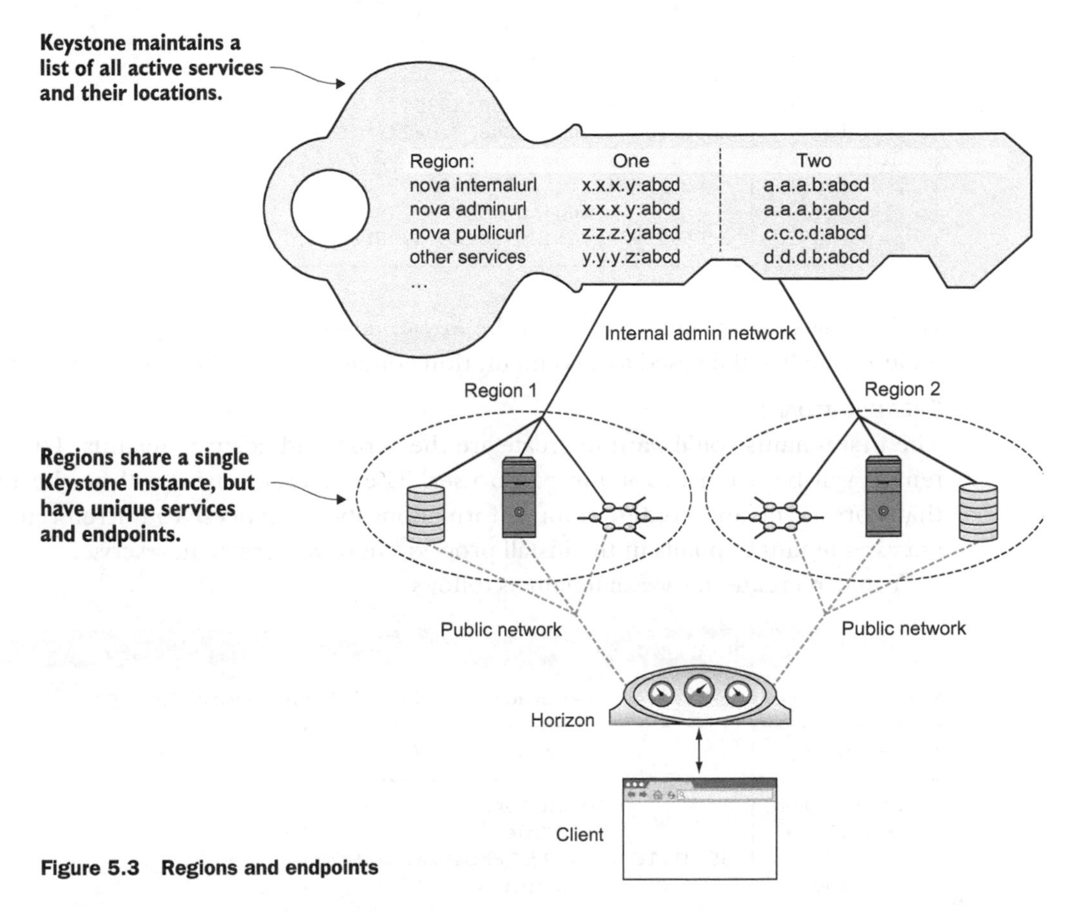

Figure 5.3 Regions and endpoints

instance. Figure 5.3 shows how OpenStack deployments are divided into regions that share a central Keystone instance.

The examples in this book are based on a single region deployment. The examples assign the name of `RegionOne` for all `region` configuration settings. The `publicurl` will correspond to the public address of the controller listed in table 5.1. The `internalurl` and `adminurl` correspond to the controller's internal address found in the same table.

You now want to create the Keystone endpoint, as follows.

Listing 5.24 Creating the Keystone endpoint

```
keystone endpoint-create \
 --region RegionOne \
 --service=keystone \
 --publicurl=http://10.33.2.50:5000/v2.0 \
 --internalurl=http://192.168.0.50:5000/v2.0 \
 --adminurl=http://192.168.0.50:35357/v2.0
```

When you run the preceding command, you'll see output like the following:

```
+-------------+----------------------------------+
|   Property  |              Value               |
+-------------+----------------------------------+
|   adminurl  |  http://192.168.0.50:35357/v2.0  |
|      id     | ad3ef29c0e2d40efb20e11eca2f2ff5d |
| internalurl |  http://192.168.0.50:5000/v2.0   |
|  publicurl  |   http://10.33.2.50:5000/v2.0    |
|    region   |            RegionOne             |
|  service_id | 8c066ff224a34d1aa354abe73708b804 |
+-------------+----------------------------------+
```

You've now created an endpoint for the Keystone service. The next step is to create a tenant, which will be used as a configuration container for additional configurations.

CREATING TENANTS

The first tenants you'll want to create are the `admin` and `service` tenants. The `admin` tenant will be a tenant for the `admin` user. The `service` tenant will be the tenant that stores user and configuration information about services. You'll reference the `service` tenant at points in the install process where you create new services.

You can create the `admin` tenant as follows.

Listing 5.25 Creating the `admin` tenant

```
$ keystone tenant-create --name=admin --description "Admin Tenant"
+-------------+----------------------------------+
|   Property  |              Value               |
+-------------+----------------------------------+
| description |           Admin Tenant           |
|   enabled   |               True               |
|      id     | 55bd141d9a29489d938bb492a1b2884c |
|     name    |              admin               |
+-------------+----------------------------------+
```

The following listing shows how you can create the `service` tenant.

Listing 5.26 Creating the `service` tenant

```
$ keystone tenant-create --name=service \
--description="Service Tenant"
+-------------+----------------------------------+
|   Property  |              Value               |
+-------------+----------------------------------+
| description |          Service Tenant          |
|   enabled   |               True               |
|      id     | b3c5ebecb36d4bb2916fecd8aed3aa1a |
|     name    |             service              |
+-------------+----------------------------------+
```

You now have an `admin` tenant and a `service` tenant. Your next step is to create users.

CREATING USERS

After tenants have been created, you'll want to create the `admin` user as follows.

Listing 5.27 Creating the admin user

```
$ keystone user-create --name=admin \
          --pass=openstack1 \
          --email=admin@testco.com
+----------+----------------------------------+
| Property |              Value               |
+----------+----------------------------------+
|  email   |        admin@testco.com          |
| enabled  |              True                |
|    id    | 8f39cacece9b4a01b51bdef57468a76e |
|   name   |             admin                |
| username |             admin                |
+----------+----------------------------------+
```

You now have an `admin` user. Once you deploy OpenStack services, you'll use this account to administer your deployment.

CREATING ROLES

Keystone roles are assigned to each user per tenant, and they specify the privileges for that user in the particular tenant. After users have been created, you'll want to create the roles for those users.

The following listing shows how you can create the `admin` role.

Listing 5.28 Creating the admin role

```
$ keystone role-create --name=admin
+----------+----------------------------------+
| Property |              Value               |
+----------+----------------------------------+
|    id    | d566b73857234f45ab1b3cb90c560da3 |
|   name   |             admin                |
+----------+----------------------------------+
```

You'll want to create a `Member` role so that you can assign users to tenants without making them administrators of the tenant. The `Member` role is also the default role used by the OpenStack Dashboard, so it must be configured.

Listing 5.29 Creating the Member role

```
$ keystone role-create --name=Member
+----------+----------------------------------+
| Property |              Value               |
+----------+----------------------------------+
|    id    | 45f75b4422774a25be07cbab055c50d8 |
|   name   |             Member               |
+----------+----------------------------------+
```

You've now created some roles, but they haven't yet been assigned to any users and tenants. The next step is to assign a role to a user in a particular tenant.

ASSIGNING ROLES

You now need to assign the user `admin` the `admin` role in the `admin` tenant. This is done by referencing the user, role, and tenant with the Keystone command `user-role-add`, as follows. This command has no output if it's successful.

Listing 5.30 Assigning the `admin` role

```
keystone user-role-add --user=admin --role=admin --tenant=admin
```

You've now assigned the Keystone role of `admin` to the `admin` user in the `admin` tenant. Don't worry, most assignments will be more descriptive than admin > admin > admin.

> **DEVELOPMENT OF OPENSTACK MANAGEMENT TOOLS** In previous OpenStack releases, you had to reference long IDs, not names, when making role assignments. Make sure you check the command-line utilities for new commands and command alternatives with each OpenStack release.

The next thing to do is verify the role that you assigned.

LISTING ROLES

To verify that the `admin` user has been assigned the appropriate roles in the `admin` tenant, you can use the keystone `user-role-list` command to list all roles of a particular user.

Listing 5.31 Verify `admin` roles in `admin` tenant

```
$ keystone user-role-list --user=admin --tenant=admin
+----------------------------------+----------------------+
|                id                |         name         |
+----------------------------------+----------------------+
| 42639ba997424e7d8fbf24353bff2a08 |        admin         |
+----------------------------------+----------------------+
```

Object IDs in OpenStack are both long and unique to each instance created, so the `user_id` and `tenant_id` information has been truncated in the displayed output.

Congratulations! You've now completed all manual steps in a Keystone deployment and you've verified its proper operation.

> **CHECK KEYSTONE SERVICE LOGS** Before you move forward, take a look at the Keystone logs (/var/log/keystone) for any errors or other obvious problems (such as trace output from failures). You can also find logs for all OpenStack services under /var/log/upstart/.

If you had no difficulty deploying OpenStack using DevStack in chapter 2, you'll likely have a greater appreciation for DevStack after installing this first component . If you did have trouble with DevStack, the benefits of this manual walk-through portion of the book should now be clear.

In the next section, you'll walk through the installation of your last shared-service component, Glance. After that, you'll move on to core component installation.

5.2.2 Deploying the Image Service (Glance)

Virtual machine images are copies of previously configured VM instances. These images can be cloned and applied to new VMs as the VMs are created. This process saves the user from having to deploy the operating system and other software when deploying VMs.

Glance is the module of OpenStack that's used to discover, deploy, and manage VM images in the OpenStack environment. By default, Glance will take advantage of RabbitMQ services, which allow OpenStack components to remotely communicate with Glance without communicating through the controller.

In the following sections, you'll manually configure the required Glance service and the Glance endpoint in Keystone. You'll also create MySQL tables and grant MySQL rights so the Glance service can use it as a central data store.

CREATING THE GLANCE DATA STORE

You now need to create the Glance database, which database will be used to store configuration and state information about images. Then you can grant the MySQL user `glance_dbu` access to the new database.

In MySQL, user creation and rights authorization functions can be completed in the same step. First, log in to the database server as `root`, as described in the subsection "Accessing the MySQL console." Then use the MySQL `GRANT` command, as follows.

Listing 5.32 Creating the database and granting access

```
CREATE DATABASE glance;
GRANT ALL ON glance.* TO 'glance_dbu'@'localhost' \
    IDENTIFIED BY 'openstack1';
```

You can check that the grants were successful:

```
mysql> SHOW GRANTS FOR 'glance_dbu'@'localhost';
+---------------------------------------------------------------------+
| Grants for glance_dbu@localhost                                           |
+---------------------------------------------------------------------+
| GRANT USAGE ON *.* TO 'glance_dbu'@'localhost' <removed password> |
| GRANT ALL PRIVILEGES ON `glance`.* TO 'glance_dbu'@'localhost'    |
+---------------------------------------------------------------------+
2 rows in set (0.00 sec)
```

To exit the MySQL shell, type `quit` and press Enter.

In the example, the `glance_dbu.* TO 'glance_dbu'@'localhost'` part means that the MySQL user `glance` is being given access to all objects under the Glance database from the localhost.

The local system user glance will be created in the subsection "Installing Glance"; the account referenced in this section is an internal MySQL account. The Glance service will be run under the local glance account, whereas the MySQL glance_dbu account will be used strictly for accessing the Glance database table you created under MySQL.

In the next sections, you'll configure the Glance user, service, and endpoint in Keystone. This will allow the Glance component to be recognized and operate in your deployment.

CONFIGURING A GLANCE KEYSTONE USER

You must create a Keystone service user account for Glance. This account will be used by the Glance service to validate tokens and authenticate and authorize other user requests. For Glance to be visible to the system, you must create a service and an endpoint in Keystone.

Create the glance user in Keystone as follows.

Listing 5.33 Creating a glance user

```
$ keystone user-create --name=glance \
           --pass="openstack1" \
           --email=glance@testco.com
+----------+----------------------------------+
| Property |              Value               |
+----------+----------------------------------+
|  email   |        glance@testco.com         |
| enabled  |               True               |
|    id    | 2ec6f7d7fbc64da090770be764d9c6a8 |
|   name   |              glance              |
| username |              glance              |
+----------+----------------------------------+
```

You can now assign the admin role to the glance user. The following listing shows how you can use the glance Keystone username, the service Keystone tenant name, and the admin Keystone role name to assign the admin role to the glance user in the service tenant.

Listing 5.34 Assigning admin role to glance user in service tenant

```
keystone user-role-add --user=glance --role-id=admin --tenant=service
```

If the command is successful, there won't be any output.

Next, check to make sure the user was created and that the proper roles were assigned.

```
keystone user-role-list --user=glance --tenant=service
+----------------------------------+-------+
|                id                | name  |
+----------------------------------+-------+
| 42639ba997424e7d8fbf24353bff2a08 | admin |
+----------------------------------+-------+
```

Information on user_id and tenant_id has been truncated in the preceding output.

You're now ready to move forward with creating the service and endpoint.

CREATING THE GLANCE SERVICE AND ENDPOINT

You can now create the service and endpoint for the Glance Image Service. Endpoint and service information is maintained by Keystone, as described in section 5.2.1. Registering the service allows it to be known by the OpenStack deployment, and registering the endpoint defines the API locations for the service.

During the service-creation process, you must provide parameters that describe your service. For instance, in the following example, notice the --type=image parameter, which specifies to Keystone that this is an image service. The name and description are really for the benefit of readability; the type is the value used for service differentiation.

Listing 5.35 Creating the Glance service

```
$ keystone service-create --name=glance --type=image \
--description="Image Service"
+-------------+----------------------------------+
|   Property  |              Value               |
+-------------+----------------------------------+
| description |          Image Service           |
|   enabled   |               True               |
|      id     | ff29dcdc693e4e55b3720a4da2771da8 |
|     name    |              glance              |
|     type    |              image               |
+-------------+----------------------------------+
```

To create the endpoint, you must provide the Glance service name you just generated, along with the region, publicurl, internalurl, and adminurl. As previously mentioned, this book assumes a single region deployment, so you'll use RegionOne for all region settings. The publicurl will correspond to the public address listed in table 5.1. The internalurl and adminurl will correspond to the OpenStack internal address of the controller in the same table. The command and its output is shown in the following listing.

Listing 5.36 Creating the Glance endpoint

```
$ keystone endpoint-create \
>   --region RegionOne \
>   --service=glance \
>   --publicurl=http://10.33.2.50:9292 \
>   --internalurl=http://192.168.0.50:9292 \
>   --adminurl=http://192.168.0.50:9292
+-------------+----------------------------------+
|   Property  |              Value               |
+-------------+----------------------------------+
|   adminurl  |     http://192.168.0.50:9292     |
|      id     | aaeaaf52c3c94b2eaf3bc33bd16db0b3 |
| internalurl |     http://192.168.0.50:9292     |
|  publicurl  |      http://10.33.2.50:9292      |
```

```
| region      |             RegionOne            |
| service_id  | ff29dcdc693e4e55b3720a4da2771da8 |
+-------------+----------------------------------+
```

Now that the Keystone configuration for Glance is complete, you can proceed with installing the Glance package.

INSTALLING GLANCE

You're now ready to install the Glance binaries on the controller. The following listing shows how that's done.

Listing 5.37 Installing Glance binaries

```
$ sudo apt-get -y install glance glance-api \
    glance-registry python-glanceclient \
    glance-common
The following extra packages will be installed:
  libgmp10 libyaml-0-2 python-amqplib python-anyjson
python-boto python-crypto python-dateutil python-glance
python-httplib2 python-json-patch python-json-pointer
  python-jsonschema python-kombu python-oslo-config
python-swiftclient python-warlock python-xattr python-yaml
...
Adding system user `glance' (UID 109) ...
Adding new user `glance' (UID 109) with group `glance' ...
...
ldconfig deferred processing now taking place
```

The files etc/glance/glance-api.conf and /etc/glance/glance-registry.conf require modifications to set the MySQL information.

The following listing shows the changes for /etc/glance/glance-api.conf

Listing 5.38 Modifying /etc/glance/glance-api.conf

```
[DEFAULT]                                    <-- 1 Configures MySQL information
rpc_backend = rabbit
rabbit_host = 192.168.0.50
rabbit_password = openstack1

[database]                                   <-- 2 Configures backend setting
#sqlite_db = /var/lib/glance/glance.sqlite
connection = mysql://glance_dbu:openstack1@localhost/glance
mysql_sql_mode = TRADITIONAL
...
```

As you can see, in /etc/glance/glance-api.conf you need to configure both the MySQL information ❶ and the rpc_backend settings ❷.

Where does Glance store data?

Glance can be configured to use several backends as data stores, including the local filesystem, Cinder, and Swift (OpenStack object storage). By default, the following parameter in /etc/glance/glance-api.conf sets the /var/lib/glance/images directory as the Glance repository:

```
# Directory that the Filesystem backend store
# writes image data to
filesystem_store_datadir = /var/lib/glance/images/
```

Now you can modify glance-registry.conf. Here you only need to make the changes to the `[database]` section.

Listing 5.39 Modifying /etc/glance/glance-registry.conf

```
[database]
#sqlite_db = /var/lib/glance/glance.sqlite
connection = mysql://glance_dbu:openstack1@localhost/glance
mysql_sql_mode = TRADITIONAL
```

In order to update the configuration, you must restart the glance-api and glance-registry services.

Listing 5.40 Restarting glance-api and glance-registry

```
$ sudo service glance-api restart
glance-api stop/waiting
glance-api start/running, process 5372

$ sudo service glance-registry restart
glance-registry stop/waiting
glance-registry start/running, process 5417
```

Now that you've configured the Glance service with the required database and account information, you need to initialize the Glance database with the following command. There will be no output if it's successful.

Listing 5.41 Initializing the data store

```
sudo glance-manage db_sync
```

The Glance module should now be initialized, and Glance should be ready to manage images.

UTF8 error

If during `db_sync` you experience the error `CRITICAL glance [-] ValueError: Tables "migrate_version" have non utf8 collation, please make sure all tables are CHARSET=utf8`, take the following action, which converts the table encoding (CHARSET) to Unicode (utf8):

```
$ mysql --user=root --password=openstack1 glance

mysql> alter table migrate_version convert to \
  character set utf8 collate utf8_unicode_ci;
Query OK, 1 row affected (0.25 sec)
Records: 1  Duplicates: 0  Warnings: 0
```

IMAGE MANAGEMENT

To test Glance, you can download a prebuilt image and register it with Glance. For testing purposes, we'll use a publicly available Ubuntu Cloud Image, which has been developed specifically to run on cloud environments like OpenStack.

The command for downloading it is shown in the following listing. You're free to use any Glance-supported image type, but keep in mind that KVM paravirtualization drivers might need to be added to any stock images.

Listing 5.42 Downloading a prebuilt image

```
wget http://cdn.download.cirros-cloud.net/0.3.2/cirros-0.3.2-x86_64-disk.img
```

UBUNTU IMAGE You can follow the instructions in listing 5.43 to add any image, like this Ubuntu one: http://uec-images.ubuntu.com/trusty/current/trusty-server-cloudimg-amd64-disk1.img.

Once the image downloads, you can use it to create a Glance image, as shown in listing 5.43. There are several image container and disk formats, and you'll want to store this image in the *qcow2* format, which is commonly used with KVM environments. The image container itself is in *OVF*, which is specified on the command line. Disk and container formats can vary based on the disk image. As of the Grizzly OpenStack release, all containers will be treated as *bare*, so if you're unsure of the container format, bare is a safe option.

Tables 5.2 and 5.3 list supported disk and container formats as documented on the OpenStack site.

Table 5.2 Disk formats

Format	Description
raw	This is an unstructured disk image format.
vhd	This is the VHD disk format, a common disk format used by virtual machine monitors from VMware, Xen, Microsoft, VirtualBox, and others.

Table 5.2 Disk formats *(continued)*

Format	Description
vmdk	This is another common disk format supported by many common VM monitors.
vdi	This is a disk format supported by VirtualBox VM monitor and the QEMU emulator.
iso	This is an archive format for the data contents of an optical disc (such as a CD-ROM).
qcow	This is a disk format supported by the QEMU emulator that can expand dynamically and that supports copy-on-write.
aki	This indicates that what is stored in Glance is an Amazon kernel image.
ari	This indicates that what is stored in Glance is an Amazon ramdisk image.
ami	This indicates that what is stored in Glance is an Amazon machine image.

Table 5.3 Container formats

Format	Description
bare	This indicates there is no container or metadata envelope for the image.
ovf	This is the OVF container format.
aki	This indicates that what is stored in Glance is an Amazon kernel image.
ari	This indicates that what is stored in Glance is an Amazon ramdisk image.
ami	This indicates that what is stored in Glance is an Amazon machine image.
ova	This indicates that what is stored in Glance is an OVA TAR file.

KEYSTONE AUTHENTICATION The previous commands you've run were authenticated through service credentials provided by environment variables set during the Keystone install. For the following commands, you'll need to either provide Keystone user credentials through the command line or set environmental variables for user authentication. For the sake of clarity, we'll use the command-line authentication option for the remainder of the book where user credentials are required.

The following listing shows the creation of a Glance image.

Listing 5.43 Creating a Glance image

```
$ glance --os-username=admin --os-password openstack1 \
> --os-tenant-name=admin \
> --os-auth-url=http://10.33.2.50:5000/v2.0  \
>  image-create \
> --name="Cirros 0.3.2" \
> --is-public=true \
> --disk-format=qcow2 \
```

```
> --container-format=bare \
> --file cirros-0.3.2-x86_64-disk.img
+------------------+--------------------------------------+
| Property         | Value                                |
+------------------+--------------------------------------+
| checksum         | 64d7c1cd2b6f60c92c14662941cb7913     |
| container_format | bare                                 |
| created_at       | 2014-09-05T14:04:09                  |
| deleted          | False                                |
| deleted_at       | None                                 |
| disk_format      | qcow2                                |
| id               | e02a73ef-ba28-453a-9fa3-fb63c1a5b15c |
| is_public        | True                                 |
| min_disk         | 0                                    |
| min_ram          | 0                                    |
| name             | Cirros 0.3.2                         |
| owner            | None                                 |
| protected        | False                                |
| size             | 13167616                             |
| status           | active                               |
| updated_at       | 2014-09-05T14:04:09                  |
| virtual_size     | None                                 |
+------------------+--------------------------------------+
```

You've now uploaded, registered, and made available an image to be served from the Glance service.

At this point aside from listing the available images, you don't have a good way to test Glance like you did Keystone, because you don't have the rest of the components to deploy a virtual machine. Unfortunately, you'll need to work your way through chapter 8 before you can fully test Glance.

GLANCE SERVICE CHECK Before you move forward, take a look at the Glance logs (/var/log/glance) for any errors or other obvious problems (such as trace output from failure). You can also find logs for all OpenStack services under /var/log/upstart/. On startup, the Glance API will likely complain of unconfigured options (sheepdog, rdb, gridfs, swift, and so on), which is fine. Keep an eye out for repeated warnings and errors that occur after startup.

Congratulations! You've completed the shared services section. You're well on your way to completing the controller configuration steps of your OpenStack deployment. In the following sections, you'll start the controller-side (server-side) configuration of other core services. Starting with the Storage service, then Network, and finally Compute, you'll complete the controller deployment.

5.3 Deploying the Block Storage (Cinder) service

Cinder is the module of OpenStack used to provide block (volume) storage to VM images in the OpenStack environment. It manages the process of provisioning remotely available storage to VMs running on compute nodes. This relationship is shown in figure 5.4, where *VM Compute* and *VM Volume* are provided by two separate

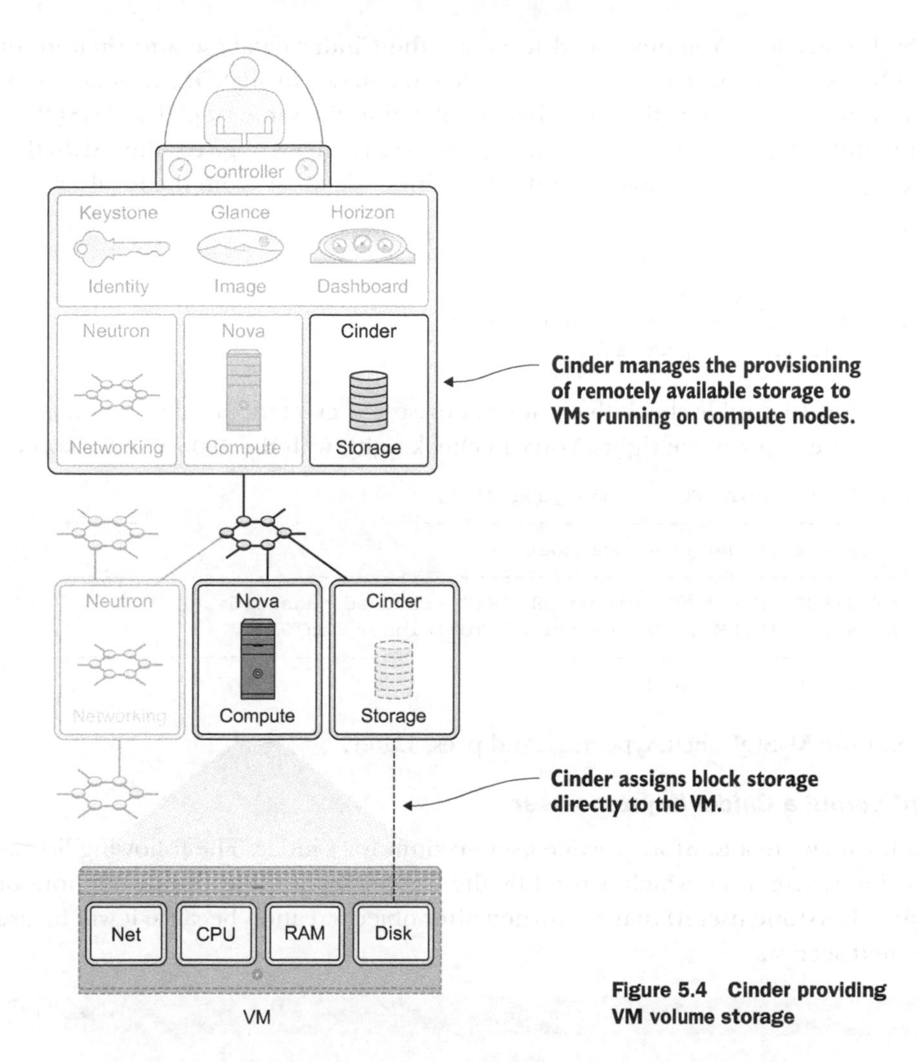

Figure 5.4 Cinder providing VM volume storage

physical resources, *Compute hardware* and *Cinder resource node*. This separation might seem strange, but for the time being just accept that the benefits of flexibility outweigh the complexity and performance drawbacks for most use cases.

By default, Cinder will take advantage of RabbitMQ services, which allow other client components, like Nova, to remotely communicate with Cinder without passing communication through the controller. You'll manually configure the required Cinder service and endpoint in Keystone. You'll also manually create a MySQL database and tables, and assign MySQL rights so that database can be used as a central Cinder data store.

5.3.1 Creating the Cinder data store

To create the Cinder data store, you first need to log in to the MySQL database instance on your controller as `root` (this was described in section "Accessing the

MySQL console"). You next need to create the Cinder database and then grant the MySQL user `cinder_dbu` access to the new database. In MySQL, user-creation and rights-authorization functions can be completed in the same step. The MySQL `GRANT` command `cinder_dbu.* TO 'cinder_dbu'@'localhost'` gives the MySQL user `cinder_dbu` access to all objects under the Cinder database from the localhost.

Listing 5.44 Create the Cinder database and grant access

```
CREATE DATABASE cinder;
GRANT ALL ON cinder.* TO 'cinder_dbu'@'%' \
    IDENTIFIED BY 'openstack1';
```

You'll want to double-check that your database was created and that the `cinder_dbu` user has the appropriate rights. You can check rights with the `SHOW GRANTS` command:

```
mysql> SHOW GRANTS FOR 'cinder_dbu'@'%';
+--------------------------------------------------------------------+
| Grants for cinder_dbu@localhost                                     |
+--------------------------------------------------------------------+
|GRANT USAGE ON *.* TO 'cinder_dbu'@'%'<removed password>          |
|GRANT ALL PRIVILEGES ON `cinder`.* TO 'cinder_dbu'@'%'              |
+--------------------------------------------------------------------+
2 rows in set (0.00 sec)
```

To exit the MySQL shell, type `quit` and press Enter.

5.3.2 Configuring a Cinder Keystone user

You must create a Keystone service user account for Cinder. The following listing creates the `cinder` user, which is used by the Cinder service. Please make a note of the `cinder` Keystone user ID that's returned after object creation because it will be used in the next section.

Listing 5.45 Creating a `cinder` user

```
$ keystone user-create --name=cinder \
            --pass="openstack1" \
            --email=cinder@testco.com
+----------+----------------------------------+
| Property |              Value               |
+----------+----------------------------------+
|  email   |        cinder@testco.com         |
| enabled  |               True               |
|    id    | 86f8b74446084fdfb44b66781cc72fa9 |
|   name   |              cinder              |
| username |              cinder              |
+----------+----------------------------------+
```

Next, you'll use the `cinder` Keystone username, the `service` Keystone tenant name, and the `admin` Keystone role name to assign the `admin` role to the `cinder` user in the `service` tenant. This command produces no output if it's successful.

Listing 5.46 Assigning `admin` role to `cinder` user in `service` tenant

```
keystone user-role-add --user=cinder --role-id=admin --tenant=service
```

You can now check to make sure the user was created and that the proper roles were assigned:

```
keystone user-role-list --user=cinder --tenant=service
+----------------------------------+-------+
|                id                | name  |
+----------------------------------+-------+
| ae2a897f8a1e4762a7f0f8da596511ce | admin |
+----------------------------------+-------+
```

Information on `user_id` and `tenant_id` has been truncated in the displayed output.

You're now ready to move forward with the service and endpoint creation.

5.3.3 Creating the Cinder service and endpoint

You can now create the service and endpoint for the Cinder service. In the following listing, you specify the type of service as a storage volume using the argument `--type=volume`.

Listing 5.47 Creating a Cinder service

```
$ keystone service-create --name=cinder --type=volume \
--description="Block Storage"
+-------------+----------------------------------+
|  Property   |              Value               |
+-------------+----------------------------------+
| description |          Block Storage           |
|   enabled   |               True               |
|     id      | 939010f014bf406693e70bfc4862e8cd |
|    name     |              cinder              |
|    type     |              volume              |
+-------------+----------------------------------+
```

To create the endpoint, you must provide the Cinder service name you just generated, along with the `region`, `publicurl`, `internalurl`, and `adminurl`. This book describes a single-region deployment, so you'll use `RegionOne` for all `region` settings. The `publicurl` will correspond to the public address in table 5.1, and the `internalurl` and `adminurl` correspond to the OpenStack internal address of the controller found in the same table. The following listing shows the endpoint creation. Make sure you enter the following information exactly as shown, including percent signs and backslashes.

Listing 5.48 Creating a Cinder endpoint

```
$ keystone endpoint-create \
>  --region RegionOne \
>  --service=cinder \
>  --publicurl=http://10.33.2.50:8776/v1/%\(tenant_id\)s \
```

```
> --internalurl=http://192.168.0.50:8776/v1/%\(tenant_id\)s \
> --adminurl=http://192.168.0.50:8776/v1/%\(tenant_id\)s
+-------------+------------------------------------------+
|   Property  |                  Value                   |
+-------------+------------------------------------------+
|   adminurl  | http://192.168.0.50:8776/v1/%(tenant_id)s |
|      id     |     2cf277bd14b94566b306ff303c2ab993     |
| internalurl | http://192.168.0.50:8776/v1/%(tenant_id)s |
|  publicurl  |  http://10.33.2.50:8776/v1/%(tenant_id)s |
|    region   |                RegionOne                 |
|  service_id |     939010f014bf406693e70bfc4862e8cd     |
+-------------+------------------------------------------+
```

The Keystone configuration for Cinder has now been completed. You can move forward with the package installation.

5.3.4 Installing Cinder

You're now ready to install the Cinder binaries on the controller.

Listing 5.49 Installing Cinder

```
$ sudo apt-get -y install cinder-api cinder-scheduler \
Processing triggers for ureadahead (0.100.0-16) ...
Setting up python-concurrent.futures (2.1.6-3) ...
Setting up python-networkx (1.8.1-0ubuntu3) ...
Setting up python-taskflow (0.1.3-0ubuntu3) ...
...
INFO migrate.versioning.api [-] 21 -> 22...
INFO migrate.versioning.api [-] done
Setting up cinder-api (1:2014.1.1-0ubuntu2) ...
cinder-api start/running, process 16558
Setting up cinder-scheduler (1:2014.1.1-0ubuntu2) ...
cinder-scheduler start/running, process 16601
```

You must now modify the main Cinder configuration file (etc/cinder/cinder.conf), providing queue, database, and Keystone information.

Listing 5.50 Modifying /etc/cinder/cinder.conf

```
[DEFAULT]
rpc_backend = rabbit
rabbit_host = 192.168.0.50
rabbit_password = openstack1

[database]
connection = mysql://cinder_dbu:openstack1@localhost/cinder

[keystone_authtoken]
auth_uri = http://192.168.0.50:35357
admin_tenant_name = service
admin_password = openstack1
auth_protocol = http
admin_user = cinder
```

In order to update the configuration, you must restart Cinder with the following two commands.

Listing 5.51 Restarting Cinder

```
sudo service cinder-scheduler restart
sudo service cinder-api restart
```

Now that you've configured the Cinder service with the required queue, database, and Keystone information, you need to initialize the Cinder database.

Listing 5.52 Initializing the data store

```
$ sudo cinder-manage db sync
INFO migrate.versioning.api [-] 0 -> 1...
INFO migrate.versioning.api [-] done
...
INFO migrate.versioning.api [-] 21 -> 22...
INFO migrate.versioning.api [-] done
```

Congratulations! The Cinder module should now be initialized, and Cinder is ready to manage block storage. Unfortunately, you won't be able to fully test the component deployment until you've configured the resource portion of this service in chapter 8 and are ready to launch a VM using your manual deployment.

> **CINDER SERVICE CHECK** Before you move forward, take a look at the Cinder logs (/var/log/cinder) for any errors or other obvious problems (such as trace output from failures). You can also find logs for all OpenStack services under /var/log/upstart/.

OK, you've installed the basic shared service and the controller-side portions of the Cinder service. Next you need to continue installing controller-side components for Networking and Compute.

5.4 Deploying the Networking (Neutron) service

OpenStack Neutron is the core of the cloud network service. Neutron APIs form the primary interface used to manage network services inside OpenStack.

Figure 5.5 shows Neutron managing both the VM network interface on the VM and the routing and switching for the network that the VM network is attached to. Simply put, Neutron manages all physical and virtual components required to connect, create, and extend networks between VMs and public network interfaces (gateways outside OpenStack networking).

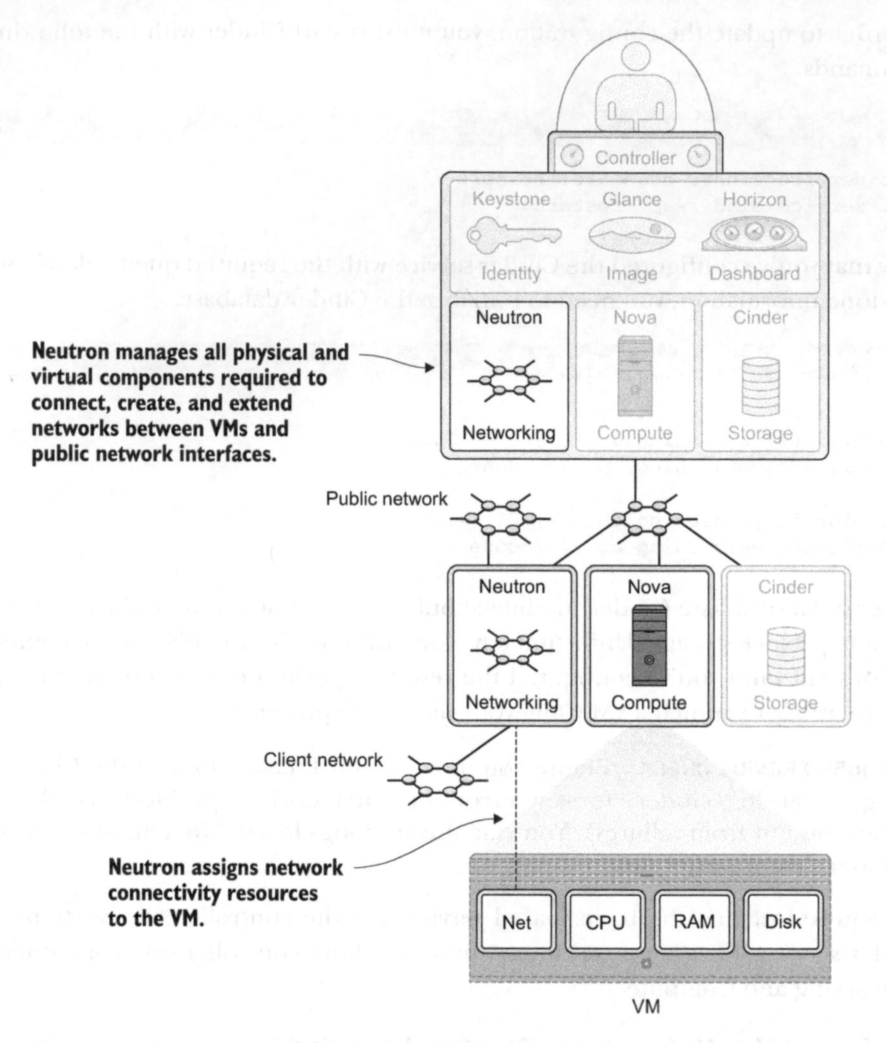

Figure 5.5 Neutron managing OpenStack Networking

5.4.1 Creating the Neutron data store

Once again, you need to log in to the MySQL console as `root` (as described in the subsection "Accessing the MySQL console"). Then you can create the Neutron database and grant the MySQL user `neutron` access to the new database.

In MySQL, the user-creation and rights-authorization functions can be completed in the same step, as shown in the following listing. The MySQL `GRANT` command `neutron.* TO 'neutron_dbu'@'localhost'` specifies that the MySQL user `neutron_dbu` has access to all objects under the Neutron database from the localhost. Likewise, the command `neutron.* TO 'neutron_dbu'@'%'` means that rights have been granted to the `neutron_dbu` user from any host.

Listing 5.53 Creating a database and granting access

```
CREATE DATABASE neutron;
GRANT ALL ON neutron.* TO 'neutron_dbu'@'localhost' IDENTIFIED BY 'openstack1';
GRANT ALL ON neutron.* TO 'neutron_dbu'@'%' IDENTIFIED BY 'openstack1';
```

You'll want to double-check that your database was created and that the neutron_dbu user has appropriate rights:

```
mysql> SHOW GRANTS FOR 'neutron_dbu'@'%';
+--------------------------------------------------------------------+
| Grants for neutron@%                                               |
+--------------------------------------------------------------------+
|GRANT USAGE ON *.* TO 'neutron_dbu'@'%'<removed password>           |
|GRANT ALL PRIVILEGES ON `neutron`.* TO 'neutron_dbu'@'%'            |
+--------------------------------------------------------------------+
2 rows in set (0.00 sec)
```

5.4.2 Configuring a Neutron Keystone user

You can now create the Keystone neutron user. Please make note of the neutron Keystone user ID that's returned after object creation.

Listing 5.54 Creating a *neutron* user

```
$ keystone user-create --name=neutron \
            --pass="openstack1" \
            --email=neutron@testco.com
+----------+----------------------------------+
| Property |              Value               |
+----------+----------------------------------+
|  email   |        neutron@testco.com        |
| enabled  |               True               |
|    id    | e817903594c843f7a79e1404a6f2a82c |
|   name   |             neutron              |
| username |             neutron              |
+----------+----------------------------------+
```

ASSIGNING A ROLE TO THE NEUTRON USER

You can now use the neutron Keystone username, service Keystone tenant name, and admin Keystone role name to assign the admin role to the neutron user. The following listing shows this command; there's no output if it's successful.

Listing 5.55 Assigning the admin role to the neutron user in the service tenant

```
keystone user-role-add \
--user=neutron \
--role=admin \
--tenant=service
```

You can now check to make sure the user has been created and that the appropriate roles have been assigned:

```
keystone user-role-list --user=neutron --tenant=service
+----------------------------------+-------+
|                id                | name  |
+----------------------------------+-------+
| 42639ba997424e7d8fbf24353bff2a08 | admin |
+----------------------------------+-------+
```

Information on the user ID and tenant ID has been truncated in the displayed output.

You're now ready to move forward with creating the service and endpoint.

CREATING THE NEUTRON SERVICE AND ENDPOINT

The next step is to create the service and endpoint for the Neutron network service. This service is specified as the network service by the parameter `--type=network`.

Listing 5.56 Creating the Neutron service

```
$ keystone service-create --name=neutron --type=network \
--description="OpenStack Networking Service"
+-------------+----------------------------------+
|   Property  |              Value               |
+-------------+----------------------------------+
| description |   OpenStack Networking Service   |
|   enabled   |               True               |
|      id     | 7d92cd9f66c34cd882b88be2f486e123 |
|     name    |             neutron              |
|     type    |             network              |
+-------------+----------------------------------+
```

To create the endpoint, you must provide the Neutron service name you just generated, along with your `region`, `publicurl`, `internalurl`, and `adminurl`. This book describes a single-region deployment, so you'll use `RegionOne` for all `region` settings. The `publicurl` will correspond to the public address found in table 5.1. The `internalurl` and `adminurl` correspond to the OpenStack internal address for the controller found in the same table. The following listing shows the endpoint creation

Listing 5.57 Creating the Neutron endpoint

```
$ keystone endpoint-create \
>  --region RegionOne \
>  --service=neutron \
>  --publicurl=http://10.33.2.50:9696 \
>  --internalurl=http://192.168.0.50:9696 \
>  --adminurl=http://192.168.0.50:9696
+-------------+----------------------------------+
|   Property  |              Value               |
+-------------+----------------------------------+
|   adminurl  |     http://192.168.0.50:9696     |
|      id     | 678fa049587a4f9b8b758c6158b67599 |
| internalurl |     http://192.168.0.50:9696     |
|  publicurl  |      http://10.33.2.50:9696      |
|    region   |            RegionOne             |
|  service_id | 7d92cd9f66c34cd882b88be2f486e123 |
+-------------+----------------------------------+
```

Keystone configuration for Neutron has now been completed. You can go ahead and install the Neutron packages.

5.4.3 Installing Neutron

In this section, you'll prepare the Neutron network service for operation. First, you install Neutron with the following command.

Listing 5.58 Install Neutron

```
$ sudo apt-get install -y neutron-server
...
Adding system user `neutron' (UID 115) ...
Adding new user `neutron' (UID 115) with group `neutron' ...
...
neutron-server start/running, process 8058
Processing triggers for ureadahead ...
```

The next step is configuration. First you must modify the /etc/neutron/neutron .conf file. You need to change the default admin information, logging verbosity, and RabbitMQ password to match your deployment parameters. You don't want to delete the entire /etc/neutron/neutron.conf file. Just replace the default values with the values specified in the following listing.

Listing 5.59 Modifying /etc/neutron/neutron.conf

```
[DEFAULT]
core_plugin = neutron.plugins.ml2.plugin.Ml2Plugin
service_plugins = router,firewall,lbaas,vpnaas,metering
allow_overlapping_ips = True
...
nova_url = http://192.168.0.50:8774/v2
nova_admin_username = admin
nova_admin_password = openstack1
nova_admin_tenant_id = 55bd141d9a29489d938bb492a1b2884c
nova_admin_auth_url = http://10.33.2.50:35357/v2.0
...
[keystone_authtoken]
auth_uri = http://10.33.2.50:5000
auth_protocol = http
admin_tenant_name = service
admin_user = neutron
admin_password = openstack1
...
[database]
connection = mysql://neutron_dbu:openstack1@localhost/neutron
```

- Configures Neutron to use the ML2 plug-in
- Enables the service plug-ins. At a minimum, the router plug-in will be required for the example deployment.
- Tells Neutron how to communicate with Nova. You can use the service tenant_id that was generated in listing 5.26.

Now that the core Neutron components are configured, you need to configure the Neutron Modular Layer 2 (ML2) plug-in. The ML2 plug-in, which combines several deprecated standalone plug-ins, is a standard framework for managing multiple OSI Layer 2 technologies commonly used in OpenStack deployments. In the walk-through examples in chapter 6, the ML2 plug-in will allow Neutron to control the Open vSwitch (virtual switch) on your compute nodes. The following configurations tell Neutron/ML2 how it should manage your Layer 2 connectivity.

In this step you'll configure the ML2 plug-in in the /etc/neutron/plugins/ml2/ml2_conf.ini file.

Listing 5.60 Modifying /etc/neutron/plugins/ml2/ml2_conf.ini

```
[ml2]
type_drivers = gre
tenant_network_types = gre
mechanism_drivers = openvswitch

[ml2_type_gre]
tunnel_id_ranges = 1:1000

[securitygroup]
firewall_driver =
 neutron.agent.linux.iptables_firewall.OVSHybridIptablesFirewallDriver
enable_security_group = True
```

The final step is to restart Neutron with your new configuration.

Listing 5.61 Restarting Neutron

```
$ sudo service neutron-server restart
neutron-server stop/waiting
neutron-server start/running, process 24590
```

You'll want to check the Neutron log to make sure the service started and is listening for requests. The primary Neutron log is the /var/log/Neutron/server.log file. In this file you should see a line containing "INFO [Neutron.service] Neutron service started, listening on 0.0.0.0:9696" if the service started successfully. If a service doesn't produce a log file, you can additionally check the service upstart log found in the /var/log/upstart/Neutron-server.log file, which should provide additional debug information.

Congratulations! You've now completed the controller-side configuration of OpenStack networking. In the next section, you'll perform the controller-side configuration of the final core component, OpenStack Compute.

5.5 Deploying the Compute (Nova) service

You could consider the OpenStack Nova component as the core of the cloud framework controller. Although each component has its own set of APIs, the Nova API forms the primary interface used to manage pools of resources. Figure 5.6 shows how Nova both manages local compute (CPU and MEM) resources and orchestrates the provisioning of secondary resources (network and storage).

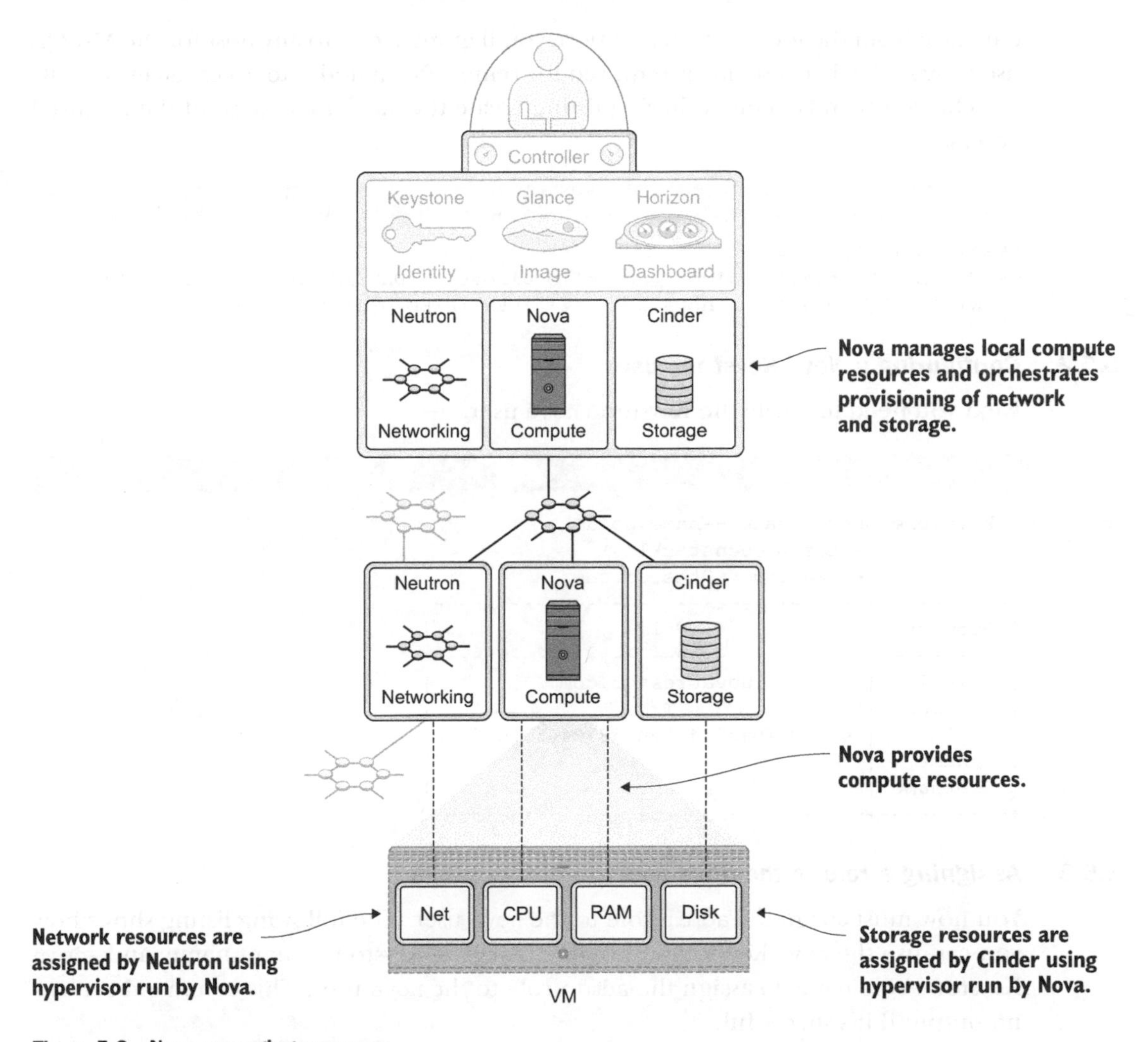
Controller
Keystone
Glance
Horizon
Identity
Image
Dashboard
Neutron
Nova
Cinder
Networking
Compute
Storage
Nova manages local compute resources and orchestrates provisioning of network and storage.
Neutron
Nova
Cinder
Networking
Compute
Storage
Nova provides compute resources.
Net
CPU
RAM
Disk
VM
Network resources are assigned by Neutron using hypervisor run by Nova.
Storage resources are assigned by Cinder using hypervisor run by Nova.

Figure 5.6 Nova managing resources

Nova supports a wide variety of hypervisors as well as bare-metal configurations. As shown in figure 5.6, Nova works with its own resource nodes along with Neutron and Cinder to bring together resources to run a virtual machine.

5.5.1 Creating the Nova data store

In this section, you'll create the Nova database and then grant the MySQL `nova_dbu` user access to the new database. Once again, you need to log in to the MySQL console as `root` to do this (see the subsection "Accessing the MySQL console" for details).

Recall that in MySQL, user-creation and rights-authorization functions can be completed in the same step. The MySQL `GRANT` command `$nova.* TO 'nova_dbu'@ 'localhost'` grants the MySQL user `nova_dbu` access to all objects under the Nova database from the localhost. In addition, you'll grant access to any host for the MySQL user nova_dbu, because this is required for remote Nova nodes to access a central DB.

The commands in the following listing create the database and grant the required access.

Listing 5.62 Create database and grant access

```
CREATE DATABASE nova;
GRANT ALL ON nova.* TO 'nova_dbu'@'localhost'  IDENTIFIED BY 'openstack1';
GRANT ALL ON nova.* TO 'nova_dbu'@'%' IDENTIFIED BY 'openstack1';
```

5.5.2 Configuring a Nova Keystone user

Next you need to create the Keystone nova user.

Listing 5.63 Creating the nova user

```
$ keystone user-create --name=nova \
            --pass="openstack1" \
            --email=nova@testco.com
+----------+----------------------------------+
| Property |              Value               |
+----------+----------------------------------+
|  email   |         nova@testco.com          |
| enabled  |               True               |
|    id    | 44fe95fbaf524c09ae633f405d9d66ca |
|   name   |               nova               |
| username |               nova               |
+----------+----------------------------------+
```

5.5.3 Assigning a role to the nova user

You now must assign the admin role to the nova user. The following listing shows how you can use the nova Keystone username, service Keystone tenant name, and admin Keystone role name to assign the admin role to the nova user. This command returns no output if it's successful.

Listing 5.64 Assigning admin role to nova user in service tenant

```
keystone user-role-add --user=nova --role=admin --tenant=service
```

You can now check to make sure that the user has been created and that appropriate roles have been assigned.

Listing 5.65 Checking role assignment

```
$ keystone user-role-list --user=nova --tenant=service
+----------------------------------+-------+
|                id                | name  |
+----------------------------------+-------+
| 42639ba997424e7d8fbf24353bff2a08 | admin |
+----------------------------------+-------+
```

Information on `user_id` and `tenant_id` has been truncated in the displayed output.

You're now ready to create the service and endpoint.

5.5.4 Creating the Nova service and endpoint

Next you need to create the service and endpoint for the Nova service. The service will be designated as a Compute service by specifying the `type=compute` parameter, as shown in the following listing.

Listing 5.66 Creating the Nova service

```
$ keystone service-create --name=nova --type=compute \
> --description="OpenStack Compute Service"
+-------------+----------------------------------+
|   Property  |              Value               |
+-------------+----------------------------------+
| description |    OpenStack Compute Service     |
|   enabled   |               True               |
|      id     | 122f7e4cbd4a48cc81018af2fd27f84c |
|     name    |               nova               |
|     type    |             compute              |
+-------------+----------------------------------+
```

To create the endpoint, you must provide the Nova service name you just generated, along with your `region`, `publicurl`, `internalurl`, and `adminurl`. As mentioned before, this book describes a single-region deployment, so you'll use `RegionOne` for all `region` settings. The `publicurl` will correspond to your public address found in table XREF _101. The `internalurl` and `adminurl` correspond to the OpenStack internal address of the controller found in the same table.

Listing 5.67 Creating the Nova endpoint

```
$ keystone endpoint-create --region RegionOne \
> --service=nova \
> --publicurl='http://10.33.2.50:8774/v2/$(tenant_id)s' \
> --internalurl='http://192.168.0.50:8774/v2/$(tenant_id)s' \
> --adminurl='http://192.168.0.50:8774/v2/$(tenant_id)s'
+-------------+--------------------------------------------+
|   Property  |                   Value                    |
+-------------+--------------------------------------------+
|   adminurl  | http://192.168.0.50:8774/v2/$(tenant_id)s  |
|      id     |      b9f064fdff014ada8c46814715082928      |
| internalurl | http://192.168.0.50:8774/v2/$(tenant_id)s  |
|  publicurl  |  http://10.33.2.50:8774/v2/$(tenant_id)s   |
|    region   |                 RegionOne                  |
|  service_id |      122f7e4cbd4a48cc81018af2fd27f84c      |
+-------------+--------------------------------------------+
```

You're now ready to install the core Nova components.

5.5.5 Installing the Nova controller

In this section, you'll prepare the Nova controller for operation by installing and configuring required packages.

Listing 5.68 Install Nova controller

```
sudo apt-get -y install nova-api nova-cert nova-conductor nova-consoleauth \
  nova-novncproxy nova-scheduler python-novaclient
...
Adding system user `nova' (UID 114) ...
Adding new user `nova' (UID 114) with group `nova' ...
...
nova-api start/running, process 28367
nova-cert start/running, process 28433
nova-conductor start/running, process 28490
nova-consoleauth start/running, process 28558
nova-novncproxy start/running, process 28664
nova-scheduler start/running, process 28710
...
Processing triggers for libc-bin ...
ldconfig deferred processing now taking place
Processing triggers for ureadahead ...
```

The next configuration is one of the most critical of the entire controller install. Because Nova pulls together several of the core and shared services, you need to provide Nova with information pertaining to your deployment. If you aren't careful, a misconfiguration of Nova could lead to component failure in the overall system, even if you have working core components.

You'll add configuration to the /etc/nova/nova.conf file, which will reference the other core OpenStack services. Add the following configuration to the existing file.

Listing 5.69 /etc/nova/nova.conf

```
[DEFAULT]
rpc_backend = rabbit
rabbit_host = 192.168.0.50
rabbit_password = openstack1

my_ip = 192.168.0.50
vncserver_listen = 0.0.0.0
vncserver_proxyclient_address = 0.0.0.0

auth_strategy=keystone
service_neutron_metadata_proxy = true
neutron_metadata_proxy_shared_secret = openstack1

network_api_class = nova.network.neutronv2.api.API
neutron_url = http://192.168.0.50:9696
neutron_auth_strategy = keystone
neutron_admin_tenant_name = service
neutron_admin_username = neutron
neutron_admin_password = openstack1
neutron_admin_auth_url =  http://192.168.0.50:35357/v2.0
```

```
linuxnet_interface_driver =
  nova.network.linux_net.LinuxOVSInterfaceDriver
firewall_driver = nova.virt.firewall.NoopFirewallDriver
security_group_api = neutron
[database]
connection = mysql://nova_dbu:openstack1@localhost/nova

[keystone_authtoken]
auth_uri = http://192.168.0.50:35357
admin_tenant_name = service
admin_password = openstack1
auth_protocol = http
admin_user = nova
```

Next you need to create the Nova tables in the database. The provided nova-manage script uses the /etc/nova/nova.conf file for configuration. You can just execute the Nova script as follows.

Listing 5.70 Executing nova-manage

```
$ sudo nova-manage db sync
INFO migrate.versioning.api [-] 215 -> 216...
...
INFO migrate.versioning.api [-] 232 -> 233...
INFO migrate.versioning.api [-] done
```

If you experience an error, please review the database settings you changed in the previous step.

Finally, you must restart all Nova services..

Listing 5.71 Restarting services

```
$ cd /usr/bin/; for i in $( ls nova-* ); \
 do sudo service $i restart; done
nova-api stop/waiting
nova-api start/running, process 5467
nova-cert stop/waiting
nova-cert start/running, process 5479
nova-conductor stop/waiting
nova-conductor start/running, process 5491
nova-consoleauth stop/waiting
nova-consoleauth start/running, process 5503
nova-novncproxy stop/waiting
nova-novncproxy start/running, process 5532
nova-scheduler stop/waiting
nova-scheduler start/running, process 5547
```

To confirm that all services are running properly, execute the `nova-manage` command and check on the status and state of each service. The status should be `enabled` and the state should show `:-)`.

Listing 5.72 Listing Nova services

```
$ sudo nova-manage service list
Binary           Host          Zone            Status      State Updated_At
nova-cert        controller    internal        enabled     :-)    2014-08-
    08 15:34:24
nova-conductor   controller    internal        enabled     :-)    2014-08-
    08 15:34:24
nova-scheduler   controller    internal        enabled     :-)    2014-08-
    08 15:34:24
nova-consoleauth controller    internal        enabled     :-)    2014-08-
    08 15:34:24
```

For each service, there's an associated log that can be found in /var/log/nova/. The format for the log is *service name*.log (for example, /var/log/nova-api.log would be the log for nova-api). If any service doesn't restart, you should check its associated log for any errors, and then check the /etc/nova/nova.conf configuration file (shown in listing 5.69). If a service doesn't produce a log file, you can additionally check the service upstart log found in the /var/log/upstart/ directory with the same log naming convention.

5.6 Deploying the Dashboard (Horizon) service

The final step of the controller install is deploying the web-based Dashboard. The Horizon module provides a graphical user interface (GUI) for both user and administration functions related to OpenStack components. This will likely be the primary interface used by end users when installing and configuring resources.

5.6.1 Installing Horizon

The Horizon install is fairly straightforward and should work well as long as the rest of the components are properly configured. Horizon makes use of the Apache web server and Python modules. During the install, modules will be added and Apache will be restarted.

Listing 5.73 Installing Horizon

```
$sudo apt-get install -y openstack-dashboard memcached python-memcache
...
Starting memcached: memcached.
Processing triggers for ureadahead ...
Processing triggers for ufw ...
Setting up apache2-mpm-worker (2.2.22-6ubuntu5) ...
 * Starting web server apache2                                     [ OK ]
Setting up apache2 (2.2.22-6ubuntu5) ...
Setting up libapache2-mod-wsgi (3.4-0ubuntu3) ...
 * Restarting web server apache2 ... waiting .                     [ OK ]
Setting up openstack-dashboard (1:2013.1.1-0ubuntu1) ...
 * Reloading web server config                                     [ OK ]
Setting up openstack-dashboard-ubuntu-theme (1:2013.1.1-0ubuntu1) ...
```

```
 * Reloading web server config                                          [ OK ]
Processing triggers for libc-bin ...
ldconfig deferred processing now taking place
```

The install process should have added the site http://10.33.2.50/horizon. If you can't reach the site, check the Apache error log found in the /var/log/apache2/error.log file for any problems that would prevent startup.

Optionally, you can remove the Ubuntu theme, which has been reported to cause problems with some modules:

```
sudo apt-get -y remove --purge openstack-dashboard-ubuntu-theme
```

5.6.2 *Accessing Horizon*

The OpenStack Dashboard should now be available at http://10.33.2.50/horizon. You can log in as the `admin` user with the password `openstack1`.

At this point you can't do much with the Dashboard because no resource nodes have been added, but it's a good idea to try logging in now to make sure that components are reported in the Dashboard. Once you're logged in to Horizon, select the Admin tab on the left toolbar. Next, click System Info and look under the Services tab, which should look something like figure 5.7.

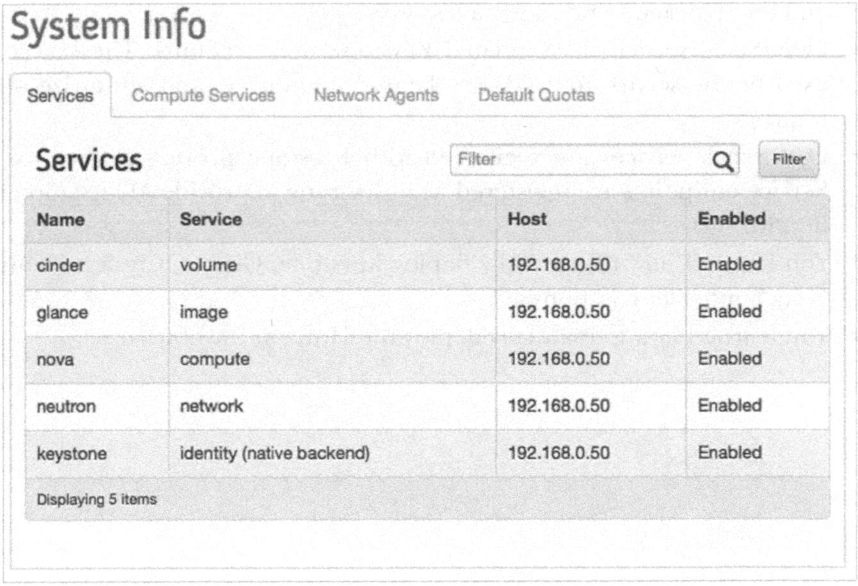

System Info

Services | Compute Services | Network Agents | Default Quotas

Services

Filter | Filter

Name	Service	Host	Enabled
cinder	volume	192.168.0.50	Enabled
glance	image	192.168.0.50	Enabled
nova	compute	192.168.0.50	Enabled
neutron	network	192.168.0.50	Enabled
keystone	identity (native backend)	192.168.0.50	Enabled

Displaying 5 items

Figure 5.7 Dashboard System Info

5.6.3 Debugging Horizon

If you experience any problems with Horizon, you can enable Horizon debugging by editing the local_settings.py file as follows.

Listing 5.74 Enabling debugging in Horizon

```
Enable Debugging Dashboard
**/usr/share/openstack-dashboard/openstack_dashboard/local/local_settings.py

#DEBUG = False
DEBUG = True
$ sudo service apache2 restart
 * Restarting web server apache2
**
```

Once the Dashboard is in debug mode, the error logs will be recorded in the Apache log: /var/log/apache2/error.log.

If you experience problems along the way, try retracing your steps, verifying services and logs along the way.

5.7 Summary

- Each OpenStack service has a related backend database that is used as a backend configuration and state data store.
- OpenStack services have related Keystone user accounts. These accounts are used by the service to validate tokens, authenticate, and authorize other user requests.
- OpenStack services are registered with Keystone providing a service catalog. Service endpoints are registered with Keystone to provide API location information for services.
- You learned how to manually deploy Keystone, Glance, Cinder, Neutron, and Nova controller components.
- You learned how to manually deploy the Horizon Dashboard.

6 Walking through a Networking deployment

This chapter covers

- Network node prerequisites
- Deploying OpenStack Networking core
- Setting up OpenStack Networking ML2 plug-in
- Configuring OpenStack Networking DHCP, Metadata, L3, and OVS agents

In chapter 5 you walked through the deployment of an OpenStack controller node, which provides the server-side management of OpenStack services. During the controller deployment, you made controller-side configurations for several OpenStack core services, including Networking, Compute, and Storage. We discussed the configurations for each core service in relation to the controller, but the services themselves weren't covered in detail.

Chapters 6 through 8 will walk you through the deployment of core OpenStack services on resource nodes. *Resource nodes* are nodes that provide a specific resource in relation to an OpenStack service. For instance, a server running OpenStack Compute (Nova) services (and all prerequisite requirements) would be considered a *compute resource node.* As you learned in chapter 2, it's possible for a specific node to provide multiple services, including Compute (Nova), Network (Neutron), and Block Storage (Cinder). But just like an exclusive node was used for the controller in chapter 5, exclusive resource nodes will be used for demonstration in chapters 6 (Networking), 7 (Block Storage), and 8 (Compute).

Take another look at the multi-node architecture introduced in chapter 5, shown in figure 6.1.

In this chapter, you'll manually deploy the Networking components in the lower right of the figure on a standalone node.

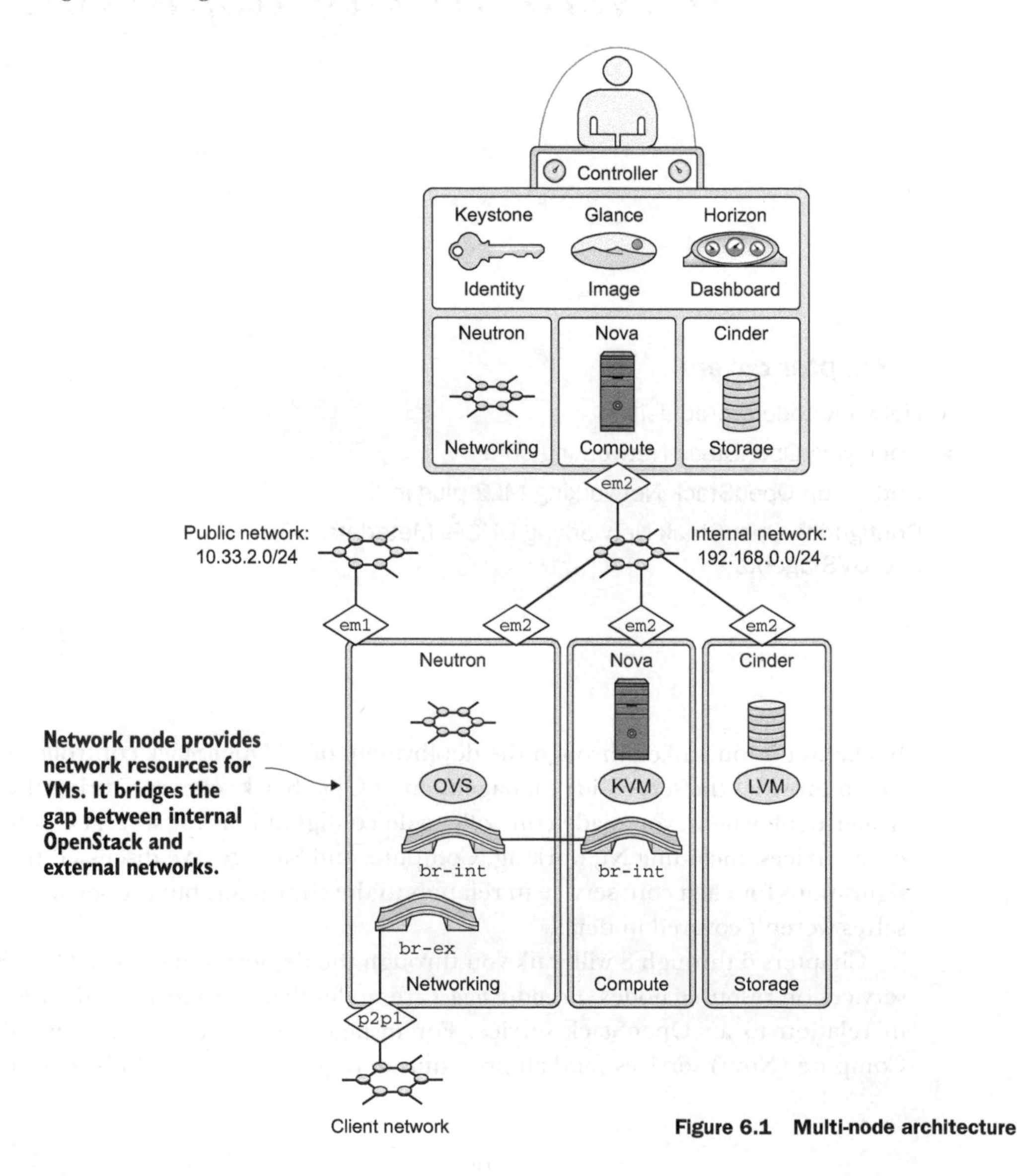

Figure 6.1 Multi-node architecture

Figure 6.2 shows your current status on your way to a working manual deployment. In this chapter, you'll first prepare the server to function as a network device. Next, you'll install and configure Neutron OSI Layer 2 (switching) components. Finally, you'll install and configure Neutron services that function on OSI Layer 3 (DHCP, Metadata, and so on). Network resources configured in this chapter will be used directly by VMs provided by OpenStack.

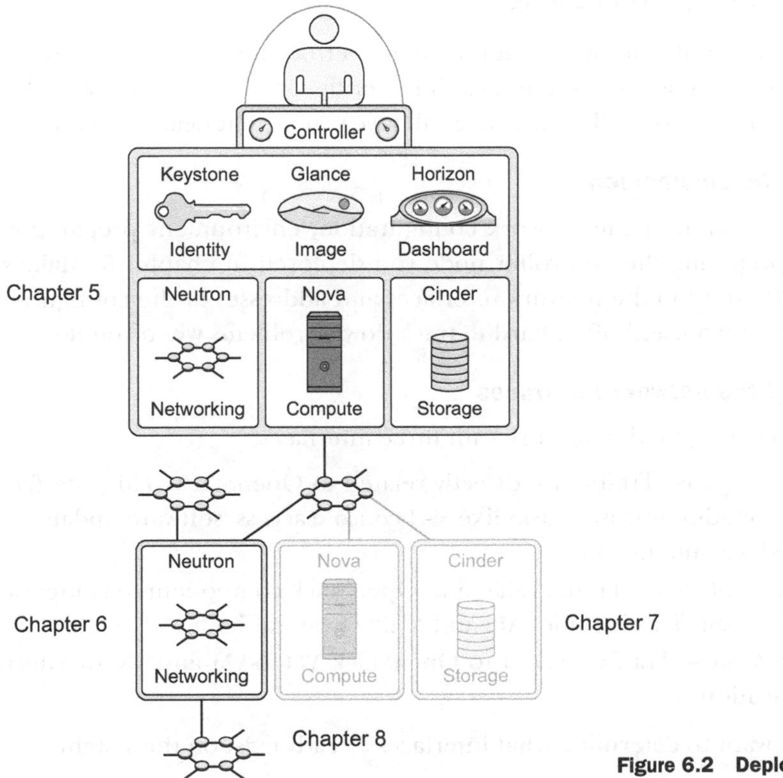

Figure 6.2 Deployment roadmap

For many people, this chapter will be the most difficult. Even if you have a deep background in traditional networking, you'll have to stop and think about how OpenStack Networking works. Overlay networks, or networks on top of other networks, are in many ways the network equivalent of the abstraction of virtual machines from bare-metal servers. This may be your first exposure to mesh/overlay/distributed networking, but these technologies are not exclusive to OpenStack. You'll learn more about overlay networks and their use in OpenStack in this chapter, but taking the time to understand the fundamental changes will be useful across many technologies.

6.1 Deploying network prerequisites

In the chapter 2 deployment, DevStack installed and configured OpenStack dependencies for you. In this chapter, you'll manually install these dependencies. Luckily, you can use a package management system to install the software: there's no compiling required, but you must still manually configure many of the components.

> **PROCEED WITH CARE** Working in a multi-node environment greatly increases deployment complexity. A small, seemingly unrelated, mistake in the configuration of one component or dependency can cause issues that are very hard to track down. Read each section carefully, making sure you understand what you're installing or configuring.

Many of the examples in this chapter include a verification step, which I highly recommend you follow. If a configuration can't be verified, retrace your steps to the last verified point and start over. This practice will save you a great deal of frustration.

6.1.1 Preparing the environment

With the exception of the network configuration, environment preparation will be similar to preparing the controller node you deployed in chapter 5. Make sure you pay close attention to the network interfaces and addresses in the configurations. It's easy to make a typo, and often hard to track down problems when you do.

6.1.2 Configuring the network interfaces

You want to configure the network with three interfaces:

- *Node interface*—Traffic not directly related to OpenStack. This interface will be used for administrative tasks like SSH console access, software updates, and even node-level monitoring.
- *Internal interface*—Traffic related to OpenStack component-to-component communication. This includes API and AMPQ type traffic.
- *VM interface*—Traffic related to OpenStack VM-to-VM and VM-to-external communication.

First, you'll want to determine what interfaces already exist on the system.

REVIEWING THE NETWORK

The following command will list the interfaces on your server.

Listing 6.1 List interfaces

```
$ ifconfig -a
em1       Link encap:Ethernet  HWaddr b8:2a:72:d5:21:c3
          inet addr:10.33.2.51  Bcast:10.33.2.255  Mask:255.255.255.0
          inet6 addr: fe80::ba2a:72ff:fed5:21c3/64 Scope:Link
          UP BROADCAST RUNNING MULTICAST  MTU:1500  Metric:1
          RX packets:9580 errors:0 dropped:0 overruns:0 frame:0
          TX packets:1357 errors:0 dropped:0 overruns:0 carrier:0
          collisions:0 txqueuelen:1000
          RX bytes:8716454 (8.7 MB)  TX bytes:183958 (183.9 KB)
          Interrupt:35
```

```
em2       Link encap:Ethernet  HWaddr b8:2a:72:d5:21:c4
          inet6 addr: fe80::ba2a:72ff:fed5:21c4/64 Scope:Link
          UP BROADCAST RUNNING MULTICAST  MTU:1500  Metric:1
          RX packets:7732 errors:0 dropped:0 overruns:0 frame:0
          TX packets:8 errors:0 dropped:0 overruns:0 carrier:0
          collisions:0 txqueuelen:1000
          RX bytes:494848 (494.8 KB)  TX bytes:680 (680.0 B)
          Interrupt:38
...
p2p1      Link encap:Ethernet  HWaddr a0:36:9f:44:e2:70
          BROADCAST MULTICAST  MTU:1500  Metric:1
          RX packets:0 errors:0 dropped:0 overruns:0 frame:0
          TX packets:0 errors:0 dropped:0 overruns:0 carrier:0
          collisions:0 txqueuelen:1000
          RX bytes:0 (0.0 B)  TX bytes:0 (0.0 B)
```

You might have configured your node interface, `em1`, during the initial installation. You'll use the `em1` interface to communicate with this node. Take a look at the two other interfaces, `em2` and `p2p1`. On the example systems used in writing this book, the `em2` interface will be used for internal OpenStack traffic and the add-on 10G adapter, whereas `p2p1` will be used for VM communication.

Next you'll review the network configuration for the example nodes, and you'll configure controller interfaces.

CONFIGURING THE NETWORK

Under Ubuntu, the interface configuration is maintained in the /etc/network/interfaces file. We'll build a working configuration based on the italicized addresses in table 6.1.

Table 6.1 Network address table

Node	Function	Interface	IP address
Controller	Pubic interface/node address	em1	10.33.2.50/24
Controller	OpenStack internal	em2	192.168.0.50/24
Network	*Node address*	*em1*	*10.33.2.51/24*
Network	*OpenStack internal*	*em2*	*192.168.0.51/24*
Network	*VM network*	*p2p1*	*None: assigned to OpenStack Networking*
Storage	Node address	em1	10.33.2.52/24
Storage	OpenStack internal	em2	192.168.0.52/24
Compute	Node address	em1	10.33.2.53/24
Compute	OpenStack internal	em2	192.168.0.53/24

In order to modify the network configuration, or any privileged configuration, you must use sudo privileges (`sudo vi /etc/network/interfaces`). Any text editor can be used in this process.

Modify your interfaces file as shown next.

Listing 6.2 Modify interface config /etc/network/interfaces

```
# The loopback network interface
auto lo
iface lo inet loopback

# The OpenStack Node Interface
auto em1                                    ❶
iface em1 inet static
        address 10.33.2.51
        netmask 255.255.255.0
        network 10.33.2.0
        broadcast 10.33.2.255
        gateway 10.33.2.1
        dns-nameservers 8.8.8.8
        dns-search testco.com

# The OpenStack Internal Interface
auto em2                                    ❷
iface em2 inet static
        address 192.168.0.51
        netmask 255.255.255.0

# The VM network interface
auto p2p1                                   ❸
iface p2p1 inet manual
```

❶ eml is the public interface used for node administration.

❷ em2 is used primarily for AMPQ and API traffic between resource nodes and the controller.

❸ p2pl virtual machine traffic between resource nodes and external networks

In your network configuration interface, `em1` will be used for node administration, such as SSH sessions to the actual server ❶. OpenStack shouldn't use this interface directly. The `em2` interface will be used primarily for AMPQ and API traffic between resource nodes and the controller ❷. The `p2p1` interface will be managed by Neutron. This interface will primarily carry virtual machine traffic between resource nodes and external networks ❸.

You should now refresh the network interfaces for which the configuration was changed. If you didn't change the settings of your primary interface, you shouldn't experience an interruption. If you changed the address of the primary interface, it's recommended you reboot the server at this point.

You can refresh the network configuration for a particular interface as shown here for interfaces `em2` and `p2p1`.

Listing 6.3 Refreshing Networking settings

```
sudo ifdown em2 && sudo ifup em2
sudo ifdown p2p1 && sudo ifup p2p1
```

The network configuration, from an operating system standpoint, should now be active. The interface will automatically be brought online based on your configuration. This process can be repeated for each interface that requires a configuration refresh. In order to confirm that the configuration was applied, you can once again check your interfaces.

Listing 6.4 Check network for updates

```
$ ifconfig -a

em1       Link encap:Ethernet  HWaddr b8:2a:72:d5:21:c3
          inet addr:10.33.2.51  Bcast:10.33.2.255  Mask:255.255.255.0
          inet6 addr: fe80::ba2a:72ff:fed5:21c3/64 Scope:Link
          UP BROADCAST RUNNING MULTICAST  MTU:1500  Metric:1
          RX packets:10159 errors:0 dropped:0 overruns:0 frame:0
          TX packets:1672 errors:0 dropped:0 overruns:0 carrier:0
          collisions:0 txqueuelen:1000
          RX bytes:8803690 (8.8 MB)  TX bytes:247972 (247.9 KB)
          Interrupt:35

em2       Link encap:Ethernet  HWaddr b8:2a:72:d5:21:c4
          inet addr:192.168.0.51  Bcast:192.168.0.255  Mask:255.255.255.0
          inet6 addr: fe80::ba2a:72ff:fed5:21c4/64 Scope:Link
          UP BROADCAST RUNNING MULTICAST  MTU:1500  Metric:1
          RX packets:7913 errors:0 dropped:0 overruns:0 frame:0
          TX packets:8 errors:0 dropped:0 overruns:0 carrier:0
          collisions:0 txqueuelen:1000
          RX bytes:506432 (506.4 KB)  TX bytes:680 (680.0 B)
          Interrupt:38
...
p2p1      Link encap:Ethernet  HWaddr a0:36:9f:44:e2:70
          inet6 addr: fe80::a236:9fff:fe44:e270/64 Scope:Link
          UP BROADCAST RUNNING MULTICAST  MTU:1500  Metric:1
          RX packets:0 errors:0 dropped:0 overruns:0 frame:0
          TX packets:8 errors:0 dropped:0 overruns:0 carrier:0
          collisions:0 txqueuelen:1000
          RX bytes:0 (0.0 B)  TX bytes:648 (648.0 B)
```

At this point you should be able to remotely access the network server, and the server should have internet access. The remainder of the install can be performed remotely using SSH or directly from the console.

6.1.3 Updating packages

The APT package index is a database of all available packages defined in the /etc/apt/sources.list file. You need to make sure your local database is synchronized with the latest packages available in the repository for your specific Linux distribution. Prior to installation, you should also upgrade any repository items, including the Linux kernel, that might be out of date.

Listing 6.5 Update and upgrade packages

```
sudo apt-get -y update
sudo apt-get -y upgrade
```

You now need to reboot the server to refresh any packages or configurations that might have changed.

Listing 6.6 Reboot server

```
sudo reboot
```

As of Ubuntu Server 14.04 (Trusty Tahr), the following OpenStack components are officially supported and included with the base distribution:

- *Nova*—Project name for OpenStack Compute; it works as an IaaS cloud fabric controller
- *Glance*—Provides services for virtual machine image, discovery, retrieval, and registration
- *Swift*—Provides highly scalable, distributed, object store services
- *Horizon*—Project name for OpenStack Dashboard; it provides a web-based admin/user GUI
- *Keystone*—Provides identity, token, catalog, and policy services for the OpenStack suite
- *Neutron*—Provides network management services for OpenStack components
- *Cinder*—Provides block storage as a service to OpenStack Compute

6.1.4 Software and configuration dependencies

In this section, you'll install a few software dependencies and make a few configuration changes in preparation for the install.

INSTALLING LINUX BRIDGE AND VLAN UTILITIES

You'll want to install the package bridge-utils, which provides a set of applications for working with network bridges on the system (OS) level. Network bridging on the OS level is critical to the operation of OpenStack Networking. For the time being, it's sufficient to think about network bridges under Linux as simply placing multiple interfaces on the same network segment (the same isolated VLAN). The default operation of Linux network bridging is to act like a switch, so you can certainly think of it this way.

In addition, you may want to install the vlan package, which provides the network subsystem the ability to work with Virtual Local Area Networks (VLANs) as defined by IEEE 802.1Q. VLANs allow you to segregate network traffic using VLAN IDs on virtual interfaces. This allows a single physical interface managed by your OS to isolate multiple networks using virtual interfaces. VLAN configuration won't be used in the examples, but you should be aware of the technology.

USING VLANS WITH NEUTRON Instructions for installing the *vlan* package are included in listing 5.114 because the vast majority of deployments will make use of IEEE 802.1Q VLANs to deliver multiple networks to Neutron nodes. But, for the sake of clarity, the examples in this book will not use VLAN interfaces. Once you understand OpenStack Networking, the adoption of VLANs on the OS level is trivial.

In summary, VLANs *isolate* traffic and interfaces, whereas Linux bridges *aggregate* traffic and interfaces.

Listing 6.7 Install vlan and bridge-utils

```
$ sudo apt-get -y install vlan bridge-utils
...
Setting up bridge-utils (1.5-6ubuntu2) ...
Setting up vlan (1.9-3ubuntu10) ...
```

You now have the ability to create VLANs and Linux bridges.

SERVER-TO-ROUTER CONFIGURATION

OpenStack manages resources for providing virtual machines. One of those resources is the network used by the virtual machine to communicate with other virtual and physical machines. For OpenStack Networking to provide network services, at least one resource node that performs the functions of a network devices (routing, switching, and so on) must exist. You want this node to act as a router and switch for network traffic.

By default, the Linux kernel isn't set to allow the routing of traffic between interfaces. The command `sysctl` is used to modify kernel parameters, such as those related to basic network functions. You need to make several modifications to your kernel settings using this tool.

The first modification is related to the forwarding or routing (kernel IP forwarding) of traffic between network interfaces by the Linux kernel. You want traffic arriving on one interface to be forwarded or routed to another interface if the kernel determines that the destination network can be found on another interface maintained by the kernel. Take a look at figure 6.3, which shows a server with two interfaces.

By default, the incoming packet shown in the figure will be dropped by interface `INT_0` because the address of this interface isn't the destination of the packet. But you want the server to inspect the packet's destination address, look in the server routing table, and, if a route is found, forward the packet to the appropriate interface. The `sysctl` setting `net.ipv4.ip_forward` instructing the kernel to forward traffic can be seen in listing 5.113.

In addition to enabling kernel IP forwarding, you also have to make a few other less-common kernel configuration changes. In the world of networking, there's something called *asymmetric routing*, where outgoing and incoming traffic paths/routes are not the same. There are legitimate reasons to do such things (such as terrestrial

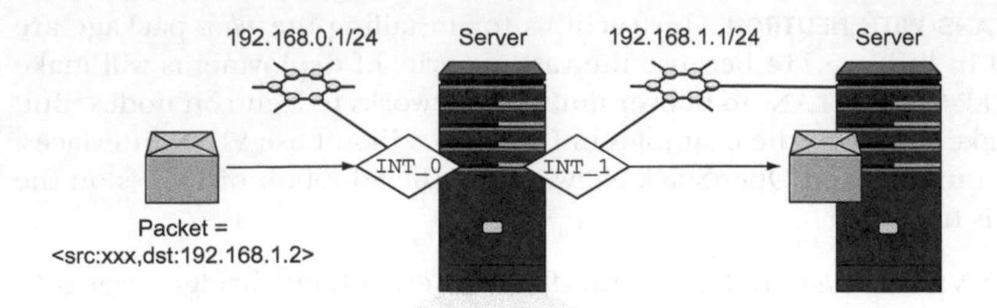

Route table

Destination	Use Iface
192.168.0.0	INT_0
192.168.1.0	INT_1

Figure 6.3 Linux IP routing

upload and satellite download; see www.google.com/patents/US6038594), but more often than not this ability was exploited by distributed denial of service (DDOS) attacks. RFC 3704, "Ingress Filtering for Multihomed Networks," also known as *reverse-path filtering*, was introduced to limit the impact of these DDOS attacks. By default, if the Linux kernel can't determine the source route of a packet, it will be dropped. OpenStack Networking is a complex platform that encompasses many layers of network resources, where the network resources themselves don't have a complete picture of the network. You must configure the kernel to disable reverse-path filtering, which leaves path management up to OpenStack.

The `sysctl` setting `net.ipv4.conf.all.rp_filter` that's used to disable reverse-path filtering for all existing interfaces is shown in listing 5.113. The `sysctl` setting `net.ipv4.conf.default.rp_filter` is used to disable reverse-path filtering for all future interfaces.

Apply the settings in the following listing to your OpenStack Network node.

Listing 6.8 Modify /etc/sysctl.conf

```
net.ipv4.ip_forward=1
net.ipv4.conf.all.rp_filter=0
net.ipv4.conf.default.rp_filter=0
```

To enable the `sysctl` kernel changes without restarting the server, invoke the `sysctl -p` command.

Listing 6.9 Execute the `sysctl` command

```
$ sudo sysctl -p
net.ipv4.conf.default.rp_filter = 0
net.ipv4.conf.all.rp_filter = 0
net.ipv4.ip_forward = 1
```

The interfaces should now forward IPv4 traffic, and reverse-path filtering should be disabled.

In the next section, you'll add advanced network features to your user with the Open vSwitch package.

6.1.5 *Installing Open vSwitch*

OpenStack Networking takes advantage of the open source distributed virtual-switching package, Open vSwitch (OVS). OVS provides the same data-switching functions as a physical switch (L2 traffic on port A destined to port B is switched to port B), but it runs in software on servers.

> **What does a switch do?**
>
> To understand what a switch does, you must first look at an Ethernet hub (you're likely using Ethernet in some form on all of your wired and wireless devices). "What is a hub?" you ask.
>
> Circa early 1990s, there were several competing OSI Layer 1 (physical) Ethernet topologies. One such topology, IEEE 10Base2, worked (and looked) much like the cable TV in your house, where you could take a single cable and add network connections by splicing in T connectors (think *splitters*). Another common topology was 10BaseT (RJ45 connector twisted pair), which is the grandfather of what most of us think of as "Ethernet" today. The good thing about 10BaseT was that you could extend the network without interrupting network service; the bad thing was that this physical topology required a device to terminate the cable segments together. This device was called a *hub*, and it also operated at the OSI Layer 1 (physical) level. If data was transmitted by a device on port A, it would be physically transmitted to all other ports on the hub.
>
> Aside from the obvious security concerns related to transmitting all data to all ports, the operation of a hub wouldn't scale. Imagine thousands of devices connected to hundreds of interconnected switches. All traffic was flooded to all ports. To solve this issue, network switches were developed. Manufacturers of Network Interface Cards (NICs) assigned a unique Ethernet Hardware Address (EHA) to every card. Switches kept track of the EHA addresses, commonly known as Media Access Control (MAC) addresses, on each port of the switch. If a packet with the destination MAC=xyz was transmitted to port A, and the switch had a record of xyz on port B, the packet was transmitted (switched) to port B. Switches operate on OSI Layer 2 (Link Layer) and switch traffic based on MAC destinations.

The examples in this book, from a network-switching standpoint, make exclusive use of the OVS switching platform.

At this point you have a server that can act like a basic network router (via IP kernel forwarding) and a basic switch (via Linux network bridging). You'll now add advanced switching capabilities to your server by installing OVS. OVS could be the topic of an entire book, but it's sufficient to say that the switching features provided by OVS rival offerings provided by standalone network vendors.

OVS is not a strict OpenStack network dependency

Without a doubt, OVS is used often with OpenStack Networking. But it's not implicitly required by the framework. The following diagram, first introduced in chapter 4, shows where OVS fits into the OpenStack Network architecture.

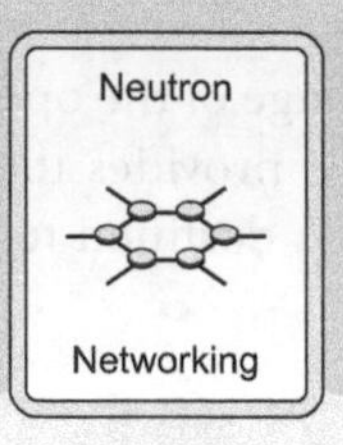

OVS is an L2 mechanism

You could use basic Linux bridging (the previously discussed virtual switch) or even a physical switch instead of OVS, as long as it's supported by a vendor-specific Neutron plug-in or module.

You can turn your server into an advanced switch with the following OVS install instructions.

Listing 6.10 Install OVS

```
$ sudo apt-get -y install openvswitch-switch
...
Setting up openvswitch-common ...
Setting up openvswitch-switch ...
openvswitch-switch start/running
```

The Open vSwitch install process will install a new OVS kernel module. In addition, the OVS kernel module will reference and load additional kernel models (GRE, VXLAN, and so on) as necessary to build network overlays.

What is a network overlay?

For a minute, forget what you know about traditional networking. Forget the concept of servers on the same switch (VLAN/network) being on the same "network." Imagine that you have a way to place any VM on any network, regardless of its physical location or underlying network topology. This is the value proposition for overlay networks.

At this point it's sufficient to think about an overlay network as a fully meshed virtual private network (VPN) between all participating endpoints (all servers being on the same L2 network segment regardless of location). To create a network such as this, you'll need technologies to tunnel traffic between endpoints. GRE, VXLAN, and other protocols provide the tunneling transports used by overlay networks. As usual, OpenStack simply manages these components. A network overlay is simply a method of extending L2 networks between hosts "overlaid" on top of other networks.

KNOW THY KERNEL Ubuntu 14.04 LTS is the first Ubuntu release to ship with kernel support for OVS overlay networking technologies (GRE, VXLAN, and the like). In previous versions, additional steps had to be taken to build appropriate kernel modules. If you're using another version of Ubuntu or another distribution altogether, make absolutely sure OVS kernel modules are loaded as shown in listing 5.116.

You want to be absolutely sure the Open vSwitch kernel modules were loaded. You can use the `lsmod` command in the following listing to confirm the presence of OVS kernel modules.

Listing 6.11 Verify OVS kernel modules

```
$ sudo lsmod | grep openvswitch
Module                       Size   Used by
openvswitch             66901  0
gre                     13796  1 openvswitch
vxlan                   37619  1 openvswitch
libcrc32c               12644  1 openvswitch
```

The output of the *lsmod* command should now show several resident modules related to OVS:

- `openvswitch`—This is the OVS module itself, which provides the interface between the kernel and OVS services.
- `gre`—Designated as "used by" the `openvswitch` module, it enables GRE functionality on the kernel level.
- `vxlan`—Just like the GRE module, `vxlan` is used to provide VXLAN functions on the kernel level.
- `libcrc32c`—Provides kernel-level support for cyclic redundancy check (CRC) algorithms, including hardware offloading using Intel's CRC32C CPU instructions. Hardware offloading is important for the high-performance calculation of network flow hashes and other CRC functions common to network headers and data frames.

Having GRE and VXLAN support on the kernel level means that the transports used to create overlay networks are understood by the system kernel, and by relation the Linux network subsystem.

> **No modules? DKMS to the rescue!**
>
> Dynamic Kernel Module Support (DKMS) was developed to make it easier to provide kernel-level drivers outside of the mainline kernel. DKMS has historically been used by OVS to provide kernel drivers for things such as overlay network devices (such as GRE and VXLAN), that were not included directly in the Linux kernel. The kernel that ships with Ubuntu 14.04 includes support for overlay devices built into the kernel, but depending on your distribution and release, you might not have a kernel with built-in support for the required network overlay technologies.
>
> The following command will deploy the appropriate dependencies and build the OVS `datapath` module using the DKMS framework:
>
> ```
> sudo apt-get -y install openvswitch-datapath-dkms
> ```
>
> Only run this command if the modules couldn't be validated as shown in listing 5.116.

If you think the kernel module should have loaded, but you still don't see it, restart the system and see if it loads on restart. Additionally, you can try to load the kernel module with the command `modprobe openvswitch`. Check the kernel log, /var/log/kern/log, for any errors related to loading OVS kernel modules. OVS won't function for your purposes without the appropriate resident kernel modules.

6.1.6 *Configuring Open vSwitch*

You now need to add an internal `br-int` bridge and an external `br-ex` OVS bridge.

The `br-int` bridge interface will be used for communication within Neutron-managed networks. Virtual machines communicating within internal OpenStack Neutron-created networks will use this bridge for communication. This interface shouldn't be confused with the internal interface on the operating system level.

Listing 6.12 Configure internal OVS bridge

```
sudo ovs-vsctl add-br br-int
```

Now that `br-int` has been created, create the external bridge interface, `br-ex`. The external bridge interface will be used to bridge OVS-managed internal Neutron networks with physical external networks.

Listing 6.13 Configure external OVS bridge

```
sudo ovs-vsctl add-br br-ex
```

You'll also want to confirm that the bridges were successfully added to OVS and that they're visible to the underlying networking subsystem. You can do that with the following commands.

Listing 6.14 Verify OVS configuration

```
$ sudo ovs-vsctl show
8cff16ee-40a7-40fa-b4aa-fd6f1f864560
    Bridge br-int
        Port br-int
            Interface br-int
                type: internal
    Bridge br-ex
        Port br-ex
            Interface br-ex
                type: internal
    ovs_version: "2.0.2"
```

Listing 6.15 Verify OVS OS integration

```
$ ifconfig -a
br-ex     Link encap:Ethernet  HWaddr d6:0c:1d:a8:56:4f          ❶ br-ex bridge
          BROADCAST MULTICAST  MTU:1500  Metric:1
          RX packets:0 errors:0 dropped:0 overruns:0 frame:0
          TX packets:0 errors:0 dropped:0 overruns:0 carrier:0
          collisions:0 txqueuelen:0
          RX bytes:0 (0.0 B)  TX bytes:0 (0.0 B)

br-int    Link encap:Ethernet  HWaddr e2:d9:b2:e2:00:4f          ❷ br-int bridge
          BROADCAST MULTICAST  MTU:1500  Metric:1
          RX packets:0 errors:0 dropped:0 overruns:0 frame:0
          TX packets:0 errors:0 dropped:0 overruns:0 carrier:0
          collisions:0 txqueuelen:0
          RX bytes:0 (0.0 B)  TX bytes:0 (0.0 B)
...
em1       Link encap:Ethernet  HWaddr b8:2a:72:d5:21:c3
          inet addr:10.33.2.51  Bcast:10.33.2.255  Mask:255.255.255.0
          inet6 addr: fe80::ba2a:72ff:fed5:21c3/64 Scope:Link
          UP BROADCAST RUNNING MULTICAST  MTU:1500  Metric:1
          RX packets:13483 errors:0 dropped:0 overruns:0 frame:0
          TX packets:2763 errors:0 dropped:0 overruns:0 carrier:0
          collisions:0 txqueuelen:1000
          RX bytes:12625608 (12.6 MB)  TX bytes:424893 (424.8 KB)
          Interrupt:35
...
ovs-system Link encap:Ethernet  HWaddr 96:90:8d:92:19:ab         ❸ ovs-system interface
          BROADCAST MULTICAST  MTU:1500  Metric:1
          RX packets:0 errors:0 dropped:0 overruns:0 frame:0
          TX packets:0 errors:0 dropped:0 overruns:0 carrier:0
          collisions:0 txqueuelen:0
          RX bytes:0 (0.0 B)  TX bytes:0 (0.0 B)
```

Notice the addition of the `br-ex` ❶ and `br-int` ❷ bridges in your interface list. The new bridges will be used by OVS and the Neutron OVS module for internal and external traffic. In addition, the `ovs-system` interface ❸ was added. This is the OVS datapath interface, but you won't have to worry about working with this interface; it's simply an artifact of Linux kernel integration. Nevertheless, the presence of this interface is an indication that the OVS kernel modules are active.

At this point you have an operational OVS deployment and two bridges. The br-int (internal) bridge will be used by Neutron to attach virtual interfaces to the network bridge. These tap interfaces will be used as endpoints for the Generic Routing Encapsulation (GRE) tunnels. GRE tunnels are used to create point-to-point network connections (think VPN) between endpoints over the Internet Protocol (IP), and Neutron will configure GRE tunnels between compute and network nodes using OVS. These tunnels will provide a mesh of virtual networks between all possible resource locations and network drains in the topology. This mesh provides the functional equivalence of a single isolated OSI L2 network for the virtual machines on the same virtual network. The internal bridge won't need to be associated with a physical interface or be placed in an OS-level "UP" state to work.

The br-ex (external) bridge will be used to connect the OVS bridges and Neutron-derived virtual interfaces to the physical network. You must associate the external bridge with your VM interface as follows.

Listing 6.16 Add interface p2p1 (VM) to bridge br-ex

```
sudo ovs-vsctl add-port br-ex p2p1
sudo ovs-vsctl br-set-external-id br-ex bridge-id br-ex
```

Now check that the p2p1 interface was added to the br-ex bridge.

Listing 6.17 Verify OVS configuration

```
$ sudo ovs-vsctl show
8cff16ee-40a7-40fa-b4aa-fd6f1f864560
    Bridge br-int
        Port br-int
            Interface br-int
                type: internal
    Bridge br-ex                          <-- 1 br-ex bridge
        Port br-ex
            Interface br-ex
                type: internal
        Port "p2p1"
            Interface "p2p1"              <-- 2 p2pl interface
    ovs_version: "2.0.1"
```

Notice the p2p1 interface ❷ listed as a port on the br-ex bridge ❶. This means that the p2p1 interface is virtually connected to the OVS br-ex bridge interface.

Currently the br-ex and br-int bridges aren't connected. Neutron will configure ports on both the internal and external bridges, including taps between the two. Neutron will do all of the OVS configuration from this point forward.

6.2 Installing Neutron

In this section, you'll prepare the Neutron ML2 plug-in, L3 agent, DHCP agent, and Metadata agent for operation. The ML2 plug-in is installed on every physical node where Neutron interacts with OVS.

You'll install the ML2 plug-in and agent on all compute and network nodes. The ML2 plug-in will be used to build Layer 2 (data link layer, Ethernet layer, and so on) configurations and tunnels between network endpoints managed by OpenStack. You can think of these tunnels as virtual network cables connecting separate switches or VMs together.

The L3, Metadata, and DHCP agents are only installed on the network nodes. The L3 agent will provide Layer 3 routing of IP traffic on the established L2 network. Similarly, the Metadata and DHCP agents provide L3 services on the L2 network.

The agents and plug-in provide the following services:

- *ML2 plug-in*—The ML2 plug-in is the link between Neutron and OSI L2 services. The plug-in manages local ports and taps, and it generates remote connections over GRE tunnels. This agent will be installed on network and compute nodes. The plug-in will be configured to work with OVS.
- *L3 agent*—This agent provides Layer 3 routing services and is deployed on network nodes.
- *DHCP agent*—This agent provides DHCP services for Neutron-managed networks using DNSmasq. Normally this agent will be installed on a network node.
- *Metadata agent*—This agent provides cloud-init services for booting VMs and is typically installed on the network node.

6.2.1 Installing Neutron components

You're now ready to install Neutron software as follows.

Listing 6.18 Install Neutron components

```
$ sudo apt-get -y install neutron-plugin-ml2 \
neutron-plugin-openvswitch-agent neutron-l3-agent \
neutron-dhcp-agent
...
Adding system user `neutron' (UID 109) ...
Adding new user `neutron' (UID 109) with group `neutron' ...
...
Setting up neutron-dhcp-agent  ...
neutron-dhcp-agent start/running, process 14910
Setting up neutron-l3-agent  ...
neutron-l3-agent start/running, process 14955
Setting up neutron-plugin-ml2 ...
Setting up neutron-plugin-openvswitch-agent  ...
neutron-plugin-openvswitch-agent start/running, process 14994
```

Neutron plug-ins and agents should now be installed. You can continue on with the Neutron configuration.

6.2.2 Configuring Neutron

The next step is configuration. First, you must modify the /etc/neutron/neutron .conf file to define the service authentication, management communication, core network plug-in, and service strategies. In addition, you'll provide configuration and credentials to allow the Neutron client instance to communicate with the Neutron controller, which you deployed in chapter 5. Modify your neutron.conf file based on the values shown below. If any of these values doesn't exist, add it.

Listing 6.19 Modify /etc/neutron/neutron.conf

```
[DEFAULT]
verbose = True
auth_strategy = keystone

rpc_backend = neutron.openstack.common.rpc.impl_kombu
rabbit_host = 192.168.0.50
rabbit_password = openstack1

core_plugin = neutron.plugins.ml2.plugin.Ml2Plugin
allow_overlapping_ips = True
service_plugins = router,firewall,lbaas,vpnaas,metering

nova_url = http://127.0.0.1:8774/v2
nova_admin_username = admin
nova_admin_password = openstack1
nova_admin_tenant_id = b3c5ebecb36d4bb2916fecd8aed3aa1a
nova_admin_auth_url = http://10.33.2.50:35357/v2.0

[keystone_authtoken]
auth_url =  http://10.33.2.50:35357/v2.0
admin_tenant_name = service
admin_password = openstack1
auth_protocol = http
admin_user = neutron

[database]
connection = mysql://neutron_dbu:openstack1@192.168.0.50/neutron
```

Now that the core Neutron components are configured, you must configure the Neutron agents, which will allow Neutron to control network services.

6.2.3 Configuring the Neutron ML2 plug-in

The Neutron OVS agent allows Neutron to control the OVS switch.

This configuration can be made in the /etc/neutron/plugins/ml2/ml2_conf.ini file. The following listing provides the database information, along with ML2-specific switch configuration.

Listing 6.20 Modify /etc/neutron/plugins/ml2/ml2_conf.ini

```
[ml2]
type_drivers = gre
tenant_network_types = gre
mechanism_drivers = openvswitch
```

```
[ml2_type_gre]
tunnel_id_ranges = 1:1000

[ovs]
local_ip = 192.168.0.51
tunnel_type = gre
enable_tunneling = True

[securitygroup]
firewall_driver =
neutron.agent.linux.iptables_firewall.OVSHybridIptablesFirewallDriver
enable_security_group = True
```

Your Neutron ML2 plug-in configuration is now complete. Clear the log file, and then restart the service:

```
sudo rm /var/log/neutron/openvswitch-agent.log
sudo service neutron-plugin-openvswitch-agent restart
```

Your Neutron ML2 plug-in agent log should now look something like the following:

```
Logging enabled!
Connected to AMQP server on 192.168.0.50:5672
Agent initialized
successfully, now running...
```

You now have OSI L2 Neutron integration using OVS. In the next section, you'll configure the OSI L3 Neutron services.

6.2.4 Configuring the Neutron L3 agent

Next, you need to configure the Neutron L3 agent. This agent provides L3 services, such as routing, for VMs. The L3 agent will be configured to use Linux namespaces.

What is Linux namespace isolation?

There's a feature built into the Linux kernel called *namespace isolation*. This feature allows you to separate processes and resources into multiple namespaces so that they don't interfere with each other. This is done internally by assigning namespace identifiers to each process and resource. From a network perspective, namespaces can be used to isolate network interfaces, firewall rules, routing tables, and so on. This is the underlying way in which multiple tenant networks, residing on the same Linux server, can have the same address ranges.

Go ahead and configure your L3 agent.

Listing 6.21 Modify /etc/neutron/l3_agent.ini

```
[DEFAULT]
interface_driver = neutron.agent.linux.interface.OVSInterfaceDriver
use_namespaces = True
verbose = True
```

The L3 agent is now configured and will use Linux namespaces.

Clear the log file, and then restart the service:

```
sudo rm /var/log/neutron/l3-agent.log
sudo service neutron-l3-agent restart
```

Your Neutron L3 agent log should look something like the following:

```
Logging enabled!
Connected to AMQP server on 192.168.0.50:5672
L3 agent started
```

6.2.5 Configuring the Neutron DHCP agent

You'll next want to configure the DHCP agent, which provides DHCP services for VM images. Modify your dhcp_agent.ini as shown in the following listing.

Listing 6.22 Modify /etc/neutron/dhcp_agent.ini

```
[DEFAULT]
...
interface_driver = neutron.agent.linux.interface.OVSInterfaceDriver
dhcp_driver = neutron.agent.linux.dhcp.Dnsmasq
use_namespaces = True
...
```

The DHCP agent is now configured and will use Linux namespaces. Clear the log file, and then restart the service:

```
sudo rm /var/log/neutron/dhcp-agent.log
sudo service neutron-dhcp-agent restart
```

Your Neutron DHCP agent log should look something like this:

```
Logging enabled!
Connected to AMQP server on 192.168.0.50:5672
DHCP agent started Synchronizing state
Synchronizing state complete
```

6.2.6 Configuring the Neutron Metadata agent

You'll next want to configure the Metadata agent, which provides environmental information to VM images. Cloud-init, which was originally created by Amazon for E2 services, is used to inject system-level settings on VM startup. To use Metadata services, you must use an image with a cloud-init–compatible agent installed and enabled.

Cloud-init is supported in most modern Linux distributions. Either download an image that has cloud-init preinstalled or install the package from your distribution.

Modify your metadata_agent.ini file to include the following information.

Listing 6.23 Modify /etc/neutron/metadata_agent.ini

```
[DEFAULT]
auth_url =  http://10.33.2.50:35357/v2.0
auth_region = RegionOne
admin_tenant_name = service
admin_password = openstack1
auth_protocol = http
admin_user = neutron
nova_metadata_ip = 192.168.0.50
metadata_proxy_shared_secret = openstack1
```

The Neutron Metadata agent is now configured and will use Linux namespaces. Clear the log file, and then restart the service:

```
sudo rm /var/log/neutron/metadata-agent.log
sudo service neutron-metadata-agent restart
```

Your Neutron Metadata agent log should look something like this:

```
Logging enabled!
(11074) wsgi starting up on http:///:v/
Connected to AMQP server on 192.168.0.50:5672
```

6.2.7 Restarting and verifying Neutron agents

It's a good idea at this point to restart all Neutron services, as shown in the following listing. Alternatively, you could simply restart the server.

Listing 6.24 Restart Neutron agents

```
$ cd /etc/init.d/; for i in $( ls neutron-* ); \
do sudo service $i restart; done
neutron-dhcp-agent stop/waiting
neutron-dhcp-agent start/running, process 16259
neutron-l3-agent stop/waiting
neutron-l3-agent start/running, process 16273
neutron-metadata-agent stop/waiting
neutron-metadata-agent start/running, process 16283
neutron-ovs-cleanup stop/waiting
neutron-ovs-cleanup start/running
```

You'll want to check the Neutron logs to make sure each service started successfully and is listening for requests. The logs can be found in the /var/log/neutron or /var/log/upstart/neutron-* directory.

Review the logs, checking for connections to the AMQP (RabbitMQ) server, and ensure there are no errors. The log files should exist even if they're empty. Ensure that there are no errors about unsupported OVS tunnels in the file /var/log/neutron/openvswitch-agent.log. If you experience such errors, restart the operating system and see if reloading the kernel modules and OVS takes care of the problem.

If you continue to experience problems starting Neutron services, you can increase the verbosity of the services through the /etc/neutron/neutron.conf file or the corresponding agent file.

6.2.8 Creating Neutron networks

In chapter 3 you were introduced to OpenStack Networking. This section reviews items presented in that chapter as they relate to the components you've deployed in this chapter.

Before you start creating networks using OpenStack Networking, you need to recall the basic differences between traditional "flat" networks, typically used for virtual and physical machines, and how OpenStack Networking works.

The term *flat* in *flat network* alludes to the absence of a virtual routing tier; the VM has direct access to a network, just as if you plugged a physical device into a physical network switch. Figure 6.4 shows an example of a flat network connected to a physical router.

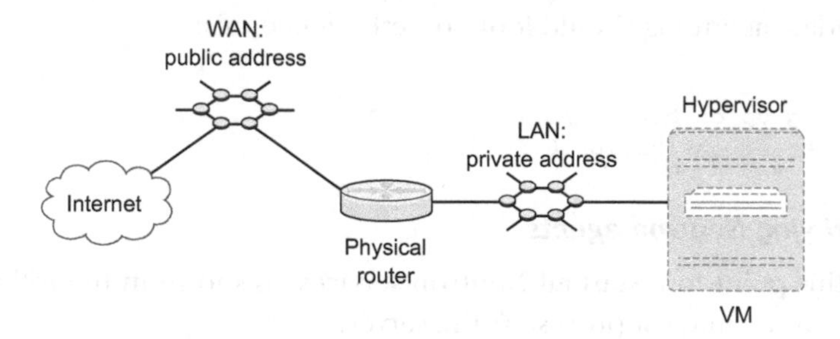

Figure 6.4 Traditional flat network

In this type of deployment, all network services (DHCP, load balancing, routing, and so on) beyond simple switching (OSI Model, Layer 2) must be provided outside of the virtual environment. For most systems administrators, this type of configuration will be very familiar, but this is not how we're going to demonstrate the power of OpenStack. You can make OpenStack Networking behave like a traditional flat network, but this approach will limit the benefits of the OpenStack framework.

In this section, you'll build a tenant network from scratch. Figure 6.5 illustrates an OpenStack tenant network, with virtual isolation from the physical external network.

Set your environment variables

The configurations in the following subsections require OpenStack authentication. In the previous examples, command-line arguments were provided for credentials. For the sake of simplicity, though, the following examples will use environment variables instead of command-line arguments.

To set your environment variables, execute the following commands in your shell:

```
$ export OS_USERNAME=admin
$ export OS_PASSWORD=openstack1
$ export OS_TENANT_NAME=admin
$ export OS_AUTH_URL=http://10.33.2.50:5000/v2.0
```

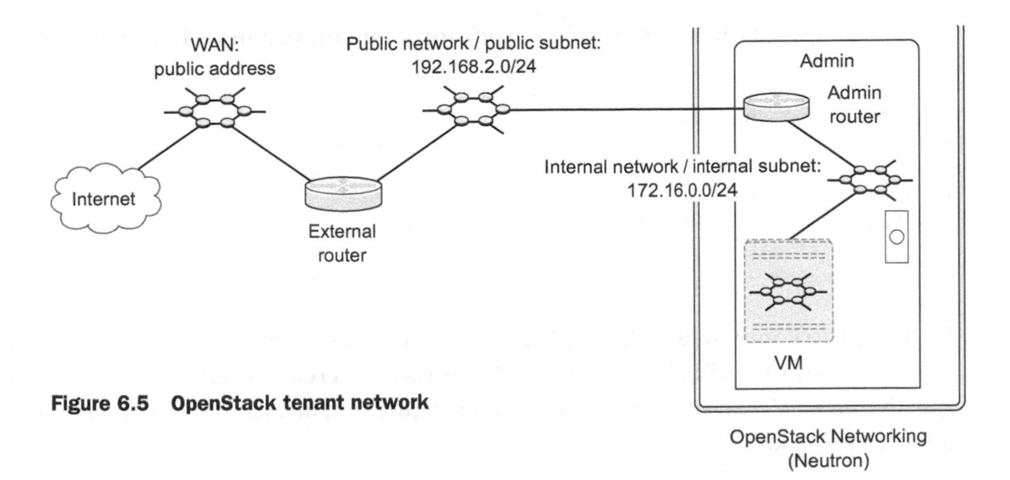

Figure 6.5 OpenStack tenant network

Network (Neutron) console

Neutron commands can be entered through the Neutron console (which is like a command line for a network router or switch) or directly through the CLI. The console is very handy if you know what you're doing, and it's a natural choice for those familiar with the Neutron command set. For the sake of clarity, however, this book demonstrates each action as a separate command, using CLI commands.

The distinction between the Neutron console and the Neutron CLI will be made clear in the following subsections. There are many things you can do with the Neutron CLI and console that you can't do in the Dashboard. Although the demonstrations will be executed using the CLI, you'll still need to know how to access the Neutron console. As you can see from the following, it's quite simple. Using the `neutron` command without arguments will take you to the console. All of the subcommands will be listed using the command shown in the following listing.

Listing 6.25 Access Neutron console

```
devstack@devstack:~/devstack$ neutron
(neutron) help

Shell commands (type help <topic>):
===================================
...
(neutron)
```

You now have the ability to access the interactive Neutron console. Any CLI configurations can be made either in the console or directly on the command line.

In the next subsection, you'll create a new network.

Internal networks

The first step you'll take in providing a tenant-based network is to configure the internal network. The internal network is used directly by instances in your tenant. The internal network works on the ISO Layer 2, so for the network types, this is the virtual equivalent of providing a network switch to be used exclusively for a particular tenant.

In order to create an internal network for a tenant, you must first determine your tenant ID:

```
$ keystone tenant-list
+----------------------------------+---------+---------+
|                id                |   name  | enabled |
+----------------------------------+---------+---------+
| 55bd141d9a29489d938bb492a1b2884c |  admin  |   True  |
| b3c5ebecb36d4bb2916fecd8aed3aa1a | service |   True  |
+----------------------------------+---------+---------+
```

By using the commands in listing 6.26, you can create a new network for your tenant. First, you tell OpenStack Networking (Neutron) to create a new network. Then you specify the `admintenant-id` on the command line. Finally, you specify the name of the tenant network.

Listing 6.26 Create internal network

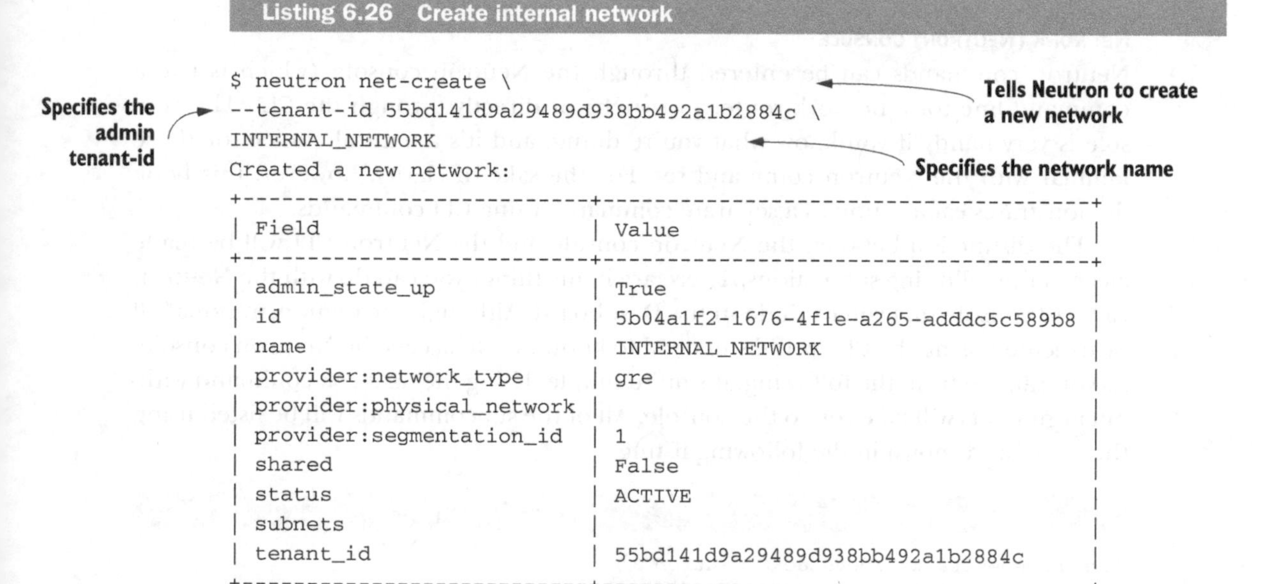

```
$ neutron net-create \
--tenant-id 55bd141d9a29489d938bb492a1b2884c \
INTERNAL_NETWORK
Created a new network:
+---------------------------+--------------------------------------+
| Field                     | Value                                |
+---------------------------+--------------------------------------+
| admin_state_up            | True                                 |
| id                        | 5b04a1f2-1676-4f1e-a265-adddc5c589b8 |
| name                      | INTERNAL_NETWORK                     |
| provider:network_type     | gre                                  |
| provider:physical_network |                                      |
| provider:segmentation_id  | 1                                    |
| shared                    | False                                |
| status                    | ACTIVE                               |
| subnets                   |                                      |
| tenant_id                 | 55bd141d9a29489d938bb492a1b2884c     |
+---------------------------+--------------------------------------+
```

Tells Neutron to create a new network
Specifies the admin tenant-id
Specifies the network name

Figure 6.6 illustrates the `INTERNAL_NETWORK` you created for your tenant. The figure shows the network you just created connected to a VM (if one was in the tenant).

You've now created an internal network. In the next subsection, you'll create an internal subnet for this network.

INTERNAL SUBNETS

In the previous subsection, you created an internal network. The internal network you created inside your tenant is completely isolated from other tenants. This will be a strange concept to those who work with physical servers, or even those who generally expose their virtual machines directly to physical networks. Most people are used to connecting their servers to the network, and network services are provided on a data center or enterprise level. We don't typically think about networking and computation being controlled under the same framework.

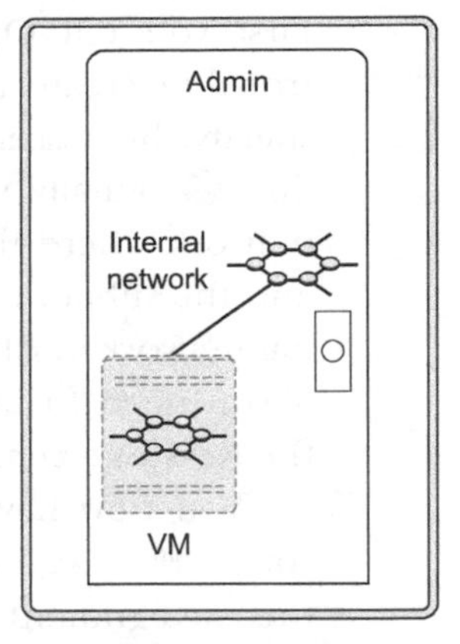

Figure 6.6 Created internal network

As previously mentioned, OpenStack can be configured to work in a flat network configuration. But there are many advantages to letting OpenStack manage the network stack. In this subsection, you'll create a subnet for your tenant. This can be thought of as an ISO Layer 3 (L3) provisioning of the tenant. You might be thinking to yourself, "What are you talking about? You can't just provision L3 services on the network!" or "I already have L3 services centralized in my data center. I don't want OpenStack to do this for me!" By the end of this section, or perhaps by the end of the book, you'll have your own answers to these questions. For the time being, just trust that OpenStack offers benefits that are either enriched by these features or that are not possible without them.

What does it mean to create a new subnet for a specific network? Basically, you describe the network you want to work with, and then you describe the address ranges you plan to use on that network. In this case, you'll assign the new subnet to the `ADMIN_NETWORK`, in the `ADMIN` tenant. You must also provide an address range for the subnet. You can use your own address range as long as it doesn't exist in the tenant or a shared tenant. One of the interesting things about OpenStack is that through the use of Linux namespaces, you could use the same address range for every internal subnet in every tenant.

Enter the command in the following listing.

Listing 6.27 Creating an internal subnet for the network

```
$ neutron subnet-create \                                      ❶ Creates new subnet
--tenant-id 55bd141d9a29489d938bb492a1b2884c \                 ❷ Specifies admin tenant-id
INTERNAL_NETWORK 172.16.0.0/24                                 ❸ Specifies network name and subnet range
Created a new subnet:
+------------------+------------------------------------------------+
| Field            | Value                                          |
+------------------+------------------------------------------------+
| allocation_pools | {"start": "172.16.0.2", "end": "172.16.0.254"} |
| cidr             | 172.16.0.0/24                                  |
| dns_nameservers  |                                                |
| enable_dhcp      | True                                           |
| gateway_ip       | 172.16.0.1                                     |
| host_routes      |                                                |
| id               | eb0c84d3-ea66-437f-9d1a-9defe8cccd06           |
| ip_version       | 4                                              |
| name             |                                                |
| network_id       | 5b04a1f2-1676-4f1e-a265-adddc5c589b8           |
| tenant_id        | 55bd141d9a29489d938bb492a1b2884c               |
+------------------+------------------------------------------------+
```

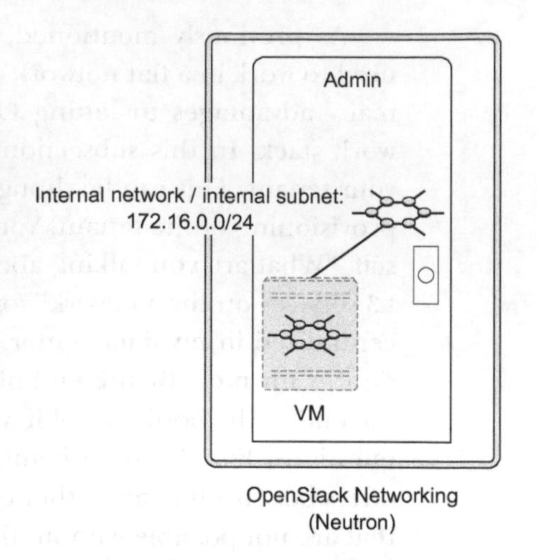

Figure 6.7 Created internal subnet

First you tell OpenStack Networking (Neutron) to create a new subnet ❶. Then you specify the `admintenant-id` on the command line ❷. Finally you specify the name of the network where the subnet should be created and the subnet range to be used on the internal network in CIDR notation ❸. Don't forget, if you need to find the `admintenant-id`, use the Keystone `tenant-id` command.

You now have a new subnet assigned to your `INTERNAL_NETWORK`. Figure 6.7 illustrates the assignment of the subnet to the `INTERNAL_NETWORK`. Unfortunately, this subnet is still isolated, but you're one step closer to connecting your private network to a public network.

In the next subsection, you'll add a router to the subnet you just created. Make a note of your `subnet-id`—it will be needed in the following sections.

> **CIDR NOTATION** As previously mentioned, CIDR is a compact way to represent subnets. For internal subnets, it's common to use a private class C address range. One of the most commonly used private ranges for internal or private networks is 192.168.0.0/24, which provides the range 192.168.0.1–192.168.0.254.

ROUTERS

Routers, put simply, route traffic between interfaces. In this case, you have an isolated network on your tenant and you want to be able to communicate with other tenant networks or networks outside of OpenStack. The following listing shows you how to create a new tenant router.

Listing 6.28 Create router

```
$ neutron router-create \                                    ❶ Creates new router
--tenant-id 55bd141d9a29489d938bb492a1b2884c \               ❷ Specifies admin tenant-id
ADMIN_ROUTER                                                 ❸ Specifies router name
Created a new router:
+-----------------------+--------------------------------------+
| Field                 | Value                                |
+-----------------------+--------------------------------------+
| admin_state_up        | True                                 |
| external_gateway_info |                                      |
| id                    | 5d7f2acd-cfc4-41bd-b5be-ba6d8e04f1e9 |
| name                  | ADMIN_ROUTER                         |
| status                | ACTIVE                               |
| tenant_id             | 55bd141d9a29489d938bb492a1b2884c     |
+-----------------------+--------------------------------------+
```

First, you tell OpenStack Networking (Neutron) to create a new router ❶. Then, you specify the admintenant-id on the command line ❷. Finally, you specify the name of the router ❸.

Figure 6.8 illustrates the router you created in your tenant.

Now you have a new router, but your tenant router and subnet aren't connected. The next listing shows how to connect your subnet to your router.

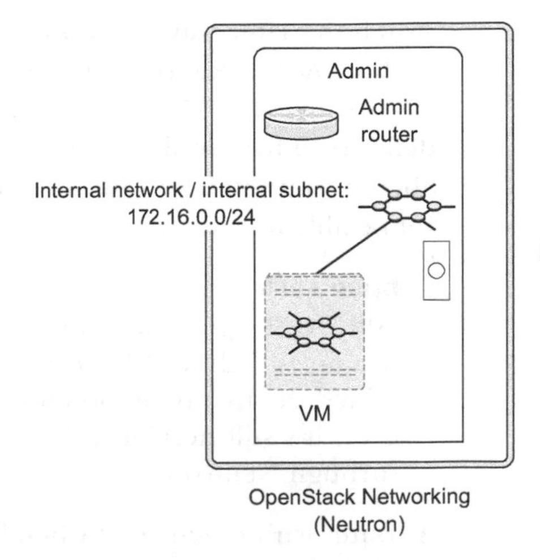

Figure 6.8 Created internal router

Listing 6.29 Adding router to internal subnet

```
$ neutron router-interface-add \                    ❶ Adds internal subnet
5d7f2acd-cfc4-41bd-b5be-ba6d8e04f1e9 \              ❷ Specifies router-id
eb0c84d3-ea66-437f-9d1a-9defe8cccd06                ❸ Specifies subnet-id
Added interface 54f0f944-06ce-4c04-861c-c059bc38fe59
    to router 5d7f2acd-cfc4-41bd-b5be-ba6d8e04f1e9.
```

First, you tell OpenStack Networking (Neutron) to add an internal subnet to your router ❶. Then, you specify the router-id of the router ❷. Finally, you specify the subnet-id of the subnet ❸.

If you need to look up Neutron-associated object IDs, you can access the Neutron console by running the Neutron CLI application without arguments: neutron. Once in the Neutron console, you can use the help command to navigate through the commands.

Figure 6.9 illustrates your router, ADMIN_ROUTER, connected to your internal network, INTERNAL_NETWORK.

The process of adding a router to a subnet will actually create a *port* on the local virtual switch. You can think of a port on a virtual

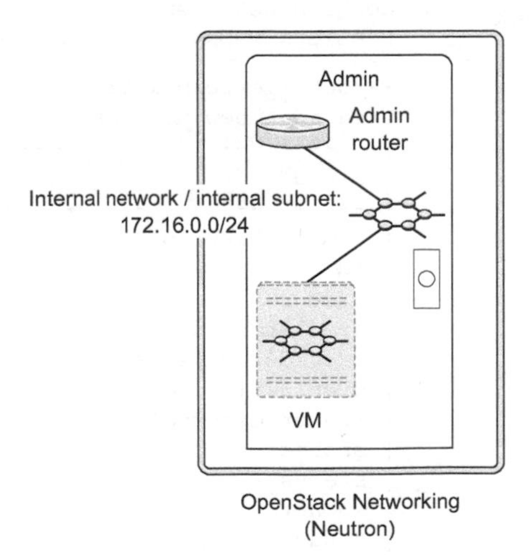

Figure 6.9 Router connected router to internal network

switch the same way you'd think of a port on a physical switch. In this case, the device is the ADMIN_ROUTER, the network is INTERNAL_NETWORK, and the subnet is 172.16.0.0/24. The router will use the address specified during subnet creation (it defaults to first available address). When you create an instance (VM), you should be able to communicate with the router address on the 172.16.0.1 address, but you won't yet be able to route packets to external networks.

> **DHCP AGENT** In past versions of OpenStack Networking, you had to manually add DHCP agents to your network. The DHCP agent is used to provide your instances with an IP address. In current versions, the agent is automatically added the first time you create an instance. In advanced configurations, however, it's still helpful to know that agents (of all kinds) can be manipulated through Neutron.

A router isn't much good when it's only connected to one network, so your next step is to create a public network that can be connected to the router you just created.

EXTERNAL NETWORK

In the subsection "Internal networks," you created a network that was specifically for your tenant. Here you'll create a public network that can be used by multiple tenants. This public network can be attached to a private router and will function as a network gateway for the internal network created in the previous section.

Only the admin user can create external networks. If a tenant isn't specified, the new external network will be created in the admin tenant. Create a new external network as shown in the next listing.

Listing 6.30 Create external network

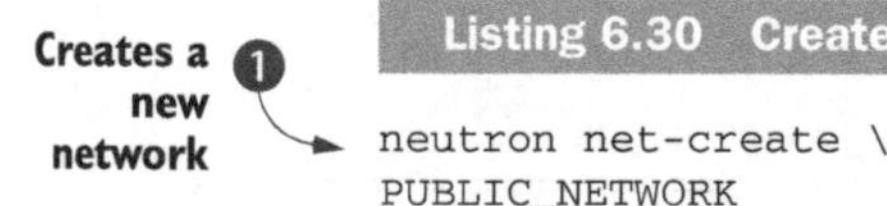


```
neutron net-create \
PUBLIC_NETWORK                                    ❷ Specifies network name

--router:external=True                            ❸ Designates as external network
Created a new network:
+---------------------------+--------------------------------------+
| Field                     | Value                                |
+---------------------------+--------------------------------------+
| admin_state_up            | True                                 |
| id                        | 64d44339-15a4-4231-95cc-ee04bffbc459 |
| name                      | PUBLIC_NETWORK                       |
| provider:network_type     | gre                                  |
| provider:physical_network |                                      |
| provider:segmentation_id  | 2                                    |
| router:external           | True                                 |
| shared                    | False                                |
| status                    | ACTIVE                               |
| subnets                   |                                      |
| tenant_id                 | 55bd141d9a29489d938bb492a1b2884c     |
+---------------------------+--------------------------------------+
```

First, you tell Neutron to create a new network ❶ and you specify the network name ❷. Then, you designate this network as an external network ❸.

You now have a network designated as an external network. As shown in figure 6.10, this network will reside in the admin tenant. Before you can use this network as a gateway for your tenant router (as shown in the subsection "Routers") you must first add a subnet to the external network you just created. That's what you'll do next.

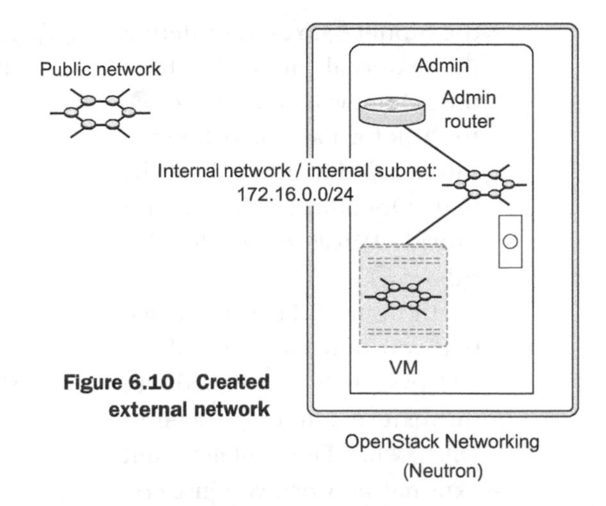

Figure 6.10 Created external network

EXTERNAL SUBNET

You must now create an external subnet, as shown in the following listing.

Listing 6.31 Create external subnet

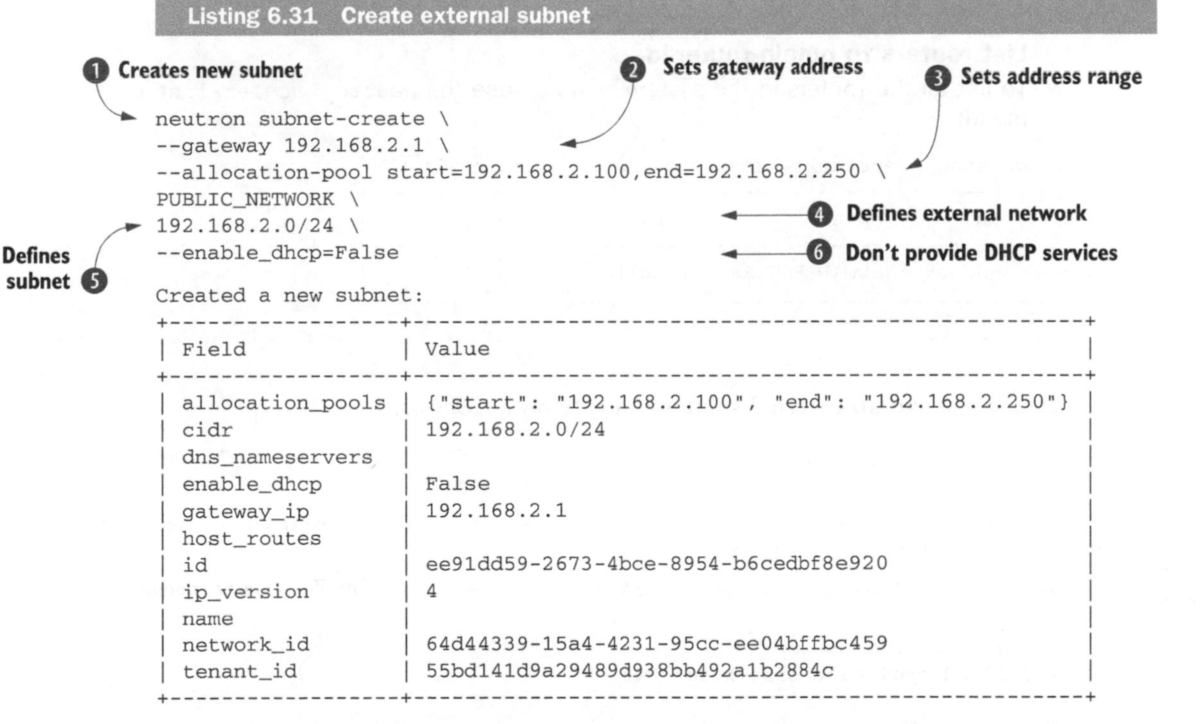

```
neutron subnet-create \                                         ❶ Creates new subnet
--gateway 192.168.2.1 \                                         ❷ Sets gateway address
--allocation-pool start=192.168.2.100,end=192.168.2.250 \       ❸ Sets address range
PUBLIC_NETWORK \                                                ❹ Defines external network
192.168.2.0/24 \                                                ❺ Defines subnet
--enable_dhcp=False                                             ❻ Don't provide DHCP services

Created a new subnet:
+-------------------+------------------------------------------------------+
| Field             | Value                                                |
+-------------------+------------------------------------------------------+
| allocation_pools  | {"start": "192.168.2.100", "end": "192.168.2.250"}   |
| cidr              | 192.168.2.0/24                                       |
| dns_nameservers   |                                                      |
| enable_dhcp       | False                                                |
| gateway_ip        | 192.168.2.1                                          |
| host_routes       |                                                      |
| id                | ee91dd59-2673-4bce-8954-b6cedbf8e920                 |
| ip_version        | 4                                                    |
| name              |                                                      |
| network_id        | 64d44339-15a4-4231-95cc-ee04bffbc459                 |
| tenant_id         | 55bd141d9a29489d938bb492a1b2884c                     |
+-------------------+------------------------------------------------------+
```

You first tell Neutron to create a new subnet ❶. You set the gateway address to the first available address ❷ and then define the range of addresses available for allocation in the subnet ❸. You then define the external network where the subnet will be assigned ❹. In CIDR format, you define the subnet ❺. Finally, you specify that OpenStack should not provide DHCP services for this subnet ❻.

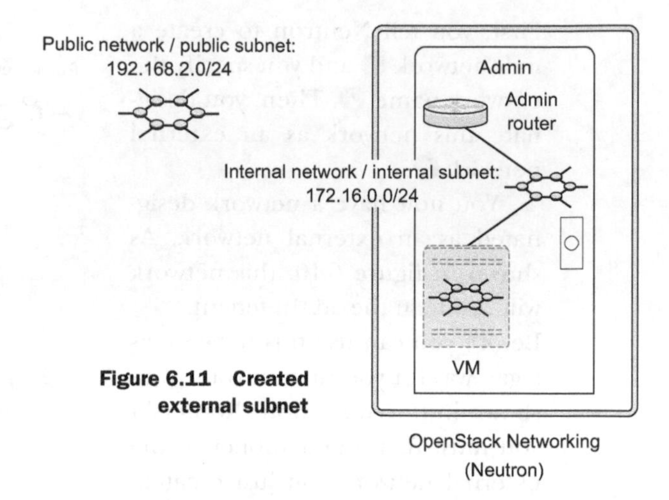

Figure 6.11 Created external subnet

In figure 6.11, you can see that you now have the subnet 192.168.2.0/24 assigned to the external network `PUBLIC_NETWORK`. The subnet and external network you just created can now be used by an OpenStack Networking router as a gateway network. In the next step, you'll assign your newly created external network as the gateway address of your internal network.

List routers to obtain router-id

To list all the routers in the system, you can use the `neutron router-list` command:

```
devstack@devstack:~/devstack$ neutron router-list
+--------+---------------+-----------------------+
| id     | name          | external_gateway_info |
+--------+---------------+-----------------------+
| 5d..e9 | ADMIN_ROUTER  | null                  |
+--------+---------------+-----------------------+
```

You can assign an external subnet as a gateway as follows.

Listing 6.32 Add new external network as router gateway

```
neutron router-gateway-set \                        ◄── Uses router-gateway-set command
5d7f2acd-cfc4-41bd-b5be-ba6d8e04f1e9 \              ◄── Specifies router-id
64d44339-15a4-4231-95cc-ee04bffbc459                ◄── Specifies external-network-id

Set gateway for router
15d7f2acd-cfc4-41bd-b5be-ba6d8e04f1e9
```

Figure 6.12 illustrates the assignment of the `PUBLIC_NETWORK` network as the gateway for the `ADMIN_ROUTER` in the `ADMIN` tenant. You can confirm this setting by running the command `neutron router-show <router-id>`, where the `<router-id>` is the ID of the `ADMIN_ROUTER`. The command will return the `external_gateway_info`, which lists

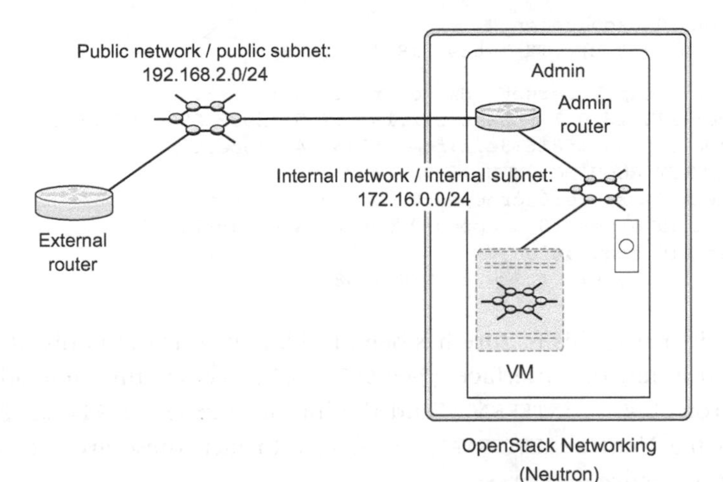

Figure 6.12 Assigned public network as router gateway

the currently assigned gateway network. Optionally, you can log in to the OpenStack Dashboard and look at your tenant network.

6.2.9 Relating Linux, OVS, and Neutron

At this point you should have a working Neutron environment and even a functioning network or two. But something will inevitably break and you'll need to troubleshoot the problem. Naturally, you'll turn up the log level in the suspected Neutron component. If you're lucky, there will be an obvious error. If you're not so lucky, there could be a problem with the underlying systems that Neutron depends on. Throughout this chapter, those dependencies and component relations have been explained. In many cases, you've created networks that make use of Linux namespaces, which you might not be used to working with. Now you'll work with Linux namespaces to relate the components you created on the network and systems layers.

Start by looking at the Linux network namespaces on the Neutron node:

```
$ sudo ip netns list
qrouter-5d7f2acd-cfc4-41bd-b5be-ba6d8e04f1e9
```

This result suggests you should look at the namespace `qrouter-5d7f2acd-cfc4-41bd-b5be-ba6d8e04f1e9`. Referencing the namespace, you'll display all network interface adapters:

```
sudo ip netns exec qrouter-5d7f2acd-cfc4-41bd-b5be-ba6d8e04f1e9\
 ifconfig -a

qg-896674d7-52 Link encap:Ethernet  HWaddr fa:16:3e:3b:fd:28
          inet addr:192.168.2.100  Bcast:192.168.2.255  Mask:255.255.255.0
          inet6 addr: fe80::f816:3eff:fe3b:fd28/64 Scope:Link
          UP BROADCAST RUNNING  MTU:1500  Metric:1
          RX packets:0 errors:0 dropped:0 overruns:0 frame:0
          TX packets:9 errors:0 dropped:0 overruns:0 carrier:0
```

❶ Interface qg-896674d7-52

```
          collisions:0 txqueuelen:0
          RX bytes:0 (0.0 B)  TX bytes:738 (738.0 B)

qr-54f0f944-06 Link encap:Ethernet  HWaddr fa:16:3e:e7:f3:35
          inet addr:172.16.0.1  Bcast:172.16.0.255  Mask:255.255.255.0
          inet6 addr: fe80::f816:3eff:fee7:f335/64 Scope:Link
          UP BROADCAST RUNNING  MTU:1500  Metric:1
          RX packets:0 errors:0 dropped:0 overruns:0 frame:0
          TX packets:9 errors:0 dropped:0 overruns:0 carrier:0
          collisions:0 txqueuelen:0
          RX bytes:0 (0.0 B)  TX bytes:738 (738.0 B)
```

❷ Interface qr-54f0f944-06

Whether you knew it or not, this feature has been lurking in your Linux distribution for some time. Notice that the interface `qg-896674d7-52` ❶ has the same address range as the Neutron `PUBLIC_INTERFACE`, and the interface `qr-54f0f944-06` ❷ has the same range as the Neutron `INTERNAL_INTERFACE`. In fact, these are the router interfaces for their respective networks.

> **Working with Linux network namespaces**
>
> To work with network namespaces, you must preface each command with `ip netns <function> <namespace_id>`:
>
> ```
> sudo ip netns <function> <namespace_id> <command>
> ```
>
> For more information about `ip netns`, consult the online man pages (which list it as "ip-netns"): http://man7.org/linux/man-pages/man8/ip-netns.8.html.

OK. You have some interfaces in a namespace, and these interfaces are related to the router interfaces you created earlier in the chapter. At some point, you'll want to communicate either between VM instances on OpenStack Neutron networks or with networks external to OpenStack Neutron. This is where OVS comes in.

Take a look at your OVS instance:

```
$ sudo ovs-vsctl show
    Bridge br-int
...
        Port "qr-54f0f944-06"
            tag: 1
            Interface "qr-54f0f944-06"
                type: internal
...
    Bridge br-ex
        Port br-ex
            Interface br-ex
                type: internal
        Port "p2p1"
            Interface "p2p1"
        Port "qg-896674d7-52"
            Interface "qg-896674d7-52"
                type: internal
...
```

Some things have been added to OVS since you last saw it in listing 6.44. Notice that the interface `qr-54f0f944-06` shows up as `Port "qr-54f0f944-06"` on the internal bridge, `br-int`. Likewise, the interface `qg-896674d7-52` shows up as `Port "qg-896674d7-52"` on the external bridge, `be-ex`.

What does this mean? The external interface of the router in your configuration is on the same bridge, `br-ex`, as the physical interface, `p2p1`. This means that the OpenStack Neutron network `PUBLIC_NETWORK` will use the physical interface `br-ex` to communicate with networks external to OpenStack.

Now that all of the pieces are tied together, you can move on to the next section, where you can graphically admire your newly created networks.

6.2.10 *Checking Horizon*

In chapter 5 you deployed the OpenStack Dashboard. The Dashboard should now be available at http://<controller address>/horizon.

It's a good idea to log in at this point to make sure that components are reported in the Dashboard. Log in as `admin` with the password `openstack1`. Once logged in to Horizon, select the Admin tab on the left toolbar. Next, click System Info and look under the Network Agents tab, which should look similar to figure 6.13. If you followed the instructions in the previous sections, your network should be visible in the Dashboard.

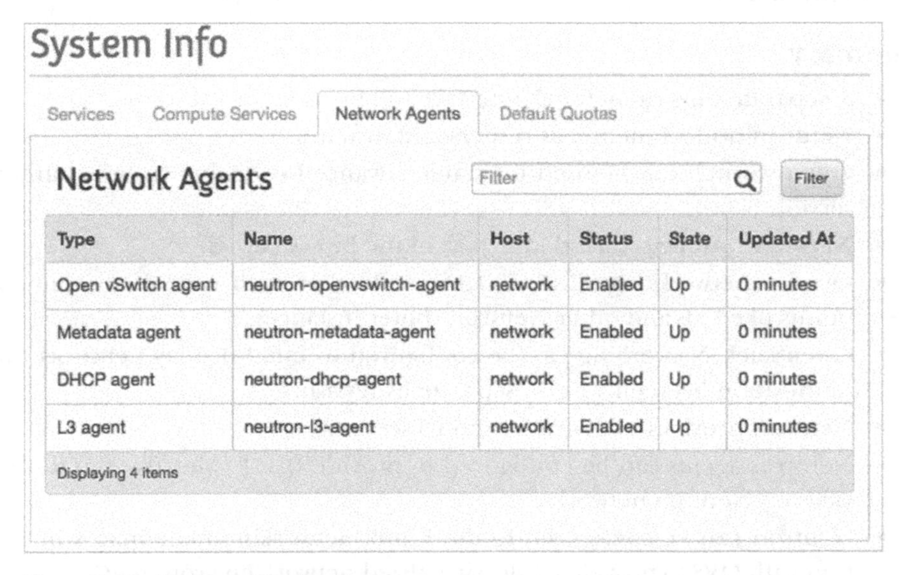

Figure 6.13 Dashboard System Info

Now make sure you're in the `admin` tenant and select the Project tab on the left toolbar. Next, click Network and then Network Topology. Your Network Topology screen should look like figure 6.14.

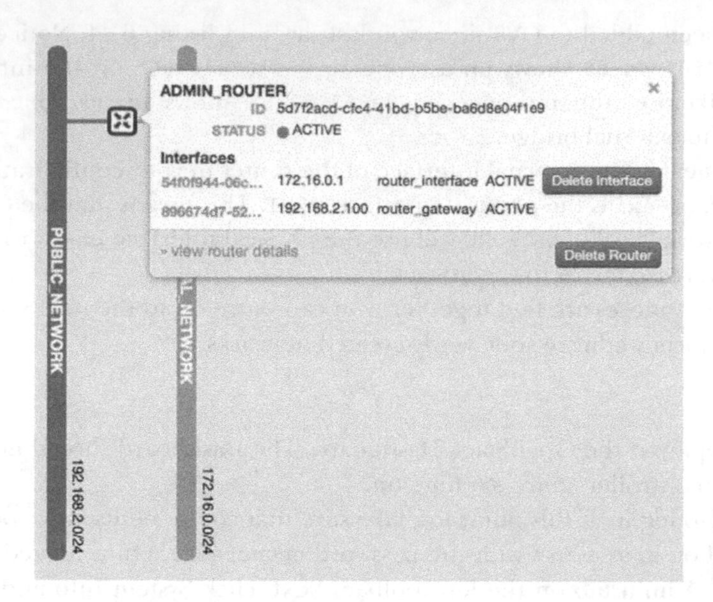

Figure 6.14 Network topology of PUBLIC/INTERNAL/ADMIN network

The figure shows your public network, tenant router, and tenant network in relation to your tenant. If you've made it to this screen, you've successfully manually deployed your network node.

6.3 Summary

- A separate physical network interface will be used for VM traffic.
- Neutron nodes function as routers and switches.
- Open vSwitch can be used to enable advanced switching features on a typical server.
- Network routing is included as part of the Linux kernel.
- Overlay networks use GRE, VXLAN, and other such tunnels to connect endpoints like VMs and other Neutron router instances.
- OpenStack Networking can be configured to build overlay networks for communication between VMs on separate hypervisors.
- Neutron provides both OSI L2 and L3 services.
- Neutron agents can be configured to provide DHCP, Metadata, and other services on Neutron networks.
- Neutron can be configured to use Linux networking namespaces in conjunction with OVS to provide a fully virtualized network environment.
- Internally, all tenants can use the same network IP ranges without conflict, because they're separated by using Linux namespaces.
- Neutron routers are used to route traffic between internal and external Neutron networks.

7 Walking through a Block Storage deployment

This chapter covers

- Storage node prerequisites
- Understanding Logical Volume Manager (LVM)
- Deploying OpenStack Block Storage
- Managing LVM storage with OpenStack Block Storage
- Testing OpenStack Block Storage

In chapter 5 you walked through the deployment of an OpenStack controller node, which provides the server-side management of OpenStack services. During the controller deployment, you set up controller-side configurations for several OpenStack core services including Networking, Compute, and Storage. The configurations for each core service were discussed in relation to the controller, but the services themselves weren't discussed in detail.

In chapter 6 you deployed your first standalone resource node. That node will provide OpenStack Networking services for the deployment. In this chapter, you'll deploy another standalone resource node that will provide OpenStack Block Storage services for the deployment.

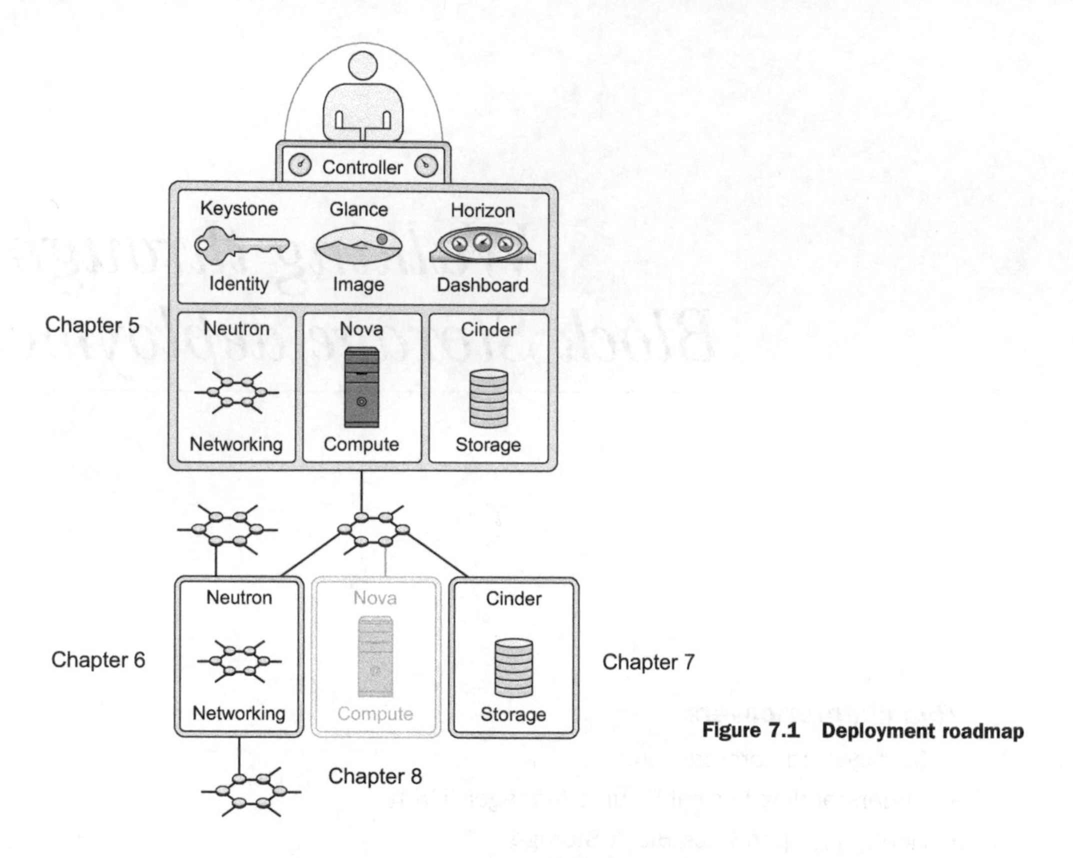

Figure 7.1 Deployment roadmap

You're now halfway through your manual OpenStack deployment, as shown in figure 7.1.

Take another look at the multi-node architecture introduced in chapter 5, shown in figure 7.2. In this chapter, you'll deploy the OpenStack Block Storage components at the lower left of the figure. You'll deploy them manually on a standalone node. If you've worked in virtual environments before, this chapter is unlikely to introduce any fundamental concepts that will seem strange to you.

First, you'll prepare the server to function as a storage device. In this process, you'll take the raw storage capacity found on a physical disk and prepare it so that it can be managed by a system-level volume manager. Then, you'll configure OpenStack Block Storage to manage your storage resources, using this volume manager. The storage resources configured in this chapter will be used directly by VMs provided by OpenStack. Storage resources used directly by VMs are generally referred to as *VM volumes.*

In the multi-node example presented in this book, you'll use storage provided by the same server that's acting as the OpenStack Block Storage node, but as you learned in chapter 4, this doesn't have to be the case. Instead of using storage on the storage node (such as physically attached disks), the storage node could be used to manage storage provided by a vendor-storage system (SAN, Ceph, and the like). For the sake of

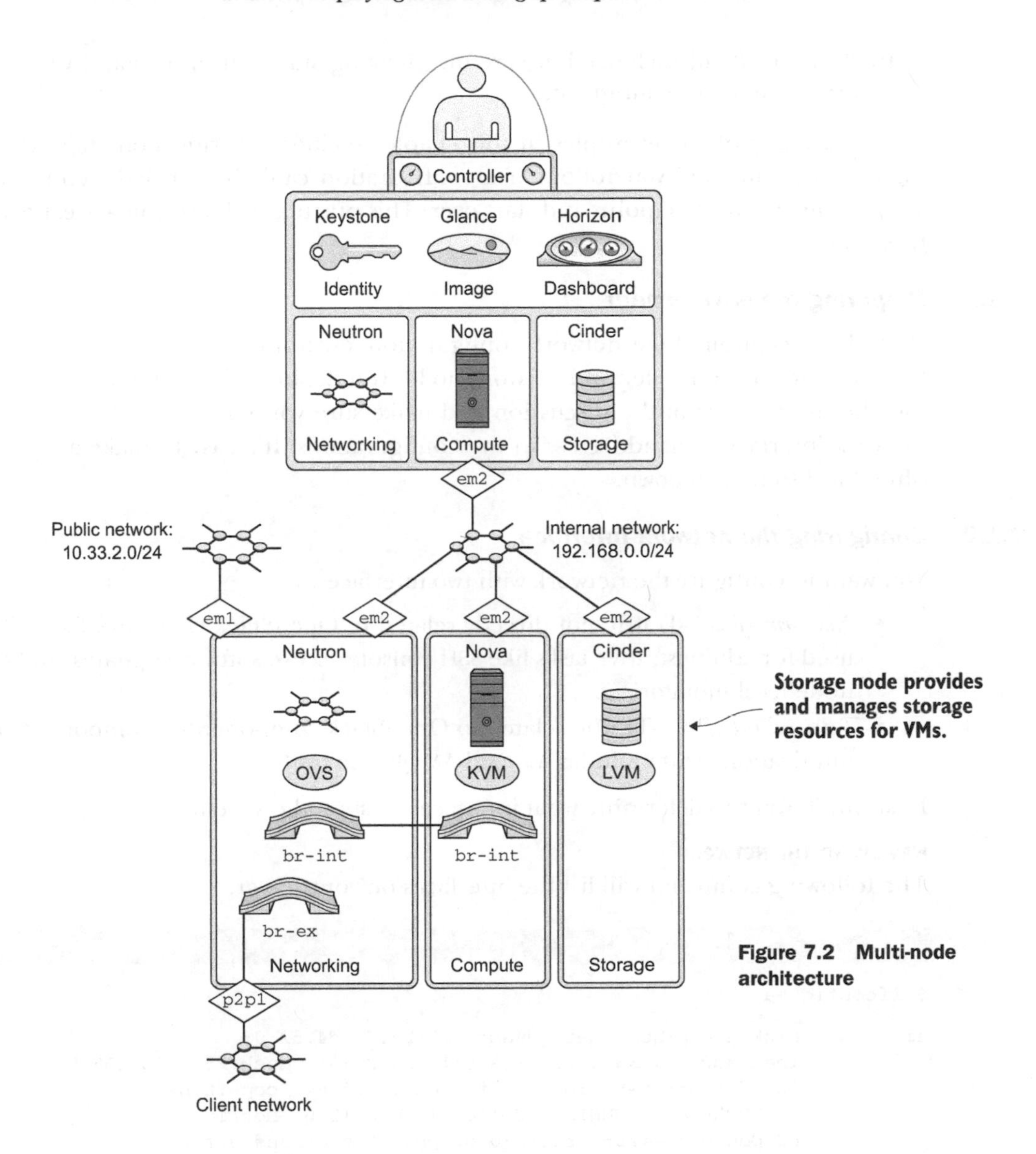

Figure 7.2 Multi-node architecture

simplicity, the examples in this book will use storage physically located on the storage node, and no vendor-storage system will be introduced.

Let's get started!

7.1 Deploying Block Storage prerequisites

As is true of all chapters in this part of the book, this chapter walks you through manually installing and configuring dependencies and core OpenStack packages.

> **PROCEED WITH CARE** Working in a multi-node environment greatly increases deployment complexity. A small, seemingly unrelated, mistake in the configuration of one component or dependency can cause issues that are very hard to

track down. Read each section carefully, making sure you understand what you're installing or configuring.

As usual, many of the examples in this chapter include a verification step, which I highly recommended you follow. If a configuration can't be verified, retrace your steps to the last verified point and start over. This practice will save you a great deal of frustration.

7.1.1 Preparing the environment

With the exception of the network configuration, environment preparation here will be similar to the controller and network nodes you deployed in chapters 5 and 6. Follow the instructions in the discussion and make sure you pay close attention to the network interfaces and addresses in the configurations. It's easy to make a typo and often hard to track it down.

7.1.2 Configuring the network interface

You want to configure the network with two interfaces:

- *Node interface*—Traffic not directly related to OpenStack. This interface will be used for administrative tasks like SSH console access, software updates, and even node-level monitoring.
- *Internal interface*—Traffic related to OpenStack component-to-component communication. This includes API and AMQP type traffic.

First, you'll want to determine what interfaces exist on the system.

REVIEWING THE NETWORK

The following command will list the interfaces on your server.

Listing 7.1 List interfaces

```
$ ifconfig -a

em1       Link encap:Ethernet  HWaddr b8:2a:72:d4:52:0f
          inet addr:10.33.2.62  Bcast:10.33.2.255  Mask:255.255.255.0
          inet6 addr: fe80::ba2a:72ff:fed4:520f/64 Scope:Link
          UP BROADCAST RUNNING MULTICAST  MTU:1500  Metric:1
          RX packets:44205 errors:0 dropped:0 overruns:0 frame:0
          TX packets:7863 errors:0 dropped:0 overruns:0 carrier:0
          collisions:0 txqueuelen:1000
          RX bytes:55103938 (55.1 MB)  TX bytes:832282 (832.2 KB)
          Interrupt:35

em2       Link encap:Ethernet  HWaddr b8:2a:72:d4:52:10
          BROADCAST MULTICAST  MTU:1500  Metric:1
          RX packets:0 errors:0 dropped:0 overruns:0 frame:0
          TX packets:0 errors:0 dropped:0 overruns:0 carrier:0
          collisions:0 txqueuelen:1000
          RX bytes:0 (0.0 B)  TX bytes:0 (0.0 B)
          Interrupt:38
```

You might have configured your node interface, em1, during the initial installation. You'll use the em1 interface to communicate with this node. Take a look at the other interface, em2. On the example systems used in this book, interface em2 will be used for internal OpenStack traffic.

Next you'll review the network configuration for the example nodes, and you'll configure controller interfaces.

CONFIGURING THE NETWORK

Under Ubuntu, the interface configuration is maintained in the /etc/network/interfaces file. We'll build a working configuration based on the addresses italicized in table 7.1.

Table 7.1 Network address table

Node	Function	Interface	IP address
controller	Public interface/node address	em1	10.33.2.50/24
controller	OpenStack internal	em2	192.168.0.50/24
network	Node address	em1	10.33.2.51/24
network	OpenStack internal	em2	192.168.0.51/24
network	VM network	p2p1	None: assigned to OpenStack Networking
storage	*Node address*	*em1*	*10.33.2.52/24*
storage	*OpenStack internal*	*em2*	*192.168.0.52/24*
compute	Node address	em1	10.33.2.53/24
compute	OpenStack internal	em2	192.168.0.53/24

In order to modify the network configuration, or any privileged configuration, you must use sudo privileges (sudo vi /etc/network/interfaces). Any text editor can be used in this process.

Modify your interfaces file as shown here.

Listing 7.2 Modify interface file config /etc/network/interfaces

```
# The loopback network interface
auto lo
iface lo inet loopback
# The OpenStack Node Interface
auto em1                                    ❶ em1 interface
iface em1 inet static
        address 10.33.2.62
        netmask 255.255.255.0
        network 10.33.2.0
        broadcast 10.33.2.255
```

```
        gateway 10.33.2.1
        dns-nameservers 8.8.8.8
        dns-search testco.com
# The OpenStack Internal Interface
auto em2                                    ❷ em2 interface
iface em2 inet static
        address 192.168.0.62
        netmask 255.255.255.0
```

In your network configuration, the em1 interface ❶ will be used for node administration, such as SSH sessions to the actual server. OpenStack shouldn't use this interface directly. The em2 interface ❷ will be used primarily for AMQP and API traffic between resource nodes and the controller.

> **Why have a storage interface?**
>
> In practice, you want your storage traffic and VM network traffic isolated for many reasons, not the least of which is performance. Unlike a physical server with a local disk, the local volumes of these virtual machines are on other servers, and they use the network to communicate. Although networks can be used for storage traffic, variations in network performance (latency, loss, and so forth) have far more impact on storage networks than on VM networks. For example, a small delay in retrieving an API response between OpenStack components might be negligible, but the same delay could be crippling for an operating system attempting to retrieve pages swapped from RAM to disk storage from a network-attached block volume.
>
> For the sake of simplicity, the examples in this chapter don't isolate storage and internal OpenStack traffic. In a production environment, this would not be recommended.

You should now refresh the network interfaces for which the configuration was changed. If you didn't change the settings of your primary interface, you shouldn't experience an interruption. If you changed the address of the primary interface, it's recommended that you reboot the server at this point.

If you've changed your network configuration, you can refresh the settings for a particular interface, as shown here for interface em2.

Listing 7.3 Refreshing Networking settings

```
sudo ifdown em2 && sudo ifup em2
```

The network configuration, from an operating system standpoint, should now be active. The interface will automatically be brought online based on your configuration. This process can be repeated for each interface that requires a configuration refresh.

In order to confirm that the configuration was applied, you'll once again need to check your interfaces.

Listing 7.4 Check network for updates

```
$ ifconfig -a

em1       Link encap:Ethernet  HWaddr b8:2a:72:d4:52:0f
          inet addr:10.33.2.62  Bcast:10.33.2.255  Mask:255.255.255.0
          inet6 addr: fe80::ba2a:72ff:fed4:520f/64 Scope:Link
          UP BROADCAST RUNNING MULTICAST  MTU:1500  Metric:1
          RX packets:44490 errors:0 dropped:0 overruns:0 frame:0
          TX packets:8023 errors:0 dropped:0 overruns:0 carrier:0
          collisions:0 txqueuelen:1000
          RX bytes:55134915 (55.1 MB)  TX bytes:863478 (863.4 KB)
          Interrupt:35

em2       Link encap:Ethernet  HWaddr b8:2a:72:d4:52:10
          inet addr:192.168.0.62  Bcast:192.168.0.255  Mask:255.255.255.0
          inet6 addr: fe80::ba2a:72ff:fed4:5210/64 Scope:Link
          UP BROADCAST RUNNING MULTICAST  MTU:1500  Metric:1
          RX packets:1 errors:0 dropped:0 overruns:0 frame:0
          TX packets:6 errors:0 dropped:0 overruns:0 carrier:0
          collisions:0 txqueuelen:1000
          RX bytes:64 (64.0 B)  TX bytes:532 (532.0 B)
          Interrupt:38
```

At this point you should be able to remotely access the network server, and the server should have internet access. The remainder of the installation can be performed remotely using SSH or directly from the console.

7.1.3 Updating packages

As was mentioned in chapters 5 and 6, the APT package index is a database of all available packages defined by a remote list found in the /etc/apt/sources.list file. You'll want to make sure the local database is synchronized with the latest packages available in the repository for your specific Linux distribution. Prior to installation, you should also upgrade any repository items, including the Linux kernel, that might be out of date.

Listing 7.5 Update and upgrade packages

```
sudo apt-get -y update
sudo apt-get -y upgrade
```

You now need to reboot the server to refresh any packages or configurations that might have changed.

Listing 7.6 Reboot server

```
sudo reboot
```

7.1.4 Installing and configuring the Logical Volume Manager

In this section, you'll install a few software dependencies and make a few configuration changes in preparation for the OpenStack component install.

The Logical Volume Manager (LVM) is a volume manager for the Linux kernel. A volume manager is simply a management layer that provides an abstraction between the partitioning of physical devices and logical devices as they appear on the system level. One can think of LVM in the same way as a hardware RAID adapter. LVM sits in between the Linux kernel and the storage devices, providing a software layer for managing storage volumes. This software layer provides several key benefits over using storage devices directly:

- *Resize volumes*—Physical and virtual volumes can be used to expand and contract an LVM volume.
- *Snapshots*—LVM can be used to create read/write snapshots (clones or copies) of volumes.
- *Thin provisioning*—LVM volumes can be created that report a specific volume size on the system level, but the storage isn't actually allocated until it's used. Thin provisioning is a common technique for over-provisioning storage resources.
- *Cache-enabled volumes*—LVM volumes can be created that utilize SSD (fast) storage for caching of slower volumes.

LVM IS NOT A STRICT OPENSTACK BLOCK STORAGE DEPENDENCY In the examples, you'll be using raw storage (physical disk) attached to the same physical node that's running OpenStack Block Storage (Cinder). Cinder will be configured to use a specific LVM volume, `cinder-volumes`. If Cinder was configured to manage storage volumes on vendor-provided devices, you wouldn't have to configure LVM.

INSTALLING LVM

If you're using the Ubuntu 14.04 operating system, you were given the option to manage your system disks using LVM during the install process. If you chose to use LVM during this process, the following step is unnecessary, but harmless. However, if LVM wasn't installed, you should install it now.

Listing 7.7 Install LVM

```
$ sudo apt-get install lvm2
Reading package lists... Done
Building dependency tree
Reading state information... Done
The following extra packages will be installed:
  libdevmapper-event1.02.1 watershed
The following NEW packages will be installed:
  libdevmapper-event1.02.1 lvm2 watershed
...
```

You now have the LVM tools installed and are ready to create LVM volumes.

USING LVM

In this section, you'll identify physical devices you can use for storage and you'll create an LVM volume to be used by Cinder.

> **Devices used in the examples**
>
> I expect you'll be using dedicated nodes (physical or virtual) for the chapters 5 through 8 walkthroughs. If you have access to a resource configuration that matches the reference architecture used in these chapters, things should just work. On the other hand, if resources are limited (one disk, one network adapter, and so on) you can make modifications to match your working environment.
>
> The following examples work as well with one disk as they do with several. The examples can even be adapted to work with disk partitions instead of an entire disk.

Most modern Linux distributions use udev (a dynamic device manager) for kernel device management. On a system using udev, like our reference Ubuntu 14.04 system, disk devices are listed and arranged by hardware path under /dev/disk/by-path. This directory is shown in the following listing.

Listing 7.8 List disk devices

```
$ ls -la /dev/disk/by-path
pci-0000:03:00.0-scsi-0:2:2:0 -> ../../sda
pci-0000:03:00.0-scsi-0:2:2:0-part1 -> ../../sda1
pci-0000:03:00.0-scsi-0:2:2:0-part2 -> ../../sda2
pci-0000:03:00.0-scsi-0:2:2:0-part3 -> ../../sda3
pci-0000:03:00.0-scsi-0:2:3:0 -> ../../sdb
pci-0000:03:00.0-scsi-0:2:4:0 -> ../../sdc
pci-0000:03:00.0-scsi-0:2:5:0 -> ../../sdd
```

The listing shows four physical disk devices: sda, sdb, sdc, and sdd. You can see in the listing that the sda volume has three partitions: sda1, sda2, and sda3. Although it isn't apparent from the listing, the physical device sda is being used as a system volume on the reference storage node. The remaining devices, sdb, sdc, and sdd, will be used to create an LVM volume.

Now that you know which disk devices you're targeting, you can start working with LVM.

> **Know your devices or risk data loss**
>
> Make sure you can identify the storage devices (drives) that you wish to target both inside and outside the operating system. Most servers come with storage adapters that are used to connect physical hard disks to the server. Storage adapters can be used to simply present a raw disk to the operating system, or they can be used to present multiple disks as a logical volume. As far as the operating system is concerned, a physical disk and a logical disk look exactly the same (memory addressable storage),

(continued)
because the storage adapter abstracts this relationship. Storage adapters will report SCSI ID information for physical and logical (such as RAID sets) disks. Listing 7.9 shows how you can match the SCSI ID, `scsi-0:2:2:0`, to the device mapping, `sda`, to ensure you're working with the appropriate disks.

LVM relationships and commands

This is not a book on LVM, but you should have a basic understanding of LVM components and commands before you start creating volumes.

LVM is divided into three functional components:

- *Physical volume*—One or more partitions (or an entire device) on a physical drive
- *Volume group*—One or more physical volumes representing one or more logical volumes
- *Logical volume*—A volume reference contained within a volume group

There are many LVM tools and commands, but only those required for creating a volume usable by Cinder will be described here. The following commands are required to create an LVM volume for use by Cinder:

- `pvcreate <device>` is used to create physical volumes from Linux storage devices.
- `pvscan` is used to display a listing of physical volumes.
- `pvdisplay` is used to display physical volume attributes like size, state, and system-level identifiers.
- `vgcreate <name> <device>` is used to assign physical volumes to a pool of storage referenced by some `<name>`.

Physical volume operations

You'll use the `pvcreate` volume to create a physical LVM device. This process creates a volume-group descriptor at the start of the referenced disks.

The following example shows how to create an LVM physical volume from three Linux system devices: `sdb`, `sdc`, and `sdd`.

Listing 7.9 Using `pvcreate` to create a physical volume

```
$ sudo pvcreate /dev/sdb /dev/sdc /dev/sdd
  Physical volume "/dev/sdb" successfully created
  Physical volume "/dev/sdc" successfully created
  Physical volume "/dev/sdd" successfully created
```

Next, you'll want to verify that your devices have been successfully created using the `pvscan` command, as follows.

Listing 7.10 Using `pvscan` to verify physical volumes

```
$ sudo pvscan
  PV /dev/sda3    VG storage-vg       lvm2 [835.88 GiB / 24.00 MiB free]
  PV /dev/sdb                         lvm2 [4.55 TiB]
  PV /dev/sdc                         lvm2 [4.55 TiB]
  PV /dev/sdd                         lvm2 [4.55 TiB]
  Total: 4 [14.45 TiB] / in use: 1 [835.88 GiB] / in no VG: 3 [13.64 TiB]
```

The previous listing shows the physical volumes we just created, along with the `sda` volume that was shown in listing 7.9. The `sda` volume was created during the Linux installation and has already been assigned to the `storage-vg` volume group.

To get a closer look at the physical volumes, you can use the command `pvdisplay`.

Listing 7.11 Using `pvdisplay` to display physical volume attributes

```
$ sudo pvdisplay
  --- Physical volume ---
  PV Name               /dev/sda3
  VG Name               storage-vg
  PV Size               835.88 GiB / not usable 2.00 MiB
  Allocatable           yes
  PE Size               4.00 MiB
  Total PE              213986
  Free PE               6
  Allocated PE          213980
  PV UUID               XKAbeN-MI3p-kD9h-qHAS-ZXDZ-nzuh-echFIZ

  "/dev/sdb" is a new physical volume of "4.55 TiB"
  --- NEW Physical volume ---
  PV Name               /dev/sdb
  VG Name
  PV Size               4.55 TiB
  Allocatable           NO
  PE Size               0
  Total PE              0
  Free PE               0
  Allocated PE          0
  PV UUID               nUNvkZ-ggd7-8GA2-IS2n-71xr-wk0W-qdUL84
...
```

LVM physical volumes have now been created from the system-level devices `sdb`, `sdc`, and `sdd`. You can now move on to the next step, where you'll work with *volume groups*. OpenStack Block Storage (Cinder) will interface with LVM on the volume-group level.

Volume-group operations

The next step is to create a *volume group*. The volume group will simply be a pool of storage consisting of the physical volumes you created in the previous step.

Follow the steps in the next listing to create the new volume group with a group name of `cinder-volumes`.

Listing 7.12 Using vgcreate to create a volume group

```
$ sudo vgcreate cinder-volumes /dev/sdb /dev/sdc /dev/sdd
  Volume group "cinder-volumes" successfully created
```

Next, you'll want to make sure your volume group was successfully created. If you repeat the pvscan command demonstrated in listing 7.11, you'll see that the physical volumes have been assigned a volume group.

Use the command vgdisplay to list all volume groups.

Listing 7.13 Using vgdisplay to verify volume groups

```
$ sudo vgdisplay
  --- Volume group ---
  VG Name               cinder-volumes
...
  VG Size               13.64 TiB
  PE Size               4.00 MiB
  Total PE              3575037
  Alloc PE / Size       0 / 0
  Free  PE / Size       3575037 / 13.64 TiB
  VG UUID               1On40i-fPAS-EsHf-WbH7-P6M5-1U0f-TcBrX2

  --- Volume group ---
  VG Name               storage-vg
...
```

The listing shows the cinder-volumes volume group you just created, along with the storage-vg volume group that was created during the operating system installation.

OK. You've taken system-level devices and created LVM physical volumes. Next, you assigned those physical volumes to a volume group named cinder-volumes. If you've worked with LVM before, you might expect the next step to be creating a logical volume (creating a virtual volume from the cinder-volumes pool), but this isn't the case. Instead, Cinder will be configured to manage the cinder-volumes pool. Cinder will create logical volumes as needed, based on VM requirements.

Reboot the storage node. When the node comes back online, check that the volume group cinder-volumes is present, as demonstrated earlier in listing 7.14.

7.2 Deploying Cinder

Cinder provides an abstraction layer between block storage resources and Compute services (Nova). Through the Cinder API, block volumes can be managed (created, destroyed, assigned, and so on) without knowledge of the underlying resource that provides the storage.

Consider an organization that historically maintained separate groups to manage storage and compute resources. In this scenario, the storage group could expose a Cinder service to the compute group, for storage to be used by OpenStack Nova. In other words, consumers of block storage services from Cinder require no information regarding the underlying systems managing the backend storage. Cinder translates the APIs of the underlying storage system to provide both storage resources and statistical reporting related to storage. As with OpenStack Networking, underlying support for backend storage subsystems is provided through vendor-based Cinder plug-ins.

For each OpenStack release, a minimum set of features and statistical reports are required for each plug-in. If plug-ins aren't maintained between releases, and additional functions and reporting are required, they're deprecated in subsequent releases. The current list of minimum features and reports can be found in tables 7.2 and 7.3. The most current list of plug-in requirements can be found on the GitHub repository: https://github.com/openstack/cinder/blob/master/doc/source/devref/drivers.rst.

Table 7.2 Minimum features

Feature name	Description
Volume create/delete	Create or delete a volume for a VM on a backend storage system.
Volume attach/detach	Attach or detach a volume to a VM on a backend storage system.
Snapshot create/delete	Take a running snapshot of a volume on a backend storage system.
Volume from snapshot	Create a new volume from a previous snapshot on a backend storage system.
Get volume stats	Report the statistics on a specific volume.
Image to volume	Copy an image to a volume that can be used by a VM.
Volume to image	Copy a volume used by a VM to a binary image.
Clone volume	Clone one VM volume to another VM volume.
Extend volume	Extend the size of a VM volume without destroying the data on the existing volume.

Table 7.3 Minimum reporting statistics

Statistic name	Example	Description
`driver_version`	1.0a	Version of the vendor-specific driver for the reporting plug-in or driver.
`free_capacity_gb`	1000	Amount of free space in gigabytes. If unknown or infinite, the keywords *unknown* or *infinite* are reported.
`reserved_percentage`	10	Percentage of how much space is reserved, which is needed when volume extend is used.
`storage_protocol`	iSCSI	Reports the storage protocol: iSCSI, FC, NFS, and so on.
`total_capacity_gb`	102400	Amount of total capacity in gigabytes. If unknown or infinite, the keywords *unknown* or *infinite* are reported.

Table 7.3 Minimum reporting statistics *(continued)*

Statistic name	Example	Description
vendor_name	Dell	Name of the vendor that provides the backend storage system.
volume_backend_name	Equ_vol00	Name of the volume on the vendor back end. Needed for statistical reporting and troubleshooting.

In the multi-node design presented in this book, the reference implementation plug-in (LVM) will be used as the underlying storage subsystem. But there are many plug-ins available for Cinder for many vendor platforms and technologies.

In the next section, you'll continue on with the Cinder deployment process. First, you'll install Cinder components, and then you'll configure the components.

7.2.1 *Installing Cinder*

Unlike the multiple Neutron components you deployed in chapter 6, there's only one Cinder component to install and configure. You can install the `cinder-volume` component as follows.

Listing 7.14 Install Cinder component

```
$ sudo apt-get install -y cinder-volume
[sudo] password for sysop:
Reading package lists... Done
Building dependency tree
Reading state information... Done
The following extra packages will be installed:
  alembic cinder-common libboost-system1.54.0 libboost-thread1.54.0
...
Adding system user `cinder' (UID 105) ...
Adding new user `cinder' (UID 105) with group `cinder' ...
...
tgt start/running, process 3913
```

Although you only needed to reference `cinder-volume` during the install process, you can see that this service has many dependencies. One of the fundamental dependencies is the Linux Small Computer System Interface (SCSI) target framework (`tgt`).

From the relationship of OpenStack and the Linux kernel, you can think of `tgt` like Open vSwitch. Of course, on a functional level they accomplish very different tasks, but they both bridge the gap between kernel-level and user-accessible functions. Both of these frameworks are used by OpenStack plug-ins to accomplish system-level tasks, without OpenStack having to work directly with the Linux kernel. It's sufficient to think of `tgt` as a Cinder helper.

Now that the `cinder-volume` package has been installed and helper services have been started, you can proceed with your Cinder configuration.

> **What is tgt?**
>
> Linux SCSI target framework (`tgt`) simplifies the process of integrating multi-protocol SCSI targets in Linux environments. The following target drivers are supported:
>
> - iSCSI (SCSI over IP)
> - FCoE (Fibre Channel over Ethernet)
> - iSER (iSCSI over RDMA, using Infiniband)
>
> It's worth noting that a `tgt` competitor, *Linux-IO Target* (http://linux-iscsi.org/), claims that it superseded `tgt` as of the Linux 2.6.38 kernel, but OpenStack support for this framework has not been updated since the OpenStack Grizzly release.

7.2.2 Configuring Cinder

The next step is configuration. First, you must modify the /etc/cinder/cinder.conf file. You'll define the service authentication, the management communication, the storage helpers, and the name of the LVM volume group. The bold lines in the following listing are the ones that need to be added or modified.

Listing 7.15 Modify /etc/cinder/cinder.conf

```
[DEFAULT]
...
iscsi_helper = tgtadm                              ❶ iscsi_helper
volume_group = cinder-volumes                      cinder-volumes volume group
rpc_backend = cinder.openstack.common.rpc.impl_kombu
rabbit_host = 192.168.0.50
rabbit_password = openstack1

glance_host = 192.168.0.50                         ❷ Cinder and Glance communicate via 192.168.0.50.

[database]
connection = mysql://cinder_dbu:openstack1@192.168.0.50/cinder

[keystone_authtoken]
auth_uri = http://10.33.2.50:35357/v2.0
admin_tenant_name = service
admin_password = openstack1
auth_protocol = http
admin_user = cinder
```

The `iscsi_helpertgtadm` is part of the `tgt` framework discussed in section 7.2.1 ❶.

You created the `cinder-volumes` volume group in the subsection "Using LVM." Now that you know both how to create new volume groups and where to configure Cinder, you can use any name you like.

Cinder will communicate with Glance directly over the OpenStack internal network via 192.168.0.50 ❷, as defined in table 5.1. Images applied to VM volumes will be applied from Glance to Cinder directly.

You have now installed Cinder and configured it to use the LVM pool you created in an earlier section. Now continue on to the next section to verify your Cinder deployment.

7.2.3 Restarting and verifying the Cinder agents

The final step is to restart the cinder-volume service to activate your new configuration. In addition, you'll need to restart the tgt helper service.

Listing 7.16 Restart Cinder

```
sudo service cinder-volume restart                    <-- Restarts Cinder volume service
cinder-volume stop/waiting
cinder-volume start/running, process 13822

sudo service tgt restart                              <-- Restarts tgt service
tgt stop/waiting
tgt start/running, process 6955
```

You'll want to check the Cinder log to make sure the service started successfully and is listening for requests. The log can be found in the file /var/log/cinder/cinder-volume.log.

> **ERROR [-] No module named MySQLdb**
>
> If the error "[-] No module named MySQLdb" shows up in the cinder-volume.log, the MySQL interface for Python wasn't installed as a dependency on your system. Install the python-mysqldb package to correct the issue:
>
> ```
> $ sudo apt-get install python-mysqldb
> Reading package lists... Done
> Building dependency tree
> Reading state information... Done
> The following extra packages will be installed:
>   libmysqlclient18 mysql-common
> ...
> ```

Your cinder-volume log (at /var/log/upstart/cinder-volume.log) should look something like the following:

```
Starting cinder-volume node (version 2014.1.2)
Starting volume driver LVMISCSIDriver (2.0.0)
Updating volume status
Connected to AMQP server on 192.168.0.50:5672
```

Review the logs, checking for a connection to the AMQP (RabbitMQ) server, and ensure there are no errors. The log files should exist even if they're empty. If the logs look fine, move on to the next section where you'll test Cinder operations.

7.3 Testing Cinder

Although you don't yet have all the components deployed to test Cinder with a VM, you can and should test some basic Cinder functionality. This section covers the creation of a volume using both the command-line tool and the Dashboard.

7.3.1 Create a Cinder volume: command line

First, you must install the Cinder command-line tools as follows.

Listing 7.17 Install Cinder command-line tools

```
sudo apt-get install -y python-cinderclient
```

The `python-cinderclient` package provides the `cinder` command-line application. In order to use `cinder`, you must provide the application authentication credentials, including an authentication location. One option is to set credential information as part of your shell variables, and the other is to pass the information to the application through command-line arguments. In order to provide details for the volume-creation process, we'll use command-line arguments in the examples.

The following listing demonstrates how to list all Cinder volumes. Follow the example, even though you know no volumes exist at this point. This step confirms that you have a working client and service interaction.

Listing 7.18 List Cinder volumes

```
$ cinder \
--os-username admin \
--os-password openstack1 \
--os-tenant-name admin \
--os-auth-url http://10.33.2.50:35357/v2.0 \
list

+----+--------+--------------+------+-------------+----------+-------------+
| ID | Status | Display Name | Size | Volume Type | Bootable | Attached to |
+----+--------+--------------+------+-------------+----------+-------------+
+----+--------+--------------+------+-------------+----------+-------------+
```

If all worked properly, you should see output similar to that shown in the preceding listing. If you experience an error, take a look at the Cinder log (/var/log/cinder/cinder-volume.log) for any obvious problems.

If all went well, you can create a volume as shown in the following listing.

Listing 7.19 Create Cinder volume

```
$ cinder \
--os-username admin \
--os-password openstack1 \
--os-tenant-name admin \
--os-auth-url http://10.33.2.50:35357/v2.0 \
 create \
```

```
--display-name "My First Volume!" \
--display-description "Example Volume: OpenStack in Action" \
1
+-----------------------+---------------------------------------+
|       Property        |                 Value                 |
+-----------------------+---------------------------------------+
|      attachments      |                  []                   |
|   availability_zone   |                 nova                  |
|       bootable        |                 false                 |
|      created_at       |      2014-09-07T16:53:03.998340       |
|  display_description  |  Example Volume: OpenStack in Action  |
|     display_name      |           My First Volume!            |
|       encrypted       |                 False                 |
|          id           | a595d38f-5f32-48e5-903b-9559ffda06b1  |
|       metadata        |                  {}                   |
|         size          |                   1                   |
|      snapshot_id      |                 None                  |
|     source_volid      |                 None                  |
|        status         |               creating                |
|      volume_type      |                 None                  |
+-----------------------+---------------------------------------+
```

Specifies volume size in GB

During volume creation, a volume name will be generated. This volume name will match the name of the LVM logical volume provisioned on the system. Take a look in the /var/log/cinder/cinder-volume.log file, and you should see entries resembling these:

Volume name = volume-a595d38f-5f32-48e5-903b-9559ffda06b1

```
cinder.volume.flows.manager.create_volume ... _create_raw_volume
..
'volume_name': u'volume-a595d38f-5f32-48e5-903b-9559ffda06b1'
..
cinder.volume.flows.manager.create_volume ... created successfully
```

In section 5.15.13 you created the `cinder-volumes` volume group, and in section 7.2.2 you configured Cinder to use the volume group you created. As part of the Cinder volume-creation process, a logical volume was created from the `cinder-volumes` volume group. In the following listing, the logical volume command `lvdisplay` is used to display the logical volumes on the storage node.

Listing 7.20 Display logical volumes

```
$ sudo lvdisplay
  --- Logical volume ---
...
  LV Name                volume-a595d38f-5f32-48e5-903b-9559ffda06b1
  VG Name                cinder-volumes
...
  LV Size                1.00 GiB
```

In the preceding listing, the `LV Namevolume-a595d38f-5f32-48e5-903b-9559ffda06b1` matches the `volume_name` found in the Cinder volume `create` specification.

For each Cinder volume that's created, an LVM volume is created. Keeping this in mind, you can trace back Cinder volume issues through the LVM and system levels. In the next section, you'll repeat this process using the OpenStack Dashboard. Either method can be used, but OpenStack administrators will often use the command line, and end users will choose the Dashboard. It's useful to understand both processes.

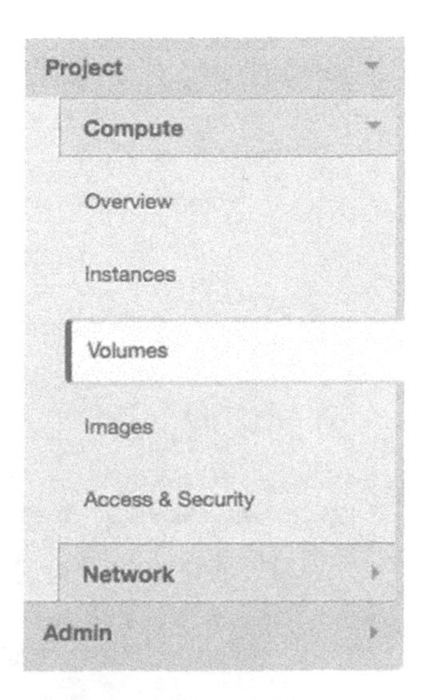

Figure 7.3 Dashboard toolbar

7.3.2 Create a Cinder volume: Dashboard

In chapter 5 you deployed the OpenStack Dashboard. The Dashboard should now be available at http://<controller address>/horizon/. Log in as `admin` with the password `openstack1`.

Once you're logged in to the Dashboard, click Volumes under the Project bar, as shown in figure 7.3.

On the Volumes & Snapshots screen, click Create Volume, shown in figure 7.4.

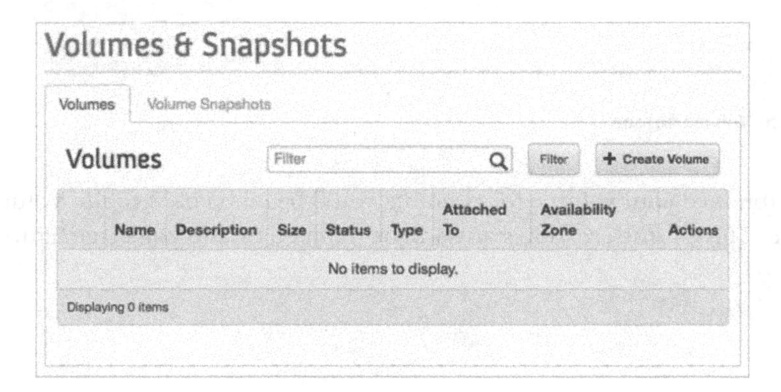

Figure 7.4 Volumes & Snapshots screen

On the Create Volume screen, specify how you'd like your image created. In chapter 5 you added the `Cirros 0.3.2` image to Glance. Unlike the previous command-line example, figure 7.5 specifies that the Glance image `Cirros 0.3.2` should be applied to the volume. Click Create Volume when you've finished describing your volume.

Create Volume

Volume Name: *
My First Volume!

Description:
Example Volume: OpenStack in Action

Type:

Size (GB): *
1

Volume Source:
Image

Use image as a source:
Cirros 0.3.2 (12.6 MB)

Availability Zone
Any Availability Zone

Description:
Volumes are block devices that can be attached to instances.

Volume Limits
Total Gigabytes (0 GB)
1,000 <django.utils.functional.__proxy__ object at 0x7f912f771f50> Available

Number of Volumes (0) 10 Available

Cancel Create Volume

Figure 7.5 Create Volume screen

Once you've submitted your volume for creation, you'll be taken back to the Volumes & Snapshots page. The volume-creation status will be updated on this page (see figure 7.6).

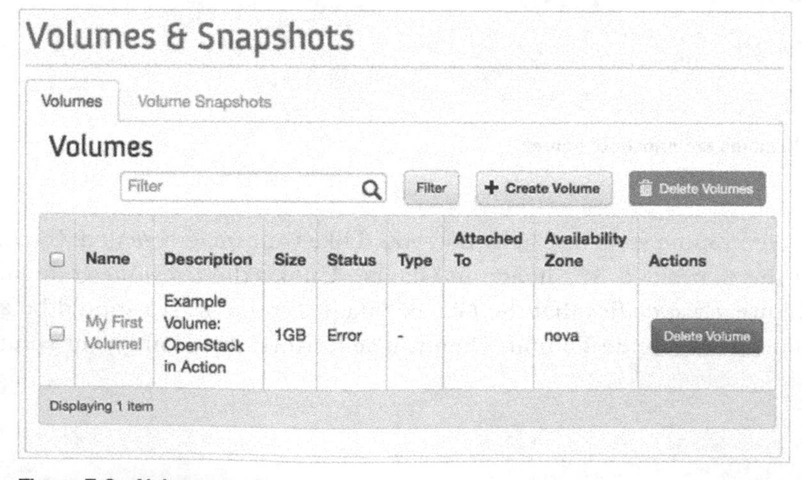

Figure 7.6 Volume-creation error

Uh-oh! What happened? With the exception of specifying an image to be applied to your volume, everything was exactly the same as in the command-line example. Where did the error come from? You'd better take a look at the /var/log/cinder/cinder-volume.log to see what's going on:

```
ERROR cinder.volume.flows.manager.create_volume
...
 is unacceptable: qemu-img is not installed and image is of type qcow2.
Only RAW images can be used if qemu-img is not installed.
```

The cinder-volume.log contains an `ERROR` that seems related to adding the image specification during volume creation. Install the `qemu-utils` package, which includes tools for image conversion.

Listing 7.21 Installing image-management tools

```
sudo apt-get install -y qemu-utils
```

The Cinder volume now has all the tools it needs to both create a volume and apply a Glance image to the volume. Go back and follow the steps in this section again. The volume-creation process with an image should now complete successfully. This procedure tests both the Cinder and Glance services.

7.4 Summary

- A separate physical network will be used for storage traffic.
- There are many vendor plug-ins for many types of storage technologies.
- Logical Volume Manager (LVM) is the reference storage plug-in interface for Cinder.
- Pools of LVM storage are used by Cinder to assign block volumes.
- The Linux SCSI target framework (`tgt`) is used as a helper by Cinder to assign block volumes.
- Glance images can be applied to volumes by Cinder during creation.

8 Walking through a Compute deployment

This chapter covers

- Installing Open vSwitch (OVS) on a compute node
- Deploying OpenStack Networking components on an OpenStack compute node
- Integrating OVS components with OpenStack Networking on a compute node
- Setting up KVM as an OpenStack Compute hypervisor
- Integrating OpenStack Compute-supporting components on a compute node

In chapter 5 you deployed an OpenStack controller node, which provides the server-side management of OpenStack services. In chapters 6 and 7, you deployed standalone resource nodes for network and storage services.

In this chapter, you'll deploy another standalone resource node, and this one will consume resources provided by the storage and network nodes. Later, in chapter 8, you'll walk through the deployment of OpenStack Compute services on a resource node. The overall multi-node architecture introduced in chapter 5 is shown again in figure 8.1, with this chapter's Compute components shown at the lower middle.

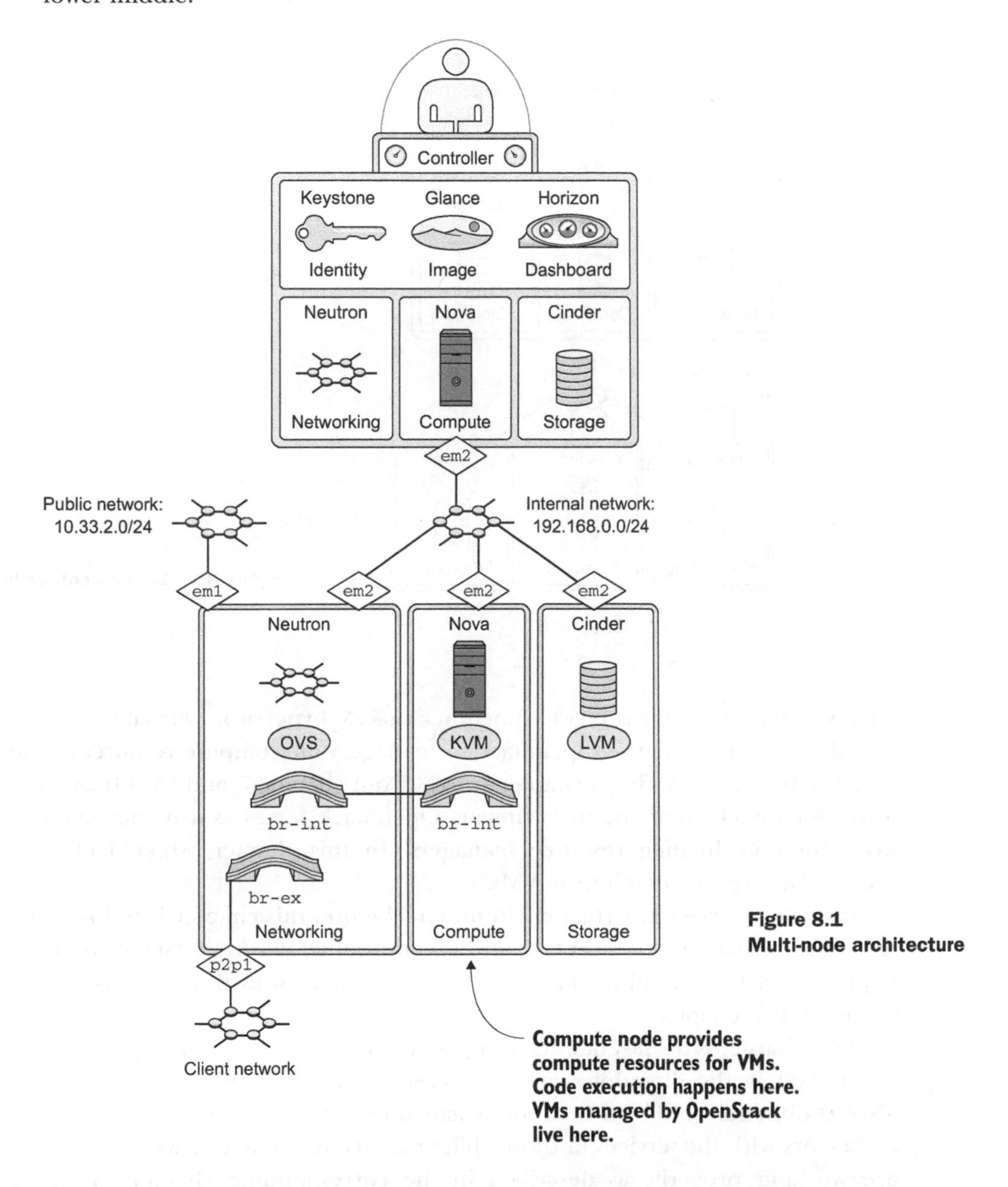

Figure 8.1 Multi-node architecture

Figure 8.2 shows your current progress on your way to a working manual deployment. In this chapter, you'll take the final step in your manual deployment—you'll add compute capabilities (hypervisor, network, and so on).

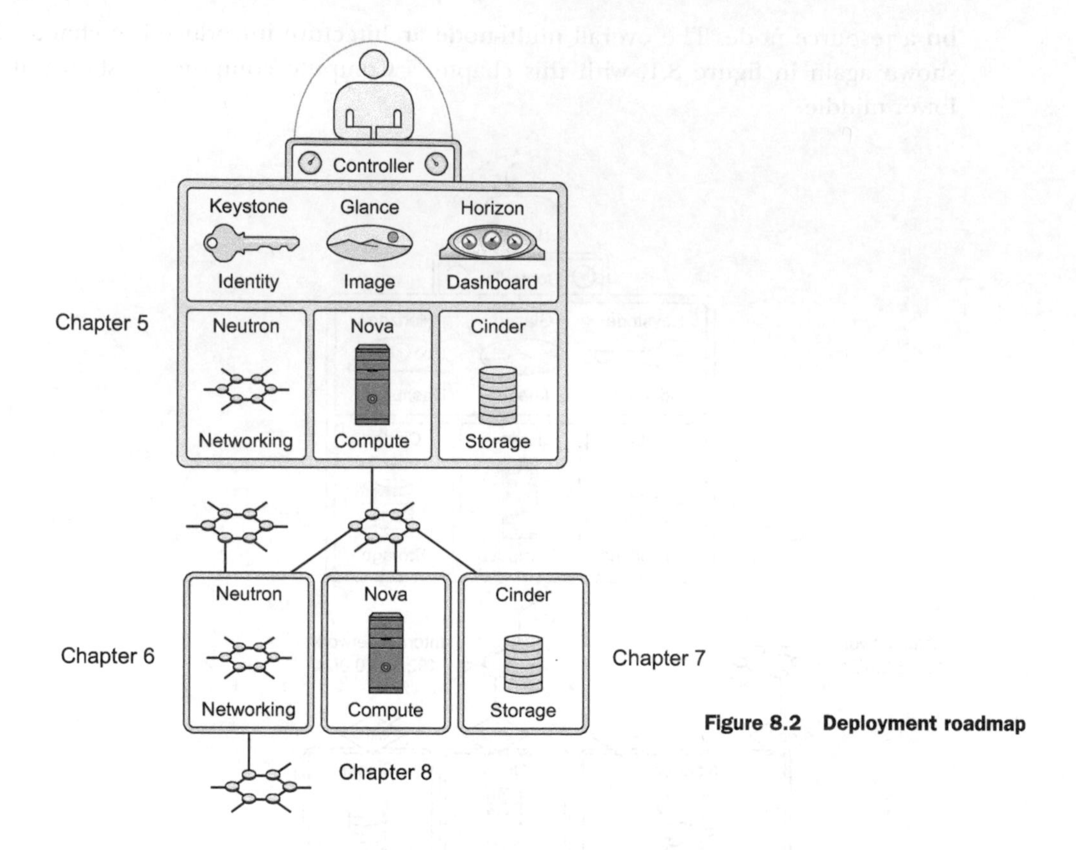

Figure 8.2 Deployment roadmap

First, you'll prepare the server to function as a KVM hypervisor (virtual machine host), and then you'll configure OpenStack to manage your compute resources. The relationship between OVS from chapter 6, LVM from chapter 7, and KVM from chapter 8 with OpenStack components is similar. OpenStack serves as a management framework for coordinating resource managers. In this chapter, OpenStack Compute relates these resources to create VMs.

If you've worked in virtual environments before, this chapter is unlikely to introduce any fundamental concepts that will seem strange to you. In fact, if you're used to deploying virtual machine clusters, you'll be familiar with the hypervisor steps presented in this chapter.

This chapter is the last one standing between you and a manual OpenStack deployment. Follow the directions carefully, keeping in mind that OpenStack Compute (Nova) depends on the services you installed in chapters 5 through 7. If you ran into any errors with the services in those chapters, take the time to make sure the services are working properly, as described in the corresponding chapters. If everything appears to be running smoothly, you can get started with this final step in your manual deployment.

8.1 Deploying Compute prerequisites

As is true of all chapters in this part of the book, you're going to manually install and configure dependencies and core OpenStack packages.

> **PROCEED WITH CARE** Working in a multi-node environment greatly increases deployment complexity. A small, seemingly unrelated mistake in the configuration of one component or dependency can cause issues that are very hard to track down. Read each section carefully, making sure you understand what you're installing or configuring.

Many of the examples in this chapter include a verification step, which I highly recommended you follow. If a configuration can't be verified, retrace your steps to the latest verified point and start over. This practice will save you a great deal of frustration.

8.1.1 Preparing the environment

With the exception of the network configuration, environment preparation will be similar to the controller nodes you deployed in chapters 5 through 7. Follow the instructions in the discussion, and make sure you pay close attention to the network interfaces and addresses in the configurations. It's easy to make a typo and often hard to track it down.

8.1.2 Configuring the network interface

You'll configure the network with two interfaces:

- *Node interface*—Traffic not directly related to OpenStack. This interface will be used for administrative tasks like SSH console access, software updates, and even node-level monitoring.
- *Internal interface*—Traffic related to OpenStack component-to-component communication. This includes API and AMQP type traffic.

Before you configure these new interfaces, though, you'll want to determine what interfaces already exist on the system.

REVIEWING THE NETWORK

To determine what interfaces exist on the system, use the following command.

Listing 8.1 List interfaces

```
$ ifconfig -a
em1       Link encap:Ethernet  HWaddr b8:2a:72:d4:ff:88
          inet addr:10.33.2.53  Bcast:10.33.2.255  Mask:255.255.255.0
          inet6 addr: fe80::ba2a:72ff:fed4:ff88/64 Scope:Link
          UP BROADCAST RUNNING MULTICAST  MTU:1500  Metric:1
          RX packets:60708 errors:0 dropped:0 overruns:0 frame:0
          TX packets:7142 errors:0 dropped:0 overruns:0 carrier:0
          collisions:0 txqueuelen:1000
          RX bytes:54254314 (54.2 MB)  TX bytes:962977 (962.9 KB)
          Interrupt:35
```

```
em2           Link encap:Ethernet  HWaddr b8:2a:72:d4:ff:89
              BROADCAST MULTICAST  MTU:1500  Metric:1
              RX packets:0 errors:0 dropped:0 overruns:0 frame:0
              TX packets:0 errors:0 dropped:0 overruns:0 carrier:0
              collisions:0 txqueuelen:1000
              RX bytes:0 (0.0 B)  TX bytes:0 (0.0 B)
              Interrupt:38
```

You might have configured your node interface, em1, during the initial installation. You'll use the em1 interface to communicate with this node. Take a look at the other interface, em2. On the example systems used in this book, the interface em2 will be used for internal OpenStack traffic.

Next, you'll review the network configuration for the example nodes and configure controller interfaces.

CONFIGURING THE NETWORK

Under Ubuntu, the interface configuration is maintained in the /etc/network/interfaces file. In this chapter, we'll build a working configuration based on the italicized addresses in table 8.1.

Table 8.1 Network address table

Node	Function	Interface	IP Address
controller	Pubic interface/node address	em1	10.33.2.50/24
controller	OpenStack internal	em2	192.168.0.50/24
network	Node address	em1	10.33.2.51/24
network	OpenStack internal	em2	192.168.0.51/24
network	VM network	p2p1	None: Assigned to OpenStack Networking
storage	Node address	em1	10.33.2.52/24
storage	OpenStack internal	em2	192.168.0.52/24
compute	*Node address*	*em1*	*10.33.2.53/24*
compute	*OpenStack internal*	*em2*	*192.168.0.53/24*

In order to modify the network configuration, or any privileged configuration, you must use sudo privileges (sudo vi /etc/network/interfaces). Any text editor can be used in this process.

Modify your interfaces file as follows.

Listing 8.2 Modify interface config /etc/network/interfaces

```
# The loopback network interface
auto lo
iface lo inet loopback
```

```
# The OpenStack Node Interface
auto em1
iface em1 inet static
        address 10.33.2.53
        netmask 255.255.255.0
        network 10.33.2.0
        broadcast 10.33.2.255
        gateway 10.33.2.1
        dns-nameservers 8.8.8.8
        dns-search testco.com

# The OpenStack Internal Interface
auto em2
iface em2 inet static
        address 192.168.0.53
        netmask 255.255.255.0
```

❶ em1 interface
❷ em2 interface

In your network configuration, the em1 interface ❶ will be used for node administration, such as SSH sessions to the host server. OpenStack shouldn't use this interface directly. The em2 interface ❷ will be used primarily for AMQP and API traffic between resource nodes and the controller.

You should now refresh the network interface settings for which the configuration was changed. If you didn't change the settings of your primary interface, you shouldn't experience an interruption. If you changed the address of the primary interface, it's recommended you reboot the server at this point.

You can refresh the network configuration for a particular interface as in the following example for interface em2.

Listing 8.3 Refreshing networking settings

```
sudo ifdown em2 && sudo ifup em2
```

The network configuration, from an operating system standpoint, should now be active. The interface will automatically be brought online based on your configuration. This process can be repeated for each interface that requires a configuration refresh.

In order to confirm that the configuration was applied, you should once again check your interfaces to see that your configuration is in place, as shown here.

Listing 8.4 Check network for updates

```
$ifconfig -a
em1       Link encap:Ethernet  HWaddr b8:2a:72:d4:ff:88
          inet addr:10.33.2.53  Bcast:10.33.2.255  Mask:255.255.255.0
          inet6 addr: fe80::ba2a:72ff:fed4:ff88/64 Scope:Link
          UP BROADCAST RUNNING MULTICAST  MTU:1500  Metric:1
          RX packets:61211 errors:0 dropped:0 overruns:0 frame:0
          TX packets:7487 errors:0 dropped:0 overruns:0 carrier:0
          collisions:0 txqueuelen:1000
          RX bytes:54305503 (54.3 MB)  TX bytes:1027531 (1.0 MB)
```

```
          Interrupt:35

em2       Link encap:Ethernet  HWaddr b8:2a:72:d4:ff:89
          inet addr:192.168.0.53  Bcast:192.168.0.255  Mask:255.255.255.0
          inet6 addr: fe80::ba2a:72ff:fed4:ff89/64 Scope:Link
          UP BROADCAST RUNNING MULTICAST  MTU:1500  Metric:1
          RX packets:4 errors:0 dropped:0 overruns:0 frame:0
          TX packets:8 errors:0 dropped:0 overruns:0 carrier:0
          collisions:0 txqueuelen:1000
          RX bytes:256 (256.0 B)  TX bytes:680 (680.0 B)
          Interrupt:38
```

At this point you should be able to remotely access the network server, and the server should have internet access. The remainder of the installation can be performed remotely using SSH or directly from the console.

8.1.3 Updating packages

As explained in previous chapters, the APT package index is a database of all available packages defined by a remote list found in the /etc/apt/sources.list file. You need to make sure the local database is synchronized with the latest packages available in the repository for your specific Linux distribution. Prior to installation, you should also upgrade any repository items, including the Linux kernel, that might be out of date.

Listing 8.5 Update and upgrade packages

```
sudo apt-get -y update
sudo apt-get -y upgrade
```

You now need to reboot the server to refresh any packages or configurations that might have changed.

Listing 8.6 Reboot server

```
sudo reboot
```

8.1.4 Software and configuration dependencies

In this section, you'll install a few software dependencies and make a few configuration changes in preparation for the OpenStack component install.

SERVER-TO-ROUTER CONFIGURATION

OpenStack manages resources used to provide virtual machines. One of those resources is the network used by virtual machines to communicate with other virtual and physical machines—you deployed OpenStack Networking in chapter 6. For OpenStack Compute to use the resources provided by OpenStack Networking, you must configure the Linux kernel to allow network traffic to be offloaded to OpenStack Networking.

The sysctl command is used to modify kernel parameters, such as those related to basic network functions. You'll need to make several modifications to your kernel settings using this tool.

In chapter 6 you had to configure the OpenStack Network node to function as a router and forward traffic between virtual and physical interfaces. For the OpenStack compute node, you don't have to make this configuration, because OpenStack Networking will be providing this service for you. But you still have to make configuration changes to allow the Linux kernel to offload traffic-forwarding decisions.

As discussed in chapter 6 (section "Server-to-router configuration"), *reverse-path filtering* was introduced to limit the impact of DDOS attacks. By default, if the Linux kernel can't determine the source route of a packet, it will be dropped. You must configure the kernel to disable reverse-path filtering, which leaves path management up to OpenStack.

Apply the following settings to your OpenStack compute node.

Listing 8.7 Modify /etc/sysctl.conf

```
net.ipv4.conf.all.rp_filter=0          ← Disables reverse-path filtering for all existing interfaces
net.ipv4.conf.default.rp_filter=0      ← Disables reverse-path filtering for all future interfaces
```

To enable the sysctl kernel changes without restarting the server, invoke the sysctl -p command.

Listing 8.8 Execute sysctl command

```
$ sudo sysctl -p
net.ipv4.conf.all.rp_filter = 0
net.ipv4.conf.default.rp_filter = 0
```

Reverse-path filtering should now be disabled on the kernel level.

In the next section, you'll add advanced network features to your node with the Open vSwitch package.

8.1.5 Installing Open vSwitch

OpenStack Compute takes advantage of the open source distributed virtual switching package Open vSwitch (OVS). OVS provides the same data-switching functions as a physical switch (L2 traffic on port A destined to port B is switched on port B), but it runs in software on servers. OVS will also be used to tunnel traffic from OpenStack compute nodes to OpenStack network nodes for routing and other L3 services. From a network-switching standpoint, the examples in this book make exclusive use of the OVS switching platform. For more on how switches work, see the sidebar "What does a switch do?" in section 5.11.9.

At this point you have a server that can act like a switch (via Linux network bridging). You'll now add advanced switching capabilities to your server through the installation of OVS. The switching features provided by OVS rival offerings provided by standalone network vendors.

You can now turn your server into an advanced switch by starting with the following command.

Listing 8.9 Install OVS

```
$ sudo apt-get -y install openvswitch-switch
...
Setting up openvswitch-common ...
Setting up openvswitch-switch ...
openvswitch-switch start/running
```

The Open vSwitch install process will install a new OVS kernel module. In addition, the OVS kernel module will reference and load additional kernel models (GRE, VXLAN, and others) as necessary to build network overlays.

You want to be absolutely sure the OVS kernel modules were loaded. Using the `lsmod` command, confirm the presence of OVS kernel modules.

Listing 8.10 Verify OVS kernel modules

```
$ sudo lsmod | grep openvswitch
Module                         Size   Used by
openvswitch                   66901  0
gre                           13796  1 openvswitch
vxlan                         37619  1 openvswitch
libcrc32c                     12644  1 openvswitch
```

The output of the `lsmod` command should show several resident modules related to OVS:

- `openvswitch` is the OVS module itself. This module provides the interface between the kernel and OVS services.
- `gre`, used by the `openvswitch` module, enables GRE functionality on the kernel level.
- `vxlan`, like the `gre` module, is used to provide VXLAN functions on the kernel level.
- `libcrc32c` provides kernel-level support for cyclic redundancy check (CRC) algorithms, including hardware offloading using Intel's CRC32C CPU instructions. Hardware offloading is important for the high-performance calculation of network flow hashes and other CRC functions common to network headers and data frames.

GRE and VXLAN support on the kernel level means that the transports used to create overlay networks are understood by the system kernel, and by extension the Linux network subsystem.

If you think the kernel module should have loaded, but you still don't see it, restart the system and see if it loads then. You should also take a look at the sidebar in section 5.11.9, "No Modules? DKMS to the Rescue!" Additionally, you can try to load the kernel module with the command modprobe openvswitch. Check the kernel log, /var/log/kern/log, for any errors related to loading OVS kernel modules. OVS won't function for your purposes without the appropriate resident kernel modules.

8.1.6 Configuring Open vSwitch

You now need to add an internal br-int OVS bridge.

The br-int bridge interface will be used for communication within Neutron-managed networks. Virtual machines communicating within internal networks created by OpenStack Neutron (not to be confused with the Internal interface on the operating-system level) will use this bridge for communication.

Listing 8.11 Configure internal OVS bridge

```
sudo ovs-vsctl add-br br-int
```

You'll also want to confirm that the bridge was successfully added to OVS and that it's visible to the underlying networking subsystem.

Listing 8.12 Show OVS configuration

```
sudo ovs-vsctl show
ff149266-a259-4baa-9744-60e7680b928d
    Bridge br-int
        Port br-int
            Interface br-int
                type: internal
    ovs_version: "2.0.2"
```

Now that you've confirmed that br-int is configured in OVS, make sure you see the bridge interface on the OS level.

Listing 8.13 Verify OVS OS integration

```
$ ifconfig -a
                                                  ❶ br-int bridge
br-int
    Link encap:Ethernet  HWaddr c6:6a:73:f4:5f:41
          BROADCAST MULTICAST  MTU:1500  Metric:1
          RX packets:0 errors:0 dropped:0 overruns:0 frame:0
          TX packets:0 errors:0 dropped:0 overruns:0 carrier:0
          collisions:0 txqueuelen:0
          RX bytes:0 (0.0 B)  TX bytes:0 (0.0 B)
...
em1       Link encap:Ethernet  HWaddr b8:2a:72:d5:21:c3
          inet addr:10.33.2.53  Bcast:10.33.2.255  Mask:255.255.255.0
          inet6 addr: fe80::ba2a:72ff:fed5:21c3/64 Scope:Link
          UP BROADCAST RUNNING MULTICAST  MTU:1500  Metric:1
          RX packets:13483 errors:0 dropped:0 overruns:0 frame:0
          TX packets:2763 errors:0 dropped:0 overruns:0 carrier:0
          collisions:0 txqueuelen:1000
```

```
          RX bytes:12625608 (12.6 MB)  TX bytes:424893 (424.8 KB)
          Interrupt:35
...
ovs-system
 Link encap:Ethernet  HWaddr 96:90:8d:92:19:ab
          BROADCAST MULTICAST  MTU:1500  Metric:1
          RX packets:0 errors:0 dropped:0 overruns:0 frame:0
          TX packets:0 errors:0 dropped:0 overruns:0 carrier:0
          collisions:0 txqueuelen:0
          RX bytes:0 (0.0 B)  TX bytes:0 (0.0 B)
```

❷ ovs-system interface

Notice the addition of the bridge `br-int` in your interface list ❶. This new bridge will be used by OVS and the Neutron OVS module for internal and external traffic. In addition, the `ovs-system` interface was added ❷. This is the OVS datapath interface, but you won't have to worry about working with this interface; it's simply an artifact of Linux kernel integration. However, the presence of this interface is an indication that the OVS kernel modules are active.

You now have an operational OVS deployment and a bridge. As explained in chapter 6, the internal `br-int` bridge will be used by Neutron to attach virtual interfaces to the network bridge; the virtual interfaces will be used as endpoints for GRE tunnels between network and compute nodes. The internal bridge will not need to be associated with a physical interface or be placed in an OS-level `UP` state to work.

Recall creating the `br-ex` OVS bridge in chapter 6. This bridge was used to interface OVS and, by relation, OpenStack Networking with a physical (external) interface and network. This step isn't needed with an OpenStack compute node because external traffic (traffic not destined for the originating node) will be sent to OpenStack Network.

You're now ready to configure the hypervisor on your OpenStack compute node.

8.2 Installing a hypervisor

As previously discussed, there are several choices for hypervisors and even containers under OpenStack. Due to its popularity, we'll use the KVM hypervisor. After the initial install, KVM will be managed by Nova, Neutron, and Cinder.

8.2.1 Verifying your host as a hypervisor platform

You first need to confirm that CPU virtualization extensions are available and enabled on your hardware. There's a nice utility called cpu-checker that will check the status of extensions that can be used by KVM. You'll want to use this utility to verify your platform as a hypervisor host.

Listing 8.14 Verify processor virtualization extensions

```
$ sudo apt-get install cpu-checker
...
Setting up cpu-checker (0.7-0ubuntu1) ...
...
```

```
$ sudo kvm-ok
INFO: /dev/kvm exists
KVM acceleration can be used
```

VIRTUAL HARDWARE EXTENSIONS The hardware assistance provided by virtualization extensions enables fully isolated virtual machines to perform at near native speeds for many workloads. Without extensions, the CPU-intensive functions of the hypervisor must be performed in software, which greatly reduces the performance of the overall system. Using hardware for OpenStack Compute that's not capable of KVM acceleration isn't recommended.

If you receive the message "INFO: Your CPU does not support KVM extensions," you can still run OpenStack, but hypervisor performance will be very poor. Virtualization extensions provide hardware assistance to the hypervisor for processor migration, priority, and memory handling.

If you receive the message "KVM acceleration can NOT be used," without the previous warning, your processor likely supports virtualization extensions, but the extensions might not be enabled in the BIOS. Confirm that your processor model supports extensions and check your BIOS setting for virtualization extensions.

Checking processor extensions

Another method for determining hardware acceleration capabilities is to check the processor extensions as reported by the Linux kernel. Use the following command:

```
egrep -c '(vmx|svm)' /proc/cpuinfo
```

If the result of this command is greater than 0, your hardware supports acceleration.

This method is listed in the OpenStack documentation.

In the next step, you'll install the KVM and Libvirt packages.

8.2.2 Using KVM

Before you install KVM, you should quickly review the components that will be installed:

- *Libvirt*—This is a management layer used to control several hypervisors from the OS and API layers.
- *QEMU (Quick Emulator)*—QEMU is a complete hardware virtualization platform (a host monitor). Complete virtualization means QEMU can emulate, in software, devices and even processors across several supported architectural platforms.
- *KVM (Kernel-based Virtual Machine)*—KVM itself doesn't perform emulation of hardware. KVM is a Linux kernel module that interfaces directly with processor-specific virtualization extensions to expose a standard \dev\kvm device. This

device is used by a host monitor, like QEMU, for hardware offloading of emulation functions.

WHAT IS KVM? When you hear someone say they're running KVM, what's really happening is that they're using KVM for virtualization-specific hardware offloading and leveraging QEMU for device emulation. When KVM extensions aren't available, QEMU will fall back to software emulation, which, although it's much slower, will work.

Unless specified, I'll refer to the Libvirt, QEMU, and KVM suite of software collectively as *KVM*.

INSTALLING THE KVM SOFTWARE

You'll now install KVM and its related packages using apt-get.

Listing 8.15 Install KVM software

```
$ sudo apt-get -y install qemu-kvm libvirt-bin
...
libvirt-bin start/running, process 13369
Setting up libvirt-bin dnsmasq configuration.
Setting up qemu-kvm (2.0.0+dfsg-2ubuntu1.3) ...
```

KVM should now be installed on the system, and the kernel modules should be loaded.

VERIFYING ACTIVE KVM KERNEL MODULES

Now that the KVM suite has been installed, you must verify that the Intel- or AMD-specific kernel module is loaded. If the KVM extension module isn't loaded, QEMU will fall back to software, and performance will be degraded.

Listing 8.16 Verify KVM acceleration

```
$sudo lsmod|grep kvm
kvm_intel              132891  0
kvm                    443165  1 kvm_intel
```

If you don't see either kvm_intel or kvm_amd listed, the processor-specific KVM extensions module wasn't loaded.

LOADING THE KVM EXTENSION MODULE

With any luck, you can skip over this section. But if the KVM module was missing from your output from the previous listing, this is what you should do.

Listing 8.17 Unload and reload KVM kernel extensions

```
$ sudo modprobe -r kvm_intel        <-- Use kvm_amd if you're
$ sudo modprobe -r kvm                  using AMD processors.
$ sudo modprobe -v kvm_intel
insmod /lib/modules/<kernel version>/kernel/arch/x86/kvm/kvm.ko
insmod /lib/modules/<kernel version>/kernel/arch/x86/kvm/kvm-intel.ko nested=1
```

Verifying KVM-accelerated QEMU environment

You can now check to make sure you have a functional KVM-accelerated QEMU environment.

Listing 8.18 Verify libvirt/qemu/kvm availability

```
$ sudo virsh --connect qemu:///system capabilities

<capabilities>

  <host>
    <uuid>44454c4c-5700-1035-8057-b8c04f583132</uuid>
    <cpu>
      <arch>x86_64</arch>
      <model>SandyBridge</model>
      <vendor>Intel</vendor>
...
<domain type='kvm'>
        <emulator>/usr/bin/kvm-spice</emulator>
...
</capabilities>
```

If you experience errors connecting, such as "Error: Failed to connect socket to /var/run/libvirt/libvirt-sock," reboot the server. If a reboot doesn't solve the problem, check the `libvirtd` log, which is found at /var/log/libvirt/libvirtd.conf. The `libvirtd` service depends on the `dubs` service, which might need to be restarted.

If problems persist, check the syslog for possible failed dependencies.

Cleaning up the KVM network

Because you'll be using OpenStack Networking (Neutron) to manage the network, you'll want to remove the default network bridge automatically created during the KVM install.

Listing 8.19 Remove KVM default virtual bridge

```
$sudo virsh net-destroy default
Network default destroyed

$ sudo virsh net-undefine default
Network default has been undefined
```

You now have a KVM hardware-accelerated QEMU environment with API-level support provided by Libvirt. OpenStack Compute (Nova) will use these software stack components to operate the Compute environment.

8.3 Installing Neutron on Compute nodes

In this section, you'll install and configure Neutron components for a compute node. The steps involved will be a subset of what you performed in section 6.2. For an OpenStack compute node, you only need to install and configure packages related to the ML2 plug-in and OVS agent. The OpenStack network nodes will take care of the rest.

8.3.1 Installing the Neutron software

Install the Neutron software packages with apt-get as follows.

Listing 8.20 Install Neutron Software

```
$ sudo apt-get -y install neutron-common \
 neutron-plugin-ml2 neutron-plugin-openvswitch-agent
...
Setting up neutron-common (1:2014.1.2-0ubuntu1.1) ...
Adding system user `neutron' (UID 108) ...
Adding new user `neutron' (UID 108) with group `neutron' ...
Not creating home directory `/var/lib/neutron'.
Setting up neutron-plugin-ml2 (1:2014.1.2-0ubuntu1.1) ...
Setting up neutron-plugin-openvswitch-agent (1:2014.1.2-0ubuntu1.1) ...
```

8.3.2 Configuring Neutron

The next step is configuration. First, you must modify the /etc/neutron/neutron .conf file. You'll define the service authentication, management communication, core network plug-in, and service strategies. In addition, you'll provide configuration and credentials to allow the Neutron client instance to communicate with the Neutron controller, which you deployed in chapter 5.

Listing 8.21 Modify /etc/neutron/neutron.conf

```
[DEFAULT]
verbose = True
auth_strategy = keystone

rpc_backend = neutron.openstack.common.rpc.impl_kombu
rabbit_host = 192.168.0.50
rabbit_password = openstack1

core_plugin = neutron.plugins.ml2.plugin.Ml2Plugin
allow_overlapping_ips = True
service_plugins = router,firewall,lbaas,vpnaas,metering

[keystone_authtoken]
auth_url =  http://10.33.2.50:35357/v2.0
admin_tenant_name = service
admin_password = openstack1
auth_protocol = http
admin_user = neutron

[database]
connection = mysql://neutron_dbu:openstack1@192.168.0.50/neutron
```

Now that the core Neutron components are configured, you must configure the Neutron ML2 plug-in, which will provide integration with OVS and L2 services.

8.3.3 Configuring the Neutron ML2 plug-in

The Neutron OVS agent allows Neutron to control the OVS switch.

The Neutron configuration can be made in the file /etc/neutron/plugins/ml2/ml2_conf.ini. We'll provide database information, along with the ML2-specific switch configuration.

Listing 8.22 Modify /etc/neutron/plugins/ml2/ml2_conf.ini

```
[ml2]
type_drivers = gre
tenant_network_types = gre
mechanism_drivers = openvswitch

[ml2_type_gre]
tunnel_id_ranges = 1:1000

[ovs]
local_ip = 192.168.0.53
tunnel_type = gre
enable_tunneling = True

[securitygroup]
firewall_driver =
neutron.agent.linux.iptables_firewall.OVSHybridIptablesFirewallDriver
enable_security_group = True
```

The Neutron ML2 plug-in configuration is now complete.

Clear the log file, and then restart the service:

```
sudo rm /var/log/neutron/openvswitch-agent.log
sudo service neutron-plugin-openvswitch-agent restart
```

Your Neutron ML2 plug-in agent log should look something like this:

```
Logging enabled!
Connected to AMQP server on 192.168.0.50:5672
Agent initialized
successfully, now running...
```

You now have OSI L2 Neutron integration using OVS. No other OpenStack Networking configuration is required for the OpenStack Compute node. In the next section, you'll install Nova-specific packages.

8.4 Installing Nova on compute nodes

In this section, you'll install and configure Nova components on a compute node. Nova components not only control the KVM hypervisor, they also pull together other OpenStack services to coordinate the resources required to launch VM instances.

8.4.1 Installing the Nova software

Install the Nova software components using `apt-get` as follows.

Listing 8.23 Install Nova Compute software

```
$ sudo -y apt-get install nova-compute-kvm
...
Adding user `nova' to group `libvirtd' ...
Adding user nova to group libvirtd
Done.
Setting up nova-compute-kvm (1:2014.1.2-0ubuntu1.1) ...
Setting up nova-compute (1:2014.1.2-0ubuntu1.1) ...
```

You now have all of the Nova software components installed on your compute node.

8.4.2 Configuring core Nova components

The next configuration is one of the most critical of the install. You'll add configuration to the /etc/nova/nova.conf file, which will reference the other core OpenStack services. Add the following configuration to the existing file.

Listing 8.24 Modify /etc/nova/nova.conf

```
[DEFAULT]
auth_strategy = keystone

rpc_backend = rabbit
rabbit_host = 192.168.0.50
rabbit_password = openstack1

my_ip = 192.168.0.5                          <- Address of compute node
vnc_enabled = True
vncserver_listen = 0.0.0.0
vncserver_proxyclient_address = 192.168.0.53   <- Address of compute node proxy
novncproxy_base_url = http://10.33.2.50:6080/vnc_auto.html

neutron_region_name = RegionOne
auth_strategy=keystone

network_api_class = nova.network.neutronv2.api.API
neutron_url = http://192.168.0.50:9696       <- URL of Neutron controller
neutron_auth_strategy = keystone
neutron_admin_tenant_name = service
neutron_admin_username = neutron
neutron_admin_password = openstack1
neutron_admin_auth_url = http://192.168.0.50:35357/v2.0
linuxnet_interface_driver = nova.network.linux_net.LinuxOVSInterfaceDriver
firewall_driver = nova.virt.firewall.NoopFirewallDriver
security_group_api = neutron

neutron_metadata_proxy_shared_secret = openstack1
service_neutron_metadata_proxy = true
glance_host = 192.168.0.50
[libvirt]
virt_type = kvm

[database]
connection = mysql://nova_dbu:openstack1@192.168.0.50/nova
```

```
[keystone_authtoken]
auth_url =  http://10.33.2.50:35357/v2.
admin_tenant_name = service
admin_password = openstack1
auth_protocol = http
admin_user = nova
```

URL of Keystone service

Your Nova configuration is now complete. Clear the log file, and then restart the service:

```
sudo rm /var/log/nova/nova-compute.log
sudo service nova-compute restart
```

Your Nova compute log should look something like the following:

```
Connected to AMQP server on 192.168.0.50:5672
Starting compute node (version 2014.1.2)
Auditing locally available compute resources
Free ram (MB): 96127
Free disk (GB): 454
Free VCPUS: 40
Compute_service record updated for compute:compute.testco.com
```

You now have a working `nova-compute` service. No other OpenStack Compute configuration is required for this node.

In the next section, you'll validate the configuration.

8.4.3 Checking Horizon

In chapter 5 you deployed the OpenStack Dashboard. The Dashboard should be available at http://<public controller address>/horizon/. Log in as `admin` with the password `openstack1` and make sure OpenStack Compute components are reported in the Dashboard.

Once you're logged in to the Dashboard, select the Admin tab on the left toolbar. Next, click System Info and look under the Compute Services tab, which should look similar to figure 8.3. Notice the addition of the `nova-compute` service on the `compute` host.

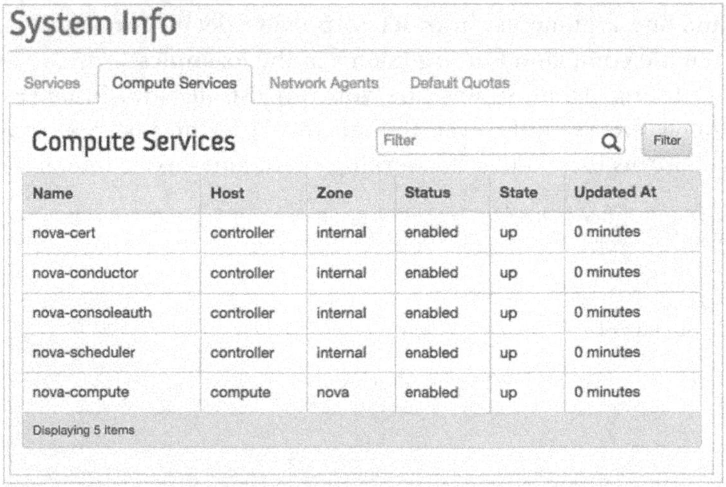

System Info

Services | Compute Services | Network Agents | Default Quotas

Compute Services

Filter | Filter

Name	Host	Zone	Status	State	Updated At
nova-cert	controller	internal	enabled	up	0 minutes
nova-conductor	controller	internal	enabled	up	0 minutes
nova-consoleauth	controller	internal	enabled	up	0 minutes
nova-scheduler	controller	internal	enabled	up	0 minutes
nova-compute	compute	nova	enabled	up	0 minutes

Displaying 5 items

Figure 8.3 Dashboard system info

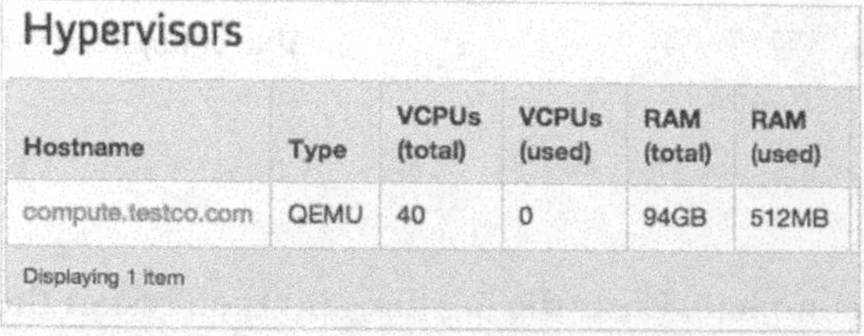

Hypervisors

Hostname	Type	VCPUs (total)	VCPUs (used)	RAM (total)	RAM (used)
compute.testco.com	QEMU	40	0	94GB	512MB
Displaying 1 item					

Figure 8.4 Dashboard hypervisor summary

Now, once again from the Admin tab, click Hypervisors. Your Hypervisors screen should look like the one in figure 8.4.

The compute node you added in the previous section should show up under the listed hypervisors. In figure 8.4, the compute node with the name `compute` is shown. The steps in this chapter can be repeated to add additional OpenStack compute nodes to your manual deployment; of course, you'll need to modify the network addresses, but other than that the process will be the same.

There are several core and many incubated OpenStack components that weren't covered in this part of the book, but with your knowledge of the framework you should be well equipped to explore additional components and perhaps even contribute your own.

If things appear to be working as expected, you can continue on to the next section, where you'll test your completed deployment.

8.5 Testing Nova

You've now installed all the OpenStack components required to create a VM. This section covers the creation of an instance using the command-line tool.

8.5.1 Creating an instance (VM): command line

In order to use `nova`, you must provide the application authentication credentials, including an authentication location. One option is to set credential information as part of your shell variables, and the other is to pass the information to the application through command-line arguments. In order to provide details for the instance creation process, we'll use command-line arguments in the examples.

The following listing demonstrates how you can list all Nova instances for the `admin` tenant. Follow the example, even though you know no instances exist at this point. This step confirms that you have a working client and service interaction.

Listing 8.25 List Nova instances

```
$ nova \
--os-username admin \
--os-password openstack1 \
--os-tenant-name admin \
--os-auth-url http://10.33.2.50:35357/v2.0 \
list
```

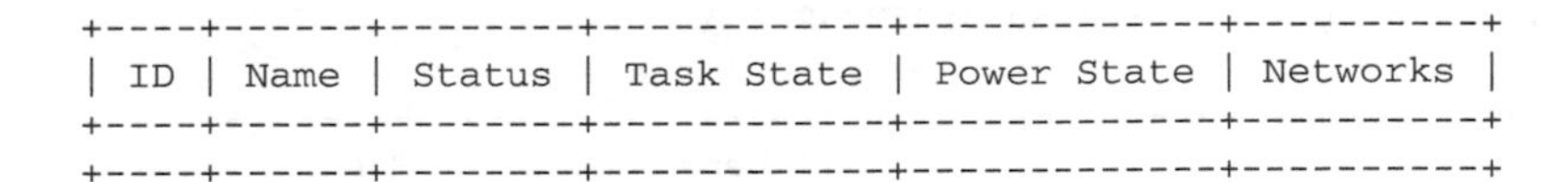

```
+----+------+--------+------------+-------------+----------+
| ID | Name | Status | Task State | Power State | Networks |
+----+------+--------+------------+-------------+----------+
+----+------+--------+------------+-------------+----------+
```

If all worked properly, you should see output similar to that shown in the listing. If you experience an error, take a look at the Nova log (/var/log/nova/nova-compute.log) for any obvious problems.

The `nova` command for creating an instance is `nova boot`. To create a Nova instance, you'll need to provide a minimum of four arguments:

- `flavor`—The size of the instance.
- `image`—The ID of an image to be applied to the volume used by the instance. This image should contain an operating system that the instance will use to boot.
- `nic net-id`—The network ID of the network you want the instance connected to.
- `<instance name>`—The name you want to use for the instance.

The next three listings show what options are available for the first three arguments. First, the Nova flavors.

Listing 8.26 List Nova flavors

```
$ nova \
--os-username admin \
--os-password openstack1 \
--os-tenant-name admin \
--os-auth-url http://10.33.2.50:35357/v2.0 \
flavor-list
+----+-----------+-----------+------+-----------+
| ID | Name      | Memory_MB | Disk | Ephemeral |
+----+-----------+-----------+------+-----------+
| 1  | m1.tiny   | 512       | 1    | 0         |
| 2  | m1.small  | 2048      | 20   | 0         |
| 3  | m1.medium | 4096      | 40   | 0         |
| 4  | m1.large  | 8192      | 80   | 0         |
| 5  | m1.xlarge | 16384     | 160  | 0         |
+----+-----------+-----------+------+-----------+
```

Select a flavor based on the defined instance size. In this example, we'll use the `m1.medium` flavor, which has a flavor ID of `3`.

Next, you can find an image to apply to the instance.

Listing 8.27 List Nova images

```
$ nova \
--os-username admin \
--os-password openstack1 \
--os-tenant-name admin \
```

```
--os-auth-url http://10.33.2.50:35357/v2.0 \
image-list
+--------------------------------------+--------------+--------+
| ID                                   | Name         | Status |
+--------------------------------------+--------------+--------+
| e02a73ef-ba28-453a-9fa3-fb63c1a5b15c | Cirros 0.3.2 | ACTIVE |
+--------------------------------------+--------------+--------+
```

Here you see only one image—the image that you uploaded in chapter 6 during the Glance installation. You'll need to reference the image ID `e02a73ef-ba28-453a-9fa3-fb63c1a5b15c` when booting your instance.

Next, you can list the Nova networks.

Listing 8.28 List Nova networks

```
$ nova \
--os-username admin \
--os-password openstack1 \
--os-tenant-name admin \
--os-auth-url http://10.33.2.50:35357/v2.0 \
net-list
+--------------------------------------+------------------+
| ID                                   | Label            |
+--------------------------------------+------------------+
| 5b04a1f2-1676-4f1e-a265-adddc5c589b8 | INTERNAL_NETWORK |
| 64d44339-15a4-4231-95cc-ee04bffbc459 | PUBLIC_NETWORK   |
+--------------------------------------+------------------+
```

Here you see two networks, the internal and public networks. For this example, we'll use the `INTERNAL_NETWORK`, which will be referenced by the network ID `5b04a1f2-1676-4f1e-a265-adddc5c589b8`.

Now that you've selected your arguments, you're ready to create an instance.

Listing 8.29 Create VM instance

```
$ nova \
--os-username admin \
--os-password openstack1 \
--os-tenant-name admin \
--os-auth-url http://10.33.2.50:35357/v2.0 \
boot \
--flavor 3 \
--image e02a73ef-ba28-453a-9fa3-fb63c1a5b15c \
--nic net-id=5b04a1f2-1676-4f1e-a265-adddc5c589b8 \
MyVM
+--------------------------------------+----------|
| Property                             | Value    |
+--------------------------------------+----------|
...
| OS-EXT-STS:vm_state                  | building |
...
| name                                 | MyVM     |
...
```

Your vm_state instance property should now be in the *building* state. List your instances again.

Listing 8.30 List Nova instances

```
$ nova \
--os-username admin \
--os-password openstack1 \
--os-tenant-name admin \
--os-auth-url http://10.33.2.50:35357/v2.0 \
list
+--------+------+--------+-------------+-----------------------------+
| ID     | Name | Status | Power State | Networks                    |
+--------+------+--------+-------------+-----------------------------+
| 82..3f | MyVM | ACTIVE |  Running    | INTERNAL_NETWORK=172.16.0.23 |
+--------+------+--------+-------------+-----------------------------+
```

With any luck, your instance will now be active with a network assigned. If your status is ERROR, or if the instance hangs in a SPAWNING state for more than a few minutes, take a look at the Nova log (/var/log/nova/nova-compute.log). If there are no apparent errors in the Nova log, start looking at the controller logs.

Assuming all went well, congratulations! You've successfully completed a manual deployment of OpenStack. You can review chapter 3 to remind yourself about basic OpenStack operations for your new deployment, creating new tenants, and creating new networks.

> **Want to try a newer version of OpenStack?**
>
> As previously mentioned, Ubuntu 14.04 uses the Icehouse release of OpenStack by default. But by using the Ubuntu CloudArchive (https://wiki.ubuntu.com/ServerTeam/CloudArchive), you can install back-ported distributions of OpenStack on older Ubuntu releases.
>
> The examples presented in this book might or might not work with newer versions of OpenStack, but now that you've gone through the deployment process yourself, you're certainly in a better position to troubleshoot upgrade-induced problems.

Through this component-level deployment, you gained a deeper understanding of the OpenStack Framework and its related dependencies. But keep in mind that the process you followed in this book was intended to teach you the framework; it doesn't serve as a best-practice reference. We'll discuss production deployments of OpenStack in the third and final part of this book.

8.6 Summary

- OpenStack Compute consumes remote network resources from Neutron, volume resources from Cinder, image resources from Glance, and hypervisor resources from local compute resources to provide VMs.
- On compute nodes, VM network traffic doesn't access external (outside OpenStack) networks directly from the compute node.
- Compute nodes function as switches for their own traffic.
- Open vSwitch can be used to enable advanced switching features on a typical server.
- OpenStack Compute uses OpenStack Networking to provide OSI L2 and L3 services.
- OpenStack Compute communicates with OpenStack Networking through overlay networks.
- OpenStack Compute VMs communicate with other OpenStack Compute VMs through overlay networks.
- Overlay networks use GRE, VXLAN, and other tunnels to connect endpoints like VMs and other OpenStack Networking services.
- Cinder is used to provide VM volume storage.
- Glance is used to provide images for VM volumes.
- KVM can be used as an OpenStack Compute hypervisor.
- Ensuring hardware acceleration support for KVM is critical for good performance.

Part 3

Building a production environment

The third and final part of this book covers topics related to deploying and utilizing OpenStack in production environments—specifically, enterprise environments where the typical systems administrator might take care of a wide variety of both infrastructure and applications. In enterprise environments, systems engineers often drive infrastructure design, deployment, and adoption. The chapters contained in this part of the book are intended to help you develop a successful OpenStack deployment for your environment.

Architecting your OpenStack

This chapter covers

- Using OpenStack to replace existing virtual server platforms
- Why you should build a private cloud
- Choices to make when building your private cloud

In the first part of this book, you dipped your toes into OpenStack through the use of DevStack. The purpose of that part was to introduce you to the hows and whys of OpenStack and to pique your interest in a deeper understanding of how things work under the covers.

In the second part of the book, you undertook a manual deployment of several core OpenStack components. Although it's important that you understand the underlying component interactions that make up OpenStack, the second part of the book isn't a blueprint for OpenStack deployment. This level of understanding builds confidence in the underlying system through low-level exposure to the components and configurations, but isn't intended to encourage you to manually install components in a production environment.

This third and final part of the book covers topics related to deploying and utilizing OpenStack in production environments, specifically enterprise environments where the typical systems administrator might take care of a wide variety of both infrastructure and applications. Often in enterprise environments, systems engineers drive infrastructure design, deployment, and adoption. The chapters in this part of the book are intended to help you develop a successful OpenStack deployment for your environment.

This chapter covers decisions—architectural, financial, and operational—that you'll need to make as you plan your deployment. This chapter isn't a cookbook, but rather a starting reference for use when developing a successful architecture. For a prescriptive approach to developing your architecture, consult the "OpenStack Architecture Design Guide" (http://docs.openstack.org/arch-design). Once you determine the type of OpenStack deployment that's right for you, the online design guide can be a valuable asset for configuration and sizing.

Many enterprise systems people will approach OpenStack from the perspective of traditional virtual and physical infrastructure platforms. We'll first discuss using OpenStack as a replacement for existing virtual server platforms, covering strategic design choices that you need to make to get the most out of your OpenStack deployment.

9.1 Replacement of existing virtual server platforms

In the 2015 Gartner report "Magic Quadrant for x86 Server Virtualization Infrastructure," it was estimated that 75% of x86 workloads were virtualized, with VMware as the predominate enterprise vendor.[1] This section covers how OpenStack can be used as either a replacement for or augmentation of your existing VM environment. In addition, the section makes a case for thinking of OpenStack as more than a replacement for a traditional virtual server platform.

Likely your traditional virtual environment was designed to provide virtual machines in a way that operationally mimics physical machines. There's also a good chance that virtualization was introduced into your environment as an infrastructure cost-savings measure for existing workloads. As workloads were moved from physical servers to their virtual equivalents through a process known as Physical to Virtual (P2V), an exact clone of the physical server was made by the P2V tool. More often than not, physical and virtual servers operated on the same networks, which allowed the P2V migration to take place without service interruption. For many environments, the process of consolidating workloads on virtual servers resulted in significant financial savings. In addition to resources being used more efficiently, new capabilities like master images and virtual machine image snapshots became part of the software deployment and upgrade processes, mitigating many types of software and hardware failures. Despite many of the new capabilities that virtual environments offered users, system

[1] See Thomas J. Bittman, Philip Dawson, Michael Warrilow, "Magic Quadrant for x86 Server Virtualization Infrastructure" (14 July 2015), www.gartner.com/technology/reprints.do?id=1-2JGMVZX&ct=150715.

administrators were still managing both the OS and application levels of virtual machines, much in the same way they had physical machines.

If your intent is to treat your virtual environment the way you do a physical environment, as described in the previous paragraph, then the benefits of OpenStack in your environment will be limited. This is to say, if your operational practice is to manually deploy virtual services through a central group without automation, you must evaluate how cloud frameworks such as OpenStack can be effectively adopted in your environment.

Suppose you've been using VMware vSphere as your server virtualization platform, and you're interested in adopting OpenStack as a VMware replacement for cost-reduction purposes. If you think of OpenStack as simply a "free" alternative to VMware, then you might be heading down the wrong path. Although in most cases OpenStack can be deployed in a way that provides feature parity with many competing virtual environments, this has to be done in a way that's compatible with your operational practices. Consider again the previous VMware replacement example, where you wanted to move all existing VMware workloads to OpenStack. Although OpenStack Storage can deal with VMDK (VMware image format) files, there's no graphical Virtual to Virtual (V2V) migration tool like what's provided by VMware, and there probably shouldn't be.

Now, consider the process most often used to build VMware-based machine images. Typically, using a desktop client, a user virtually mounts a CD or DVD from their workstation to virtual hardware and performs an install as they would with a physical machine. However, the ability to remotely mount CD and DVD images doesn't exist on the OpenStack Dashboard. One shouldn't consider these things as an indication that OpenStack is incomplete, but as an indication that it's intended to be used differently than traditional virtual server environments.

Where might OpenStack be a good replacement for VMware and other commercial hypervisors? To answer this, you must first consider strategically how you want to interact with your infrastructure resources. Table 9.1 lists the possible impact of OpenStack based on your infrastructure management strategy.

Table 9.1 OpenStack impact based on environment

Environment	Description	Impact
Siloed and manual VM	Hardware management is siloed and resources are shared. Virtual hardware is manually provisioned by IT staff to end users, where end users are responsible for OS-level operation.	Low to negative
Siloed and automated VM	Hardware management is siloed and resources are shared. Virtual infrastructure is deployed using automated methods by IT staff, where end users are responsible for application-level operation.	Medium

Table 9.1 OpenStack impact based on environment *(continued)*

Environment	Description	Impact
Application-specific backend	Hardware is dedicated and managed by the cloud framework. Infrastructure and application deployment are automated by IT staff for a specific application.	High to very high
Private cloud	Hardware is dedicated and managed by the cloud framework. Applications and standard (size and OS) VMs are provided to end users using automated self-service methods.	Very high

In the case of the *siloed and manual VM* listed in the table, there's a central group manually provisioning virtual machines without automation, and they will likely view OpenStack as either incomplete or unnecessary. Without the addition of automation, OpenStack doesn't necessarily provide them with anything that can help them do their current jobs.

The *siloed and automated VM* environment is similar to the manual environment, with the exception that at least some level of infrastructure orchestration is being used by the IT department managing the infrastructure. For instance, departments that make use of dynamic automated provisioning as part of a request workflow are considered part of this group. Much like their manual counterparts, organizations that fall into this category often evaluate OpenStack as a direct cost-saving replacement for whatever they are currently using. Although it's true that OpenStack can lead to cost savings, the operational and business processes of these organizations must change in order to take full advantage of the framework.

Now, suppose the central group manually provisioning the virtual machines is repositioned as infrastructure or application resource consultants, as in the *application-specific backend* scenario.

Suppose that through this strategic shift, not only is automation used in the infrastructure, but it's also used for application-level provisioning. Suppose further that resources are allotted to tenants, and departmental-level personnel are able to provision their own resources. That would be the *private cloud,* in which the central group is enabling departmental agility by brokering services, not prescribing them. In many respects, operating in this way allows you to change your thinking about the role of infrastructure in your environment. Automation on application and infrastructure levels removes the need for P2V and V2V tools, so there's no need to move images around. In this mode of operation, infrastructure resources are more transient and function more as an application capability than a static allocation.

The real value of OpenStack is in the automation and platform abstractions provided by the framework. The following subsections will discuss architectural considerations that you must take into account as you develop your OpenStack design.

9.1.1 Making deployment choices

If you're used to supporting virtual server platforms like VMware vSphere or Hyper-V, you might need to rethink how you purchase and support hardware. Although it's less common than in the past, physical resources in the enterprise, like network switches and central pools of storage, can be shared between physical and virtual resources. Even if a resource is exclusively assigned to a virtual server platform, you'd typically think of provisioning resources to be used by the platform, not of the platform itself managing the resources. For instance, it's common to assign a new VLAN or to make a shared logical unit number (LUN) available to a group of hypervisors. But if you need to create new VLANs or new shared LUNs, the administrators of those systems would need to go through their own provisioning process. It's also very common for the "network person" to do all network configurations, the "storage person" to do all storage assignments, and the "VM person" to tie the resources together with a physical server to produce VMs. Each person in this process has to perform manual provisioning steps along the way, often without understanding how their resource plays into the complete infrastructure.

Deploying OpenStack from slivers of shared central infrastructure is generally the wrong path to take. OpenStack detects, configures, and provisions infrastructure resources, not the other way around. Even if your shared central infrastructure provides multi-tenant (not to be confused with OpenStack tenants) operation, which would isolate OpenStack automation, you'd still need to consider the effects of making OpenStack-provisioned resources dependent on shared resources. For instance, software upgrades for reasons outside of OpenStack operations could impact services. In addition, resource utilization outside the scope of OpenStack resources could impact performance while providing no indication to OpenStack services that problems exist.

In many cases, virtual environments aren't designed to leverage the benefits of an infrastructure that can be managed programmatically. In these cases, the operational practices will have been developed around vertical management of siloed resources like compute, storage, network, load balancers, and so on. In contrast, OpenStack was designed with the complete abstraction of physical infrastructure in mind. In general, you can save yourself lots of trouble by assigning resources exclusively to OpenStack, and through plug-ins and services letting the framework manage resources, instead of the other way around.

In the following subsections, it's assumed you wish to augment or add new services using OpenStack to manage your resources. In your environment, you want to leverage the management capabilities of OpenStack, and you even want OpenStack to manage your hardware, but you want the end product to resemble what you are providing now. Specifically, you're willing to change your operational and deployment practices for the sake of efficiency, but your primary interest is to deploy VMs, just as you might be doing now with VMware or Microsoft. Section 9.2 discusses taking a more progressive approach to IaaS.

9.1.2 What kind of network are you?

If you want to take advantage of OpenStack, but you don't want OpenStack to manage L3 services such as routing, DHCP, VPN, and the like, you must evaluate your options based on management of L2 (switching) services.

For example, figure 9.1 shows a VM directly connected to a public L2 network. This example isn't specific to OpenStack and would be representative of similar network deployments for many virtual server platforms, including VMware vSphere and Microsoft Hyper-V. In this network deployment scenario, the job of the hypervisor is to direct L2 network traffic to a switch, which is typically a physical switch outside the control of the hypervisor. Unlike many of the network examples in this book, there's no concept of an "internal" or hypervisor network, because no L2 services are being provided by the virtual server platform. In this deployment type, all L3 services are provided by systems outside the hypervisor. As you might imagine, separating the majority of network services from the virtual server platform limits the benefits of the platform, but the simplification isn't without benefit. Based on your IT strategy, existing resources, and support structure, this mode of operation might be best for you.

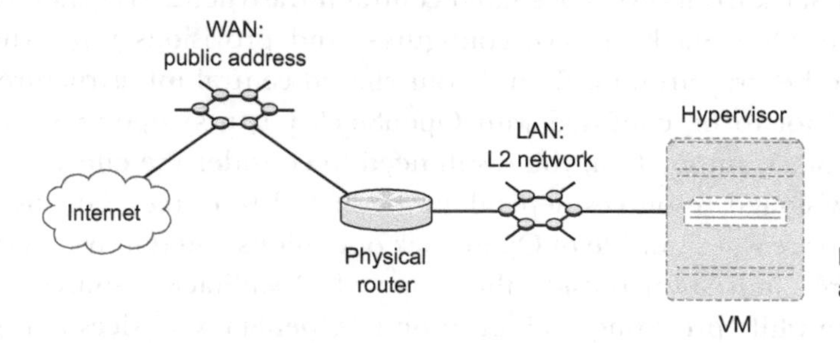

Figure 9.1 L2 network with VM and hypervisor

The majority of this book has focused on OpenStack Networking (Neutron) providing L2 and L3 services. As discussed in previous chapters, Neutron was built to manage complex networks and services within OpenStack environments, not simply to push L2 traffic to external networks. But the OpenStack Compute (Nova) project, which predates Neutron, does provide basic L2 services. If you want to limit your OpenStack deployment to L2 services only, you'll want to use Nova for networking, not Neutron.

> **Network hardware**
>
> If you're using Nova for networking, there's very little reason to worry about the integration of OpenStack with network hardware. Early on in the OpenStack project, hardware vendors were writing drivers for Nova integration in their hardware, but most development has since moved to Neutron.

A Nova network can operate in three different topologies: flat, flat DHCP, and VLAN.

In a *flat* topology, all network services are obtained externally from OpenStack. You can think of a flat topology working the same way as your office or home network connection works. When you connect your computer to the flat network, your computer relies on the existing network for services such as DHCP and DNS. In this mode of operation, OpenStack is simply connecting VMs to an existing network, just as you would a physical machine.

The *flat DHCP* topology is similar to the flat topology, with the exception that OpenStack provides a DHCP server to assign addresses to VMs.

The *VLAN* topology works in the same way as the flat topology, but it allows for *VLAN* segmentation of the network based on *VLAN IDs*. Simply put, in the flat network all VM traffic is sent to the same L2 network segment, whereas with the VLAN topology, you can assign a specific L2 network segment to a particular VM.

The next section covers choices in storage.

9.1.3 What type of storage are you?

If you're coming from a traditional virtual server platform environment, you may have no management integration between your hypervisor and storage subsystem. If you're using VMware vSphere, you'll typically attach large shared host volumes to your hypervisors, as shown in figure 9.2 SharedVolume.

In the figure, you can see a single host volume shared across all hypervisors. A shared host volume is formatted with a cluster-aware filesystem that allows hypervisor A to use the same underlying host volume to store data for VM A, as hypervisor B would for VM B. If VM B was to bc migrated to hypervisor A, no stored data would need to be transferred, because the data is already accessible by hypervisor A. In this scenario, VM volume management is done on the shared-host volume level, so from the

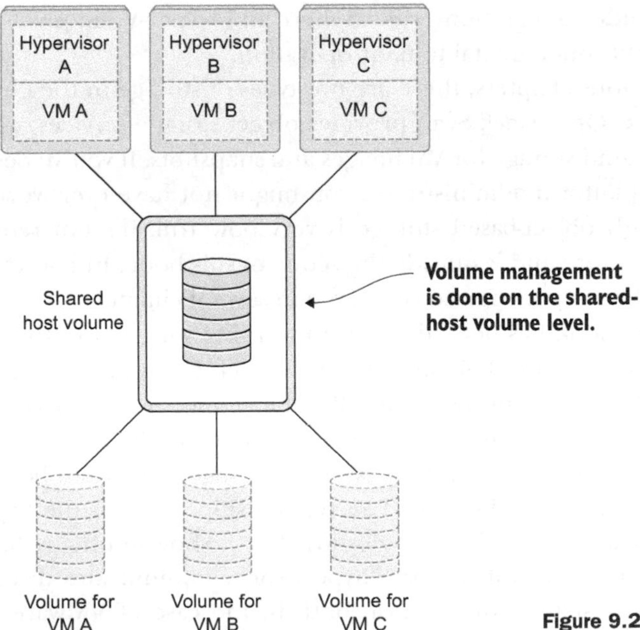

Figure 9.2 Shared volume across hypervisors

standpoint of the underlying storage subsystem, nothing is managed beyond the large shared host volume attached to the hypervisors.

In contrast, Microsoft Hyper-V promotes a "shared-nothing" model, where each hypervisor maintains its own storage and the storage for its VMs. An independent host volume model is shown in figure 9.3.

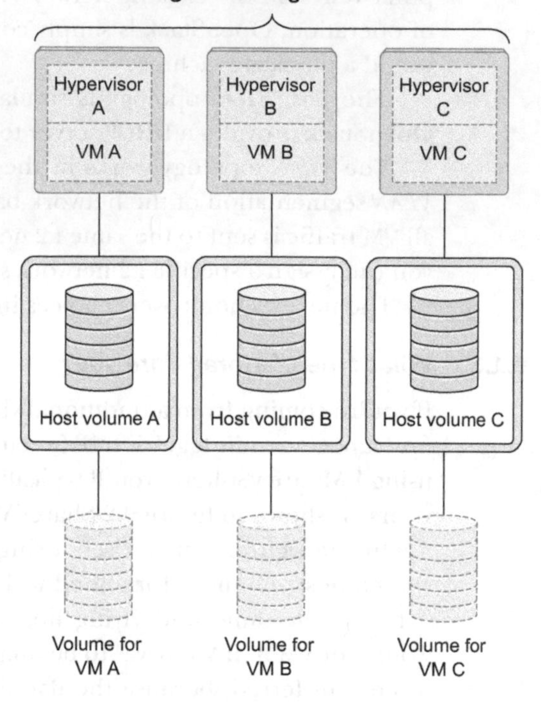

Figure 9.3 Independent host volumes

In this figure, hypervisor A uses host volume A to store data for VM A. If VM B were migrated to hypervisor A, volume information would need to be migrated to the new hypervisor. The benefit of this shared-nothing architecture is that the failure domain is reduced, but the migration costs are increased. As with the shared–host volume model, the hypervisor is managing volumes for the VMs it maintains, so the storage subsystem is still not actively managed as part of the virtual server platform. This is not to say that there are no storage vendor integrations with vSphere and Hyper-V, simply that a high level of integration is not fundamental to their operation.

As discussed in previous chapters, there are two types of storage in the OpenStack world: object and block. OpenStack Swift provides object storage services, which can be used to provide backend storage for VM images and snapshots. If you've been working as a virtual server platform administrator, you might not have ever worked with object storage. Although object-based storage is very powerful, it's not required to provide virtual infrastructure and is outside the scope of this book. In contrast, block storage is a required VM component and is covered in several chapters.

The majority of this book has been devoted to working with block storage using OpenStack Block Storage (Cinder). Using the OpenStack Compute service (Nova), it's possible to boot a VM without using Cinder. But the volume used to boot the VM is *ephemeral*, which means that when the VM is terminated, data on the VM volume is also removed. In contrast to ephemeral storage is *persistent* storage, which can be detached from a VM and reassigned to another VM. The relationship between the hypervisor, persistent VM volume, Cinder, and the storage subsystem is shown in figure 9.4.

As shown in the figure, the VM, not the hypervisor, communicates directly with an underlying storage subsystem. In contrast, both in the case of VMware vSphere and Microsoft Hyper-V, VM storage is provided by the hypervisor. This fundamental difference in operation forces a higher level of storage subsystem management for OpenStack. Where other virtual server platforms might manage VM volumes on the hypervisor level, Cinder interfaces with hardware and software storage systems to provide functions like volume creation, expansion, migration, deletion, and so on. A list of storage systems and supported functions can be found on the Cinder support matrix wiki: https://wiki.openstack.org/wikiCinderSupportMatrix.

Aside from a few corner cases, most OpenStack deployments will use Cinder to manage volume storage. But the question remains, what type of storage hardware or software platform should Cinder manage?

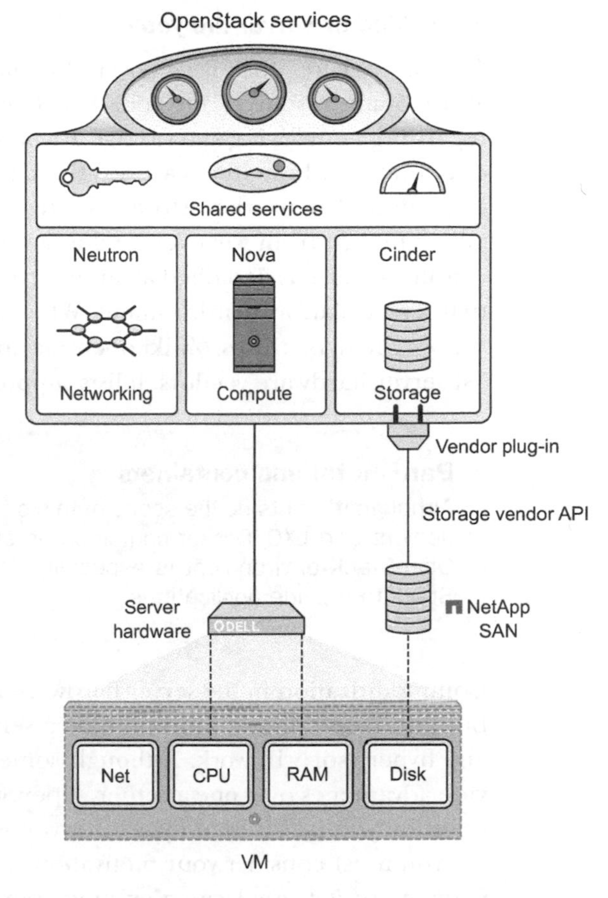

Figure 9.4 OpenStack VM volumes

The question of the underlying storage subsystem depends on the intersection of several factors, including what your current storage vendors are, whether you're willing to dedicate a storage system to OpenStack, and what your tolerance is for risk in your environment. For instance, suppose you wish to mimic the operations of vSphere or Hyper-V, and your storage is provided by a large centralized storage area network (SAN). In this case, you might not want Cinder to communicate directly with your shared central system, but you do want to use storage from this system. In this scenario, you could abstract the underlying storage subsystem by attaching independent volumes directly to compute or storage nodes, similar to what was shown in figure 9.3. You'd then use LVM to manage your independent host volume. LVM would then be managed by Cinder, thus abstracting the underlying storage subsystem. Managing an independent volume using LVM doesn't invalidate the storage model shown in figure 9.4; in fact, LVM was used as the underlying storage subsystem for Cinder in chapter 7. There are, of course, other options where centralized storage software and hardware can be used directly, but LVM is a common choice for people using shared central services.

In the next section, we'll cover choices in servers.

9.1.4 What kind of server are you?

The past few sections have covered choices you need to make in providing network and storage resources for your VM. For the most part, the choices that you make regarding networking and storage are based on your current and intended future state of operations. In neither case was there any mention of changing the underlying configuration of network or storage hardware and software in a way that was fundamentally different from what you're doing now. In fact, everything we've discussed in this chapter so far has described an architecture that can be used to mimic a traditional virtual server platform environment. When it comes to OpenStack Compute (Nova), the support matrix (https://wiki.openstack.org/wiki/HypervisorSupportMatrix) doesn't list server hardware vendors, it lists supported hypervisors.

> **Bare metal and containers**
>
> Although it's outside the scope of this book, using OpenStack to provision bare-metal servers and LXC/Docker containers is also an option. The use of containers in the OpenStack environment is especially interesting for those who'd like to use OpenStack to provide applications.

Commoditization in the server hardware market around the x86_64 instruction set all but guarantees that if you purchase a server with an Intel or AMD x86_64 processor, any hypervisor will work. Although some hardware configurations and vendors provide advantages over one another, OpenStack Compute is hardware agnostic. The real question you face is what hypervisor you want to use.

You must consider your motivations for deploying OpenStack in the first place. If your intent is to replace an existing commercial virtual server platform in a way that mimics the operation of that platform, then you likely won't want to maintain the license cost of the commercial hypervisor.

> **FREE VERSIONS OF VMWARE ESXI AND MICROSOFT HYPER-V** Recently VMware and Microsoft have released free versions of their core visualization platforms. The changes in licensing have generated a great deal of community interest in using these hypervisors with OpenStack. But there are drawbacks, including limited initial community support compared to KVM.

Based on OpenStack user surveys, KVM is the hypervisor used in the vast majority of OpenStack deployments, and most community support will be based around using KVM. In summary, you should select server hardware based on current business practices and use KVM until you have a reason to use something else.

This section covered the architectural decisions around deploying OpenStack to replace an existing virtual server environment. The next section covers the architecture for deploying a greenfield environment for an on-premise private cloud.

9.2 Why build a private cloud?

OpenStack is used in some very large public cloud services, including DreamHost DreamCompute (see www.openstack.org/marketplace/public-clouds/). These companies make use of OpenStack projects, along with their own custom integration services, to manage resources on a much larger scale than most enterprise customers. The ratio of servers to administrators varies wildly, based on the size and complexity of an organization's infrastructure and related workloads. For example, it's common to have a 30:1 or lower ratio of physical servers to administrators for small and medium-sized business, whereas a medium to large enterprise might have 500:1 virtual servers to administrators. But when you consider that the Amazons and Googles of the world achieve ratios of 10,000:1 physical servers to administrators, you can start to appreciate the infrastructure management efficiencies that large-scale providers have developed.

When an enterprise provides public cloud–like services from resources that are exclusive to the enterprise, we call this a *private cloud*. By adopting technologies and operational practices born out of large-scale providers for private clouds, enterprises are able to develop hybrid cloud strategies based on workload. So why do private clouds exist? Why aren't all workloads in the public cloud? A detailed study of IT strategy concerning public, private, and hybrid clouds is beyond the scope of this book. But this section presents several arguments for deploying a private cloud for your enterprise and for adopting a hybrid cloud strategy.

9.2.1 Public cloud economy-of-scale myth

Cloud computing is often described as the computational equivalent of the electrical power grid. Considering that the economic definition of *utility* is the ability of a commodity to satisfy human wants, it's easy to see how cloud computing gained this reputation.

But there's a fundamental flaw in this power-grid comparison. The power grid, much like cloud computing, produces a commodity that must be consumed in real time, but this is where the comparison ends. There's a vast difference between the economies of scale related to the production of the commodities. Bulk power generation from nuclear facilities are orders of magnitude more cost effective than what could be produced by a cloud of consumer-grade power generators. On the computer side, there isn't any quantum or other type of computer capable of producing computational power with a cost benefit greater than commodity clusters, so the economies of scale aren't comparable. In fact, the profit margins across commodity servers is so low that the difference between what enterprise and public cloud providers pay for the same hardware is negligible.

This is not to say that there aren't advantages for large-scale providers. For instance, a large diverse group of workloads balanced across many resources should be more efficient than its small-scale unoptimized counterpart. But enterprise customers can take advantage of the same fundamental components as public providers, and often near the same price point.

9.2.2 Global scale or tight control

Public cloud providers offer a wide variety of services beyond IaaS, but for the sake of argument, we'll restrict our evaluation of public cloud to IaaS.

Consider IaaS as providing VMs comprised of discrete components (CPU, RAM, storage, and network). Public cloud consumers are unaware of the physical infrastructure used to provide public IaaS offerings. Specifically, users have no way of knowing if they're paying for the latest and greatest or yesterday's technology. To complicate matters further, the consumer has no way to determine the level of oversubscription for a particular type of shared service. Without knowing the underlying platform and the number of shared users, there's no passive, quantitative way to measure IaaS value between public providers. Consider a case where provider A has a cost per unit of X and an oversubscription of 20:1, whereas provider B has a cost per unit of 2X and rate of 10:1. The total cost is clearly the same, but in the eyes of the customer, provider B is simply twice the cost of provider A.

In many industries, service level agreements (SLAs) are defined so consumers can evaluate the expected quality of a service provider. Typically, public cloud SLAs are based on uptime, not performance. No doubt large consumers of public cloud resources develop performance SLAs with their public providers, but this isn't common in the small business to medium enterprise space. Without a quantitative approach, you're left with qualitative evaluation based on active measurements. But although there's no shortage of benchmarks in the computing industry, an accepted workload measurement standard for cloud services has yet to emerge.

In the absence of clearly defined SLAs and verification methods, it's difficult to compare the value of public cloud offerings between providers. In addition, value comparisons can change over time as workloads change across providers. For many workloads, the benefits of global on-demand IaaS far outweigh resource performance variability. But for other workloads, the tightly controlled performance provided by a private cloud is necessary.

9.2.3 Keeping data gravity private

Dave McCrory coined the term *data gravity* to describe how applications and other services are drawn to sources of data, analogous to the attraction of physical objects in the universe proportional to their mass. Public cloud providers recognize this phenomenon and typically make it much more financially attractive to move data into their services than out. For instance, there's no charge to move data into an Amazon EC3 service, but there is a tiered pricing structure to move data out of the service. A similar pricing structure exists for Amazon EC2 instances and other IaaS provider offerings, to entice users to move data into a specific cloud provider and keep it there.

Cloud providers can exploit the data gravity phenomenon through their transfer-rate pricing structure to create cloud vendor lock-in for consumers. Consider the case where an organization determines that based on the unit price of resources (not accounting for transfers) it's cost effective to move all of its storage and related computation to a public cloud provider. Even if the majority of data was generated outside the cloud provider, there would be no transfer penalty to continuously add data to storage maintained by the public provider. Now suppose this organization wants to utilize a secondary public cloud provider for redundancy. Although the new provider might not charge transfer fees for incoming data, the existing provider will charge for outgoing data, which could greatly increase the cost.

The same holds true if you want to process information locally or if you want to take advantage of processing on a lower-cost provider. When the majority of your data is maintained in the public cloud, it's hard for services to escape the force of data gravity pulling you to your data provider.

Keeping the majority of your data in a private cloud allows you to move data in and out of public clouds as needed. For many workloads and organizations, the ability to consume services from several providers, including local resources, outweighs the benefits of pure public cloud offerings.

9.2.4 Hybrid moments

The principle of *pay as you go* is a key differentiator between public and private clouds. It's easy to understand the economic benefits of the spending 1 hour on 1000 computers versus using 1 computer for 1000 hours when timely information is essential. But at first glance, the economics of purchasing cloud services doesn't seem viable when you assume 100% service usage 24/7/365.

The idea of usage-based pricing allows for the redirection of capital that would otherwise be committed to infrastructure investment to be repositioned in other strategic investments. Large public cloud providers have a natural tolerance for load spikes for specific workloads. Given the diversity of workloads over a wide range of clients, it's unlikely that all workloads for all clients will experience a simultaneous resource peak need. Given the natural elasticity of the cloud, much of the capacity risk related to operating a private cloud is transferred to the public cloud provider. Private clouds must be built for the peak usage, regardless of peak duration, and this involves extra costs. In most cases, peak workloads exceed average workloads by a factor of 5 to 1.

The public cloud has been adopted broadly across the enterprise, but rarely exclusively. Based on enterprise surveys, public cloud services are adopted for specific strategic workloads, whereas private clouds provide services for a more diverse range of services.

The majority of IT organizations have adopted a hybrid cloud strategy, making use of both public and private cloud resources. The real economic benefits of the public cloud in the enterprise can be realized if organizations find the right balance between private and public cloud services.

For the service provider, OpenStack can provide project components that can be used to construct massively global clouds of resources. For the enterprise, the OpenStack framework can be used to deploy private cloud services. From an integration standpoint, API compatibility between public and private OpenStack-based providers allows the enterprise to optimally consume resources based on workload requirements.

9.3 Building a private cloud

This book has focused on approaching OpenStack from the enterprise perspective not as a virtual server platform replacement, but as a cloud management framework. The previous section covered the benefits of deploying a private cloud. In addition, the benefits of adopting a hybrid cloud strategy, where resources can be managed based on a common OpenStack API control set, were covered.

This section ties together what you've learned in previous chapters and prepares you for the rest of the chapters in part 3 of this book.

9.3.1 OpenStack deployment tools

The deployment tool you choose will be based on your existing vendor relationships, current operational strategy, and future cloud direction. There are three approaches you can take when deploying OpenStack.

The first approach, a manual deployment, was covered in part 2 of this book. Manual deployments offer the most flexibility but have obvious problems at scale.

The second approach is to use general orchestration tools, such as Ansible, Chef, Juju, Puppet, and Vagrant, which are used to deploy a wide range of systems and applications. Being well versed in a collection of general orchestration tools allows you to deploy not only OpenStack, but also applications using OpenStack resources. The drawback of these systems is that each tool has its particular role to play, so you end up using a wide range of general-purpose tools, which constitutes a training and operational challenge for the organization adopting this strategy.

The third approach, using a standalone OpenStack deployment and management tool, is a familiar approach for those in the enterprise. OpenStack deployment platforms like HP Helion, Mirantis Fuel, and Red Hat RDO not only provide easy-to-use tools for deploying OpenStack, they also provide their own validated OpenStack versions and deployment methods. You can think of this the same way you think of Linux distributions. Like the Linux kernel, there's one OpenStack source repository (with many branches) for community development. Enhancements and fixes recognized by Linux and OpenStack communities make their way into the respective code repositories. But as in the Linux community, vendors validate community work for specific use cases, for which they provide support. Just as you don't technically pay for the Linux kernel when you pay for a supported Linux distribution, you aren't paying for OpenStack when you purchase a commercially supported

> **Research, big data, and OpenStack**
>
> For those who deal in the area of research computing, new consolidated infrastructure management options are emerging. Traditional high-performance computing (HPC) niche vendors like Bright Computing and StackIQ are getting into the Hadoop and OpenStack game. These vendors, and many others, are adapting their HPC deployment and management platforms to provide a holistic management view across HPC, Hadoop, and OpenStack deployment.

OpenStack distribution. Commercial OpenStack vendors typically provide community supported versions of their deployment tools—one such tool, Mirantis Fuel, is covered in chapter 11.

In many IT shops, the title "systems programmer" is still used to describe roles that haven't had much to do with programming since the days of the mainframe. In some organizations, however, the systems programmer role has been reborn as part of the DevOps movement (referring to systems administrators who also write code and scripts). A systems administrator who is accustomed to manually double-clicking their way through VM and application deployments isn't likely to be comfortable with general orchestration tools that they have to code or script together. On the other hand, someone with automation experience will feel very restricted with only a standalone OpenStack deployment tool.

Based on the strategic direction of your organization, you should pick an approach that not only deploys OpenStack but that is sustainable. For some this approach will involve purchasing a commercially supported distribution and assigning existing resources to work with the supporting vendor. Other organizations will choose to develop DevOps teams that will not only be able to deploy OpenStack but also orchestrate resources and applications across private and public cloud providers.

9.3.2 Networking in your private cloud

In section 9.1.2 we discussed Nova networking. When using Nova networking, your choice of networking hardware isn't of great importance, because OpenStack does very little to manage your network. However, throughout most of this book, networking is discussed in the context of OpenStack Networking (Neutron). When using Neutron, your choice of network hardware and software is very important, because OpenStack will be managing many aspects of your network.

At the time of writing, few vendor-provided L3 (router) services exist (see https://www.openstack.org/marketplace/drivers/). Because L3 services will likely be provided by OpenStack, the focus of this discussion will be on choices related to L2.

> **Neutron Distributed Virtual Routing (DVR)**
>
> One of the many goals of the Neutron DVR subproject is to provide distributed routing with compute nodes, integration with routing hardware, and routing service migration between nodes. Although the project is fairly new, the DVR project will likely serve as the primary integration point for most advanced L3 vendor services.

From a Neutron L2 perspective, you have two choices. Your first choice is to use a community- or vendor-provided monolithic network plug-in. These plug-ins are considered monolithic because all L2 OpenStack services must be implemented by the driver, as shown in figure 9.5.

Initially, monolithic plug-ins were the only way to integrate vendor hardware and software with OpenStack Networking. Several of these plug-ins have been developed for vendor hardware, including Arista, Cisco, Melinox,

VMware, and others. The issue with this approach is that the plug-in code must be modified with subsequent OpenStack releases, even if nothing has changed on the vendor side. The effort involved in separating out vendor-specific code from OpenStack code led to the second choice in Neutron L2 networking, the modular Layer 2 (ML2) plug-in, previously covered in chapter 6 and shown in figure 9.6.

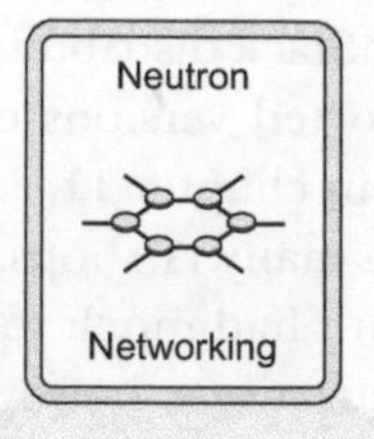

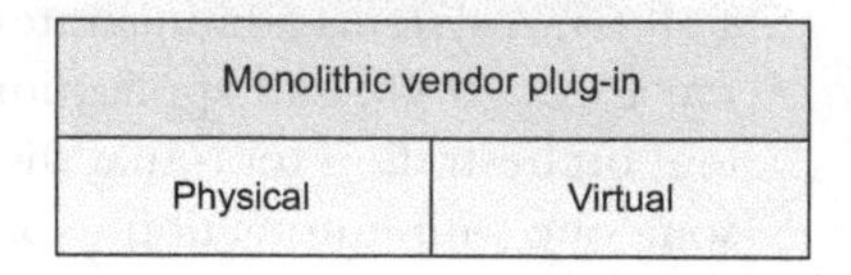

Figure 9.5 Monolithic plug-in architecture

The ML2 plug-in framework allows communities and vendors to provide L2 support much more easily than with monolithic plug-ins. The majority of vendors, even those who had previously developed monolithic plug-ins, are now adopting the ML2 plug-in by writing mechanism drivers for their specific technologies.

Based on the OpenStack user survey, Open vSwitch (OVS) is the most commonly used network driver (interface between standalone hardware and software packages and OpenStack) used in OpenStack deployments. Due to its popularity, OVS was used as a network driver in both parts 1 and 2 of this book. Specifically, using ML2 terminology, the ML2 plug-in was configured to use the GRE type driver and OVS mechanism driver. By using a combination of an overlay (GRE) type driver and a software switch (OVS), we simplified the switch hardware configuration down to simple connectivity between compute and network nodes. The hardware configuration is simple in this context, because OVS is providing virtual switching (traffic isolation on the OVS level), so you only need to worry about making sure that OVS switches on separate servers can communicate with each other.

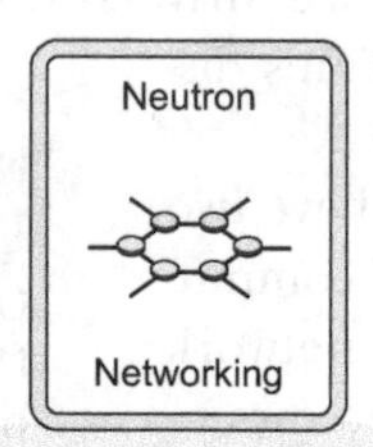

Figure 9.6 ML2 plug-in architecture

There are many benefits to using network overlays like GRE and VXLAN, including scale and flexibility. But there are performance costs related to using network overlays and software switches (OVS) in general. In chapter 11 the VLAN type driver will be used with the OVS mechanism driver. OVS is still the network driver, but instead of overlays connecting OVS instances, OVS makes use of a range of VLANS. The VLANS used by OVS in the chapter 11 example must be manually configured on the switch. In this context, some switching load is offloaded to the hardware switch and some remains in OVS.

The next progression in offloading network load from software to hardware is to use a mechanism driver that deals with all L2 operations on the hardware device (this could be a hybrid hardware and software device). In this configuration, OpenStack Networking operations are translated by the network driver into vendor-specific operations. This doesn't mean that you have to use VLAN as your type driver when using a hardware vendor mechanism driver. In fact, there are many vendor-managed types, including very powerful VXLAN type drivers, that are offloaded in hardware.

As with most things in OpenStack and technology, generalized solutions (in software) come at the cost of performance (using dedicated hardware). You must determine if the performance of software switching and overlay is acceptable, or if your private cloud can benefit from the performance gained through tight integration of OpenStack Networking and vendor hardware.

9.3.3 *Storage in your private cloud*

In chapter 7 you walked through the deployment of the OpenStack storage node using Cinder. The purpose of the storage node was to provide block storage to VMs. Just as Neutron uses network drivers to communicate with underlying software and hardware network resources, Cinder uses storage drivers (see www.openstack.org/marketplace/drivers/) to communicate with storage resources.

In chapter 7 an LVM-configured volume was managed by Cinder. In that example, an LVM storage driver was used by Cinder to interface with the underlying LVM subsystem. Where did the storage device that was used by the LVM volume come from? As discussed in section 9.1.3, the LVM volume could have been a local disk, or it could have been provided from an external source like a SAN. From the standpoint of LVM, and by relation Cinder, as long as the device shows up as a block storage device in the Linux kernel, it can be used. But this is similar to OVS using network hardware as a physical transport. Through the abstraction of the underlying storage device by LVM, you're losing many of the advanced storage features the underlying storage subsystem might offer. Just as with OVS, OpenStack is unaware of the capabilities of the underlying physical infrastructure, and storage functions are offloaded to software. Luckily, there are many Cinder storage drivers for OpenStack, including drivers for storage systems provided by Ceph, Dell, EMC, Fujitsu, Hitachi, HP, IBM, and many others. As with OpenStack Networking, integrating OpenStack Storage with hardware and software storage subsystems through the use of vendor storage drivers allows OpenStack to make use of the advanced features of the underlying system.

Based on OpenStack user surveys, the Ceph storage system is used in the majority of OpenStack deployments. Due to its popularity in the OpenStack community and its inclusion in many standalone OpenStack deployment tools, chapter 10 is dedicated to walking through a Ceph deployment.

Like vendor decisions related to OpenStack Networking, storage decisions need to be based on your current capabilities and future direction. Although it's extremely popular with the OpenStack community, building out support for a Ceph storage cluster might not be the right choice if the rest of the storage in your enterprise is EMC. Likewise, many advanced storage features previously found only in high-end arrays are now found in Cinder or are not needed because of some other aspect of private cloud operation, so purchasing a high-end array might not be necessary.

As you continue through the remaining chapters in part 3 of this book, think about the type of environment you want to construct. For some, a purpose-built system with deep vendor integration will be the best fit. For others, a flexible general-purpose deployment will be the right choice. Regardless of the path you take, make sure OpenStack is the right tool for the job and that your organization is well positioned to take advantage of the benefits of the OpenStack framework.

9.4 Summary

- If you intend to limit the use of your virtual infrastructure to what you can do with physical infrastructure, the benefits of OpenStack in your environment will also be limited.
- Systems administrators who are happy with manually provisioning infrastructure can view OpenStack as either incomplete or unnecessary.
- Systems administrators, developers, consultants, architects, and IT leadership interested in the benefits of cloud computing can view OpenStack as a disruptive technology for the enterprise.
- Users wishing to use OpenStack as a replacement for traditional virtual server infrastructure will find Nova networking comparable to their existing environment, whereas those building a private cloud will likely use Neutron networking.
- Users wishing to use OpenStack as a replacement for traditional virtual server infrastructure will find LVM-based storage comparable to their existing environment, whereas those building a private cloud will likely use Ceph or another vendor-specific directly attached VM storage system.
- By adopting technologies and operational practices born out of large-scale providers for private clouds, enterprises can develop hybrid cloud strategies for best-of-breed solutions.

10 Deploying Ceph

This chapter covers

- Preparing servers for Ceph deployment
- Deploying Ceph using the ceph-deploy tool
- Basic Ceph operations

Ceph (http://ceph.com) is an open source storage platform based on RADOS (http://ceph.com/papers/weil-rados-pdsw07.pdf) that can be used to provide block-, file-, and object-level storage services using commodity servers. Ceph works in a distributed architecture with the goal of eliminating single points of failure by replicating both user and cluster management data. So why is a chapter about Ceph included in an OpenStack book? Based on OpenStack community user surveys, Ceph is the most popular choice for OpenStack storage.[1] In chapter 7 you configured Cinder to use LVM to manage volume storage, but in a production deployment you might use a Ceph backend in place of LVM to provide storage for Cinder to manage.

Although no OpenStack book would be complete without including Ceph, its detailed design and operation is beyond the scope of this chapter. In this chapter,

[1] See "OpenStack users share how their deployments stack up," http://superuser.openstack.org/articles/openstack-users-share-how-their-deployments-stack-up.

you'll walk through a deployment of Ceph using the ceph-deploy deployment tool provided by the developers of Ceph.

You'll work with two types of nodes (commodity servers) in this chapter: *resource nodes*, which are used by Ceph to provide storage, and an *admin node*, which is used as both a Ceph client and as the environment from which you'll provision Ceph.

10.1 Preparing Ceph nodes

In the Ceph architecture, resource nodes can be further divided into nodes that are used for the operation and management of the Ceph cluster and nodes that are used to provide storage. The different types of Ceph nodes are listed in table 10.1.

Table 10.1 Ceph resource nodes

Node type	Description	Function
MON	Monitor node	Maintains a master copy of the storage cluster data map
OSD	Object-storage device node	Provides raw data storage
MDS	Metadata server node	Stores all the filesystem metadata (directories, file ownership, access modes, and so on)

The examples in this chapter are based on a Ceph cluster containing six physical servers that are used exclusively for Ceph and one shared admin server. A list of the nodes, roles, and addresses is shown in table 10.2.

Table 10.2 Ceph nodes

Node name	Node type	IP address
admin.testco.com	ADMIN	10.33.2.57
sm0.testco.com	MON/MDS	10.33.2.58
sm1.testco.com	MON/MDS	10.33.2.59
sm2.testco.com	MON/MDS	10.33.2.60
sr0.testco.com	OSD	10.33.2.61
sr1.testco.com	OSD	10.33.2.62
sr2.testco.com	OSD	10.33.2.63

THE ADMIN NODE TYPE The *admin node* is not part of the Ceph architecture. This node is simply the server that's used to automate the deployment and management of Ceph on dedicated hardware.

> **Synchronize watches!**
> Distributed systems like Ceph can't rely on a central clock, like the filesystem on a single computer can. This is important because one of the ways that distributed systems can determine a sequence of distributed events is through timestamps reported by distributed nodes. It's important that you make sure nodes participating in a Ceph cluster have a synchronized clock. In particular, the MON nodes by default must report a time within 50 ms of each other or a warning alert will be generated (this is configurable). It's recommended that you use a Network Time Protocol (NTP) service on Ceph nodes.

The first step in deploying a Ceph storage cluster is to prepare your nodes. Ceph runs on commodity hardware and software, just like the rest of the OpenStack components in this book.

First you'll configure node authentication and authorization, and then you'll deploy the Ceph software on your nodes.

10.1.1 Node authentication and authorization

On each server, you must create a user that can be used by ceph_deploy to install and configure Ceph. Create a new user using the following commands.

Listing 10.1 Create Ceph user

```
sudo useradd -d /home/cephuser -m cephuser
sudo passwd cephuser
Enter new UNIX password:
Retype new UNIX password:
passwd: password updated successfully
```

> **Setting the password without a prompt**
> Optionally, you can use the `chpasswd` command to script password updates:
> ```
> echo 'cephuser:u$block01' | sudo chpasswd
> ```

In the preceding listing, you created a user named `cephuser` with the password `u$block01`. The ceph_deploy tool will need privileged (`sudo`) access to install software on your Ceph nodes.

Typically, when you invoke privilege commands like the ones in the previous listing, you must provide a `sudo` password, but in an automated install process you don't want to type a password for each invocation of elevated privileges. In order for the `cephuser` to invoke `sudo` commands without a password, you must create a sudoers file in /etc/sudoers.d that lets the system know that the `cephuser` can run `sudo` commands without being prompted for a password. Run the commands in the following listing to create your sudoers file with appropriate privileges.

Listing 10.2 Create the sudoers file

```
echo "cephuser ALL = (root) NOPASSWD:ALL" \
 | sudo tee /etc/sudoers.d/cephuser
sudo chmod 0440 /etc/sudoers.d/cephuser
```

At this point you've created the new user `cephuser`, and your new user can now invoke sudo commands without being prompted for a sudo password.

CEPH NODE-TO-NODE AUTHENTICATION

In the previous step, you created the `cephuser`, which can be used locally on the server for which is was created. If you only have a few servers, it might not be a problem to log in to each one and run a series of scripts, but what if you're working with 10 or 100 servers? To complete the authentication and authorization steps required for automated deployment, you must configure each server to allow remote SSH-based logins for the `cephuser` without password prompts.

Just because a remote host can log in to another host without a password doesn't mean you must sacrifice security for the sake of automation. To understand how this works, you must understand the basics of SSH. Although an in-depth explanation of SSH is beyond the scope of this book, it's sufficient to consider two possible ways of identifying a remote user and a specific local user.

If you've followed the examples in this book, you're already familiar with the process of providing a password when prompted for SSH logins. But there's an alternative to password-based authentication called *key-pair* authentication, where a *public key* is shared between servers. The details are complicated—all you need to know here is that if SERVER_A shares a public key for USER_A with SERVER_B, then if USER_A exists on SERVER_B, SERVER_A can authenticate USER_A on SERVER_B without a password. This relationship is shown in figure 10.1.

Imagine SERVER_A in this scenario is your admin node, and SERVER_B is a resource node. Your admin node will be used to push automated deployment tasks to your resource nodes, which will be used to provide services. Consequently, from your admin node you'll want to access many resource nodes without using a password.

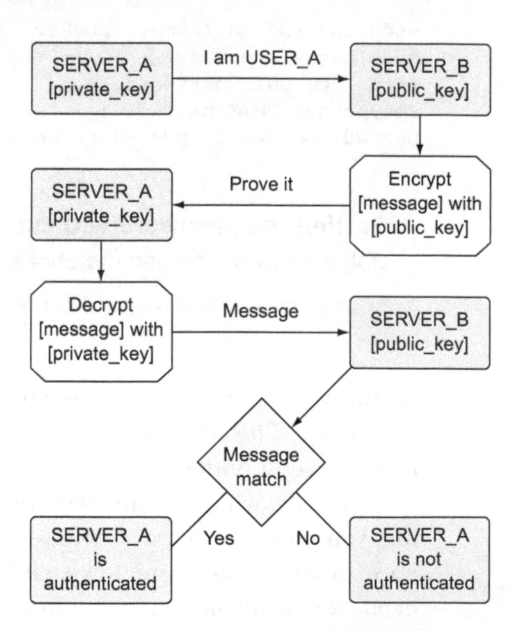

Figure 10.1 SSH key-pair exchange process

SSH KEY-PAIR AUTHENTICATION Just as a server might want to access servers without using a password, users often want this ability as well. Aside from the convenience of not having to enter a password, this form of authentication eliminates the need to store or transmit your password in clear text, and it's considered more secure (if used with key-pair and password) than password authentication alone. In fact, OpenStack provides the ability to inject key pairs into VMs as part of the instance-creation process.

From your admin node, using the `cephuser`, follow the steps in listing 10.3 to create a private/public key pair. Make sure you don't provide a passphrase when prompted, or you'll be prompted to provide this passphrase when using the key pair. You can repeat the key-pair creation process if you experience any problems.

Listing 10.3 Create private/public key pair on admin node

```
$ ssh-keygen
Generating public/private rsa key pair.
Enter file in which to save the key (/home/cephuser/.ssh/id_rsa):
Created directory '/home/cephuser/.ssh'.
Enter passphrase (empty for no passphrase):
Enter same passphrase again:
Your identification has been saved in /home/cephuser/.ssh/id_rsa.
Your public key has been saved in /home/cephuser/.ssh/id_rsa.pub.
The key fingerprint is:
90:6c:09:3d:b8:19:5e:f0:27:be:4b:00:91:34:1d:72 cephuser@admin
The key's randomart image is:
+--[ RSA 2048]----+
| .=oE=           |
|  .=+++o         |
|  .. =O..        |
|   .+o +         |
|    . . S        |
|      . .        |
|       o         |
|      . .        |
|       .         |
+-----------------+
```

Now that you have your key pair, you must distribute the public key (/home/cephuser/.ssh/id_rsa.pub) to all resource nodes. The public key must be placed in the /home/cephuser/.ssh/authorized_keys file on every resource node. Luckily, there's a tool called ssh-copy-id that will help you distribute your public key. From the admin node, follow the process shown in the next listing to distribute your public key to each of your resource nodes.

Listing 10.4 Distribute public key from admin node

```
$ ssh-copy-id cephuser@sm0.testco.com
/usr/bin/ssh-copy-id: INFO:
 attempting to log in with the new key(s),
  to filter out any that are already installed
```

```
/usr/bin/ssh-copy-id: INFO: 1 key(s) remain to be installed --
  if you are prompted now it is to install the new keys
cephuser@sm0.testco.com's password: [enter password]

Number of key(s) added: 1

Now try logging into the machine, with:
 "ssh 'cephuser@sm0.testco.com'"
and check to make sure that only the key(s) you wanted were added.
```

When logged in to your admin node as the `cephuser`, you now have the ability to remotely, yet securely, log in to your resource nodes as `cephuser`. In addition, due to the previous sudoers configuration, you can now execute privileged commands on all configured nodes.

In the next section, you'll install an automation tool used to deploy Ceph.

10.1.2 Deploying Ceph software

Ceph-deploy is a collection of scripts that can be used in the automated deployment of Ceph storage. Learning the ceph-deploy method will give you a good component-level understanding of how Ceph works, while abstracting tedious low-level repetitive tasks. Ceph-deploy, unlike more-general orchestration packages like Ubuntu Juju (covered in chapter 12), is exclusively used to build Ceph storage clusters. It's not directly used to deploy OpenStack or any other tool. (In chapter 11 you'll use a fully automated OpenStack deployment tool that deploys and configures Ceph for use with OpenStack.)

Get started by installing ceph-deploy on your admin node as show in the following listing.

Listing 10.5 Install ceph-deploy

```
$ wget -q -O- \
'https://ceph.com/git/?p=ceph.git;a=blob_plain;f=keys/release.asc' \
| sudo apt-key add -
OK

$ echo deb http://ceph.com/debian-dumpling/ $(lsb_release -sc) \
main | sudo tee /etc/apt/sources.list.d/ceph.list

$ sudo apt-get update
Hit http://ceph.com trusty InRelease
Ign http://us.archive.ubuntu.com trusty InRelease
...
Fetched 2,244 kB in 5s (423 kB/s)
Reading package lists... Done

$ sudo apt-get install ceph-deploy
Reading package lists... Done
Building dependency tree
Reading state information... Done
The following packages will be upgraded:
  ceph-deploy
...
```

You now have ceph-deploy installed on your admin node. Next, you'll configure your Ceph cluster.

10.2 Creating a Ceph cluster

In this section, you'll walk through the deployment of a Ceph cluster. This cluster can be used to provide storage resources, including object and block storage for OpenStack.

10.2.1 Creating the initial configuration

The first step in creating a Ceph cluster is to create the cluster configuration that will be used in your deployment. During this step, configuration files will be generated on the admin node.

Create your new Ceph cluster configuration as shown in the following listing. You'll want to reference all previously designated MON nodes in this step, as they maintain the master copy of the data map for the storage cluster.

Listing 10.6 Generate initial cluster configuration

```
$ ceph-deploy new sm0 sm1 sm2
[ceph_deploy.conf][DEBUG ]
 found configuration file at: /home/cephuser/.cephdeploy.conf
[ceph_deploy.cli][INFO  ]
 Invoked (1.5.21): /usr/bin/ceph-deploy new sm0
[ceph_deploy.new][DEBUG ] Creating new cluster named ceph
...
ceph_deploy.new][DEBUG ] Resolving host sm2
[ceph_deploy.new][DEBUG ] Monitor sm2 at 10.33.2.60
[ceph_deploy.new][DEBUG ] Monitor initial members are
 ['sm0', 'sm1', 'sm2']
[ceph_deploy.new][DEBUG ] Monitor addrs are
 ['10.33.2.58', '10.33.2.59', '10.33.2.60']
[ceph_deploy.new][DEBUG ] Creating a random mon key...
[ceph_deploy.new][DEBUG ]
 Writing monitor keyring to ceph.mon.keyring...
[ceph_deploy.new][DEBUG ] Writing initial config to ceph.conf...
```

LAST CHANCE TO MODIFY THE CONFIG Your initial configuration file, ceph.conf, is generated in listing 10.6. If you need to make any modifications to the configuration, do it now, because it's not trivial to change your configuration once the cluster has been created. Refer to the Ceph documentation (http://ceph.com/docs/master/rados/configuration/ceph-conf/) for configuration options that might be applicable in your deployment. In the examples, we'll use the default configuration.

The /home/cephuser/ceph.conf file will be used in the rest of your deployment. Look under the `[global]` heading in this file and make sure the initial monitor members are listed under `mon_initial_members`. Also check that the correct IP addresses were resolved to the initial members under `mon_host`.

The next step is to install the Ceph software on all the resource nodes. You'll use ceph-deploy to do this.

10.2.2 Deploying Ceph software

As the cephuser, from the admin node, follow the steps in the following listing to install the most current version of Ceph for each of your resource nodes. You can use the fully qualified domain name (such as sm0.testco.com) or the short name (sm0) for ceph-deploy commands.

Listing 10.7 Deploying Ceph software to resource nodes

```
$ ceph-deploy install admin sm0 sm1 sm2 sr0 sr1 sr2
[ceph_deploy.conf][DEBUG ]
  found configuration file at: /home/cephuser/.cephdeploy.conf
[ceph_deploy.cli][INFO  ]
  Invoked (1.5.21): /usr/bin/ceph-deploy install sm0
...
[sm0][DEBUG ]
 ceph version 0.87
```

Ceph-deploy version selection

By default, the ceph-deploy install process will use the most current version of Ceph. If you need a feature in the development branch, or you want to avoid an issue in a specific release, you can select your version using the following command arguments:

- `--release <code-name>`
- `--testing`
- `--dev <branch-or-tag>`

At this point you should have ceph-deploy installed on your admin node and the Ceph software installed on your resource nodes. You now have all the components on your physical nodes to configure and launch your Ceph cluster. Next, you'll start the process of cluster deployment.

Removing Ceph

If for some reason you want to remove Ceph from your resource nodes (perhaps you want to repurpose the hardware), you can either `uninstall` (remove the software only) or `purge` (remove the software and the configuration):

- `ceph-deploy uninstall [hostname]`
- `ceph-deploy purge [hostname]`

10.2.3 Deploying the initial configuration

Enter the command in the following listing, adjusting it for your cluster configuration. This process defines the monitor nodes within your Ceph cluster and gathers keys from the monitoring nodes.

Listing 10.8 Add monitor nodes and gather keys

```
$ ceph-deploy mon create-initial
ceph-deploy mon create sm0 sm1 sm2
...
[ceph_deploy.mon][DEBUG ]
 Deploying mon, cluster ceph hosts sm0 sm1 sm2
...
[sm0][INFO  ] monitor: mon.sm0 is running
...
[sm1][INFO  ] monitor: mon.sm1 is running
...
[sm2][INFO  ] monitor: mon.sm2 is running
```

Removing MON nodes

To gracefully remove a MON node from your cluster, use this command:

```
ceph-deploy mon destroy [hostname]
```

If you decommission a node (format the server) without completing this step, the Ceph cluster will consider the missing node to be in a failure state.

At this point you've deployed the Ceph cluster configuration and you have active MON nodes. In addition, the keys from the MON nodes should now exist on the admin node. These keys will be necessary in order to provision the OSD and MDS nodes.

Before you go any further, you should check and make sure that the MON nodes are up and running. One way to do this is to use the ceph client, which was deployed in listing 10.7. But before you can use the ceph client, you must first deploy a client configuration for your cluster.

The following listing shows how to deploy a client configuration to your admin node. This process can be repeated for any node where you wish to issue client commands using the ceph CLI.

Listing 10.9 Deploy the ceph client configuration

```
$ ceph-deploy admin admin
[ceph_deploy.conf][DEBUG ]
 found configuration file at: /home/cephuser/.cephdeploy.conf
[ceph_deploy.cli][INFO  ]
 Invoked (1.5.21): /usr/bin/ceph-deploy admin admin
[ceph_deploy.admin][DEBUG ] Pushing admin keys and conf to admin
...
```

```
[admin][DEBUG ]
write cluster configuration to /etc/ceph/{cluster}.conf
```

Now that your client configuration is in place, check on the health of your cluster using the following command.

Listing 10.10 Checking Ceph health

```
$ ceph
ceph> health
HEALTH_ERR 64 pgs stuck inactive; 64 pgs stuck unclean; no osds
```

If all went well, the ceph client will report back on the current state of the cluster. At this point you should receive a HEALTH_ERR result, because you don't have any OSD nodes. For now you just want to make sure that the ceph CLI can communicate with your cluster.

In the next section, you'll add storage to your Ceph cluster.

10.3 Adding OSD resources

At this point you have a working Ceph cluster, but no storage has been assigned to this cluster. In the following subsections, you'll walk through the process of provisioning local storage on OSD nodes and assigning those storage resources to your Ceph cluster. As you assign storage resource via OSD processes running on OSD-designated nodes, the reported storage available to the cluster will increase.

A typical Ceph OSD node is a physical server with several directly attached physical disks. Although it's technically possible to use any block device that can be formatted with a Ceph-supported filesystem (ext4, XFS, or Btrfs), direct attached disk is commonly used for financial and performance reasons. The physical disk found in an OSD node can take on several roles, as shown in table 10.3.

Table 10.3 Ceph OSD device roles

Disk type	Description
System	OS storage for server running as OSD node
Journal	Log of changes related to data resources
Data	Storage resources

First, take a look at the physical disks on your OSD nodes. Table 10.4 shows the physical disk for one of the OSD nodes in the example environment used in this book. You'll likely have a different configuration, so follow the steps carefully, making appropriate device-name substitutions where needed.

Table 10.4 Device assignments

Path	Type	Size	Usage
/dev/sda	RAID	500 GB	System
/dev/sdb	SSD	375 GB	Journal
/dev/sdc	SSD	375 GB	Journal
/dev/sdd	SSD	375 GB	Journal
/dev/sde	SSD	375 GB	Journal
/dev/sdf	SAS	1000 GB	Data
...	...	...	...
/dev/sdu	SAS	100 GB	Data

As you can see at the top of table 10.4, a single logical (physical RAID) device is being used for OS storage. In this particular system, there are four SSD devices used as journal volumes for the data drives. Journal volumes are used to temporarily store data to be replicated across OSD nodes, so the performance of journal volumes is very important. Although you don't need to have separate journal volumes, the use of dedicated SSD journal volumes has been shown to greatly improve performance. The storage in this example system will come from sixteen 1 TB data volumes.

Next, you'll start the process by identifying and clearing, or as Ceph calls it, *zapping*, your storage devices.

10.3.1 Readying OSD devices

On a single OSD node in the example system, there are 21 logical volumes, for a total of 63 in the entire example system, as shown in figure 10.2.

In a manual configuration, you'd have to log in to each node and prepare each disk independently. Luckily, ceph-deploy can be used to perform these tasks remotely.

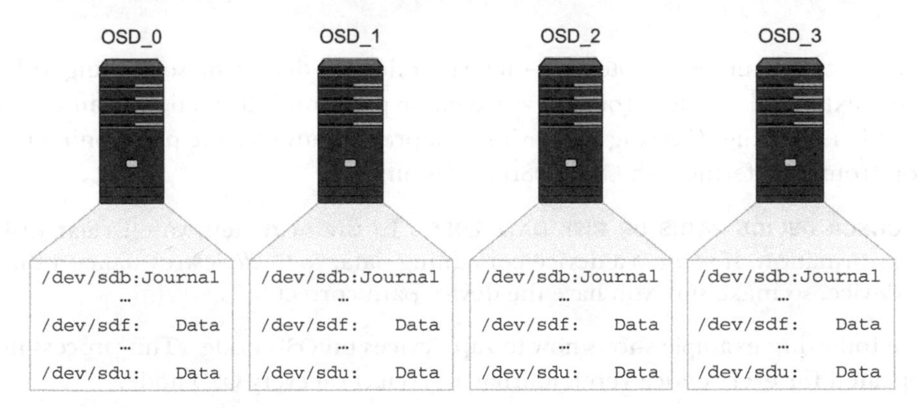

Figure 10.2 Ceph OSD nodes

The first step in this process is to identify the devices on your OSD nodes. This can be done for each OSD node as shown in the following listing. This command should be executed from the admin node.

Listing 10.11 List OSD node devices

```
$ ceph-deploy disk list sr2
...
[sr2][DEBUG ]  /dev/sda1 other, vfat, mounted on /boot/efi
[sr2][DEBUG ]  /dev/sda2 other, ext2, mounted on /boot
[sr2][DEBUG ]  /dev/sda3 other, LVM2_member
[sr2][DEBUG ] /dev/sdb other, unknown
[sr2][DEBUG ] /dev/sdc other, unknown
...
[sr2][DEBUG ] /dev/sdu other, unknown
```

Devices sdb–sdu will be used for Ceph.

Although this listing isn't very comprehensive from a decision-making standpoint, it does let you know which devices ceph-deploy can see. You'll want to make sure that all the devices listed in your device assignments table (table 10.4) are visible in ceph-deploy.

Physical-logical device mapping

Most modern servers will have a disk controller of some type. By default, most controllers require that each physical device or group of devices be configured as a logical device. This can be a time-consuming process if not automated. Check with your hardware vendor about automated hardware configuration tools, which can often be used in conjunction with dedicated hardware management cards.

The mapping of logical disk to a device path is the job of the kernel device mapper. Explaining this process is beyond the scope of this book, but it's sufficient to know that there are tools that will output all known storage devices and their attributes. For example, you can use `fdisk` with the following command:

```
sudo fdisk -l
```

Now that you can see remote disks on OSD nodes, it's time to do something with them. The next step is to clear (or *zap*) any data or partition information from the devices you plan on using. Clearing the devices will prevent any existing partitioning information from interfering with Ceph OSD provisioning.

CHECK DEVICE PATHS OR RISK DATA LOSS In the next step, you'll clear disk information. If you *zap* a device containing data, you *will* destroy data on that device, so make sure you have the device path correct.

The following example shows how to zap devices on OSD nodes. This process must be repeated for every device (both journal and data) on every OSD node.

Listing 10.12 Clear disks on OSD nodes

```
$ ceph-deploy disk zap sr0:sdb
...
[sr0][DEBUG ] zeroing last few blocks of device
...
[sr0][DEBUG ] GPT data structures destroyed!
 You may now partition the disk using fdisk
[sr0][INFO  ] Running command: sudo partprobe /dev/sdb
```

Once this process has been repeated for every disk to be used as either a journal or data device, you are ready to move on to disk preparation.

10.3.2 Creating OSDs

At this point you've identified your devices and their roles, and you've cleared the devices of any data. There are two steps remaining to add these storage resources to your Ceph cluster. First, you must prepare the OSDs, and then you must activate them.

As previously discussed, our example system will make use of four dedicated journal volumes. The failure of a journal volume can be thought of as a failure of all storage that uses that journal. For this reason, a quarter of each disk on a particular OSD server will share a journal. During OSD disk preparation, you must reference your OSD node, data disk, and journal disk, as shown here:

```
ceph-deploy osd prepare {node-name}:{disk}[:{path/to/journal}]
```

The output of the OSD disk prepare is shown in the following example. This step must be completed for every disk on every OSD.

Listing 10.13 OSD disk prepare

```
$ ceph-deploy osd prepare sr0:sdf:/dev/sdb
...
[ceph_deploy.osd][DEBUG ] Deploying osd to sr0
...
[sr0][WARNIN] DEBUG:ceph-disk:
Creating journal partition num 1 size 5120 on /dev/sdb
...
[sr0][WARNIN] DEBUG:ceph-disk:Creating xfs fs on /dev/sdf1
...
[ceph_deploy.osd][DEBUG ] Host sr0 is now ready for osd use.
```

Watching Ceph cluster activity

You can watch the activity in your Ceph cluster by running the following command:

```
$ ceph -w
```

This allows you to observe changes in the system, like the preparation and activation of disks.

By this point, you should have every disk on every OSD prepared for the cluster. In our example cluster, we have 4 OSD servers, with 48 data volumes, prepared across 12 journal volumes (as shown in figure 10.2).

ACTIVATE OSD VOLUMES

In the next and final step, you'll activate your OSD volumes for each OSD, as shown in the following listing.

Listing 10.14 OSD disk activate

```
$ ceph-deploy osd activate sr0:/dev/sdf1:/dev/sdb1
...
[sr0][WARNIN] DEBUG:ceph-disk:Starting ceph osd.0...
...
```

You should now have all OSDs activated in your Ceph cluster. Check your cluster health and stats as follows.

Listing 10.15 Ceph health and stats

```
$ ceph health
HEALTH_OK

$ ceph -s
    cluster 68d552e3-4e0a-4a9c-9852-a4075a5a99a0
     health HEALTH_OK
     monmap e1: 3 mons at
...
      pgmap v39204: 2000 pgs, 2 pools, 836 GB data, 308 kobjects
            1680 GB used, 42985 GB / 44666 GB avail
                2000 active+clean
```

At this point you're finished with the basic deployment of your Ceph cluster. In the next section, we'll cover some basic operations, including benchmarking.

Whoops, starting over

Enter the wrong value? Experience a strange failure? Want to change your design? Part of the benefit of automation is the ability to quickly and easily start the process over.

From the admin node, you can follow these steps to completely clean your environment of a ceph-deploy install:

```
ceph-deploy purge {node-name}
ceph-deploy purgedata {node-name}
ceph-deploy forgetkeys
```

The procedure to clean the environment, as described in this chapter, is as follows:

```
ceph-deploy purge admin sm0 sm1 sm2 sr0 sr1 sr2
ceph-deploy purgedata admin sm0 sm1 sm2 sr0 sr1 sr2
ceph-deploy forgetkeys
```

10.4 Basic Ceph operations

You now have a Ceph cluster, and although a fully automated OpenStack and Ceph deployment system are covered in chapter 11, you should understand a few Ceph basics. This section covers creating Ceph pools and benchmarking your Ceph cluster.

10.4.1 Ceph pools

As previously mentioned, covering Ceph completely could fill an entire book, so only minimal configuration and operation are covered here. One thing that you do need to understand is the concept of a Ceph *pool*. A pool, as the name suggests, is a user-defined grouping of storage, much like a tenant in OpenStack. A pool of storage is created with specific parameters, including *resilience type*, *placement groups*, *CRUSH rules*, and *ownership*, as described in table 10.5.

Table 10.5 Ceph pool attributes

Attribute	Description
Resilience type	The resilience type specifies how you want to prevent data loss, along with the degree to which you're willing to ensure loss doesn't occur. Two types of resilience are *replication* and *erasure coding*. The default resilience level for replication is two copies.
Placement groups	Placement groups are defined aggregations of data objects used for tracking data across OSDs. Simply put, this specifies the number of groups in which you want to place your data, across OSDs.
CRUSH rules	These rules are used to determine where and how to place distributed data. Different rules exist based on the appropriateness of placement. For example, rules used in the placement of data across a single rack of hardware might not be optimal for a pool across geographic boundaries, so different rules could be used.
Ownership	This defines the owner of a particular pool through user ID.

Now that you know the basic attributes of a Ceph pool, create one with the following command.

Listing 10.16 Create Ceph pool

```
ceph osd pool create {pool-name} {pg-num} [{pgp-num}] \
    [replicated] [crush-ruleset-name]
```

The specific command used in the example Ceph cluster is as follows:

```
$ ceph osd pool create mypool 2000 2000
pool 'mypool' created

$ ceph health
HEALTH_WARN 1959 pgs stuck inactive; 1959 pgs stuck unclean
$ ceph health
HEALTH_OK
```

In this example, a pool named "mypool" was created with 2000 placement groups. Notice in the example that two health checks were performed after the pool was created. The first check resulted in a `HEALTH_WARN`, because the placement groups were being created across OSDs. Once the placement groups were created, the cluster reported `HEALTH_OK`.

Next, we'll benchmark the performance of the cluster, using the pool you just created.

10.4.2 Benchmarking a Ceph cluster

There are many ways to benchmark a storage system based on ratios of reads and writes and data rates and sizes. Consequently, there are nearly countless configuration options that can be applied to optimize a storage system, including everything from how the Linux kernel manages low-level I/O operations to the block size of a filesystem or the distribution of data across Ceph nodes. Luckily, storage providers like Ceph do a good job of creating system-wide defaults to cover typical storage workload profiles.

You'll now benchmark your Ceph cluster using a Ceph benchmark tool. This benchmark will be on the pool level, so in terms of the Ceph architecture, this benchmark can be considered low-level and representative of core system performance. If your system is slow on this level, it will only get worse at a higher level of abstraction, where additional performance-impacting constraints could be applied.

WRITING BENCHMARKS

You'll use the Ceph-provided rados tool to perform a write benchmark on the pool you created in the previous section. The `rados` command syntax is as follows.

Listing 10.17 Ceph pool benchmark tool

```
rados -p mypool bench <seconds> write|seq|rand \
 [-t concurrent_operations] [--no-cleanup]
```

The `--no-cleanup` flag will leave data generated during the write test on your pool, which is required for the read test.

You'll want to perform the test from your admin node, because the other Ceph nodes are participating in managing your storage. Here's an example of what it looks like.

Listing 10.18 Ceph write benchmark

```
$ rados -p mypool bench 60 write --no-cleanup
 Maintaining 16 concurrent writes of 4194304
 bytes for up to 60 seconds or 0 objects
...
Total writes made:        16263
Write size:               4194304
Bandwidth (MB/sec):       1083.597      <-- Average write bandwidth for 60 sec

Stddev Bandwidth:         146.055
Max bandwidth (MB/sec):   1164
```

```
Min bandwidth (MB/sec): 0
Average Latency:        0.0590456
Stddev Latency:         0.0187798
Max latency:            0.462985
Min latency:            0.024006
```

Average write latency for 60 sec

In this example, the average write bandwidth was 1,083 MB/sec, out of the theoretical maximum bandwidth (10 gigabit Ethernet) of 1,250 MB/sec.

READING BENCHMARKS

You can now test random reads as shown in the following listing. Keep in mind that if you didn't specify the --no-cleanup command in the previous step, you'll receive an error.

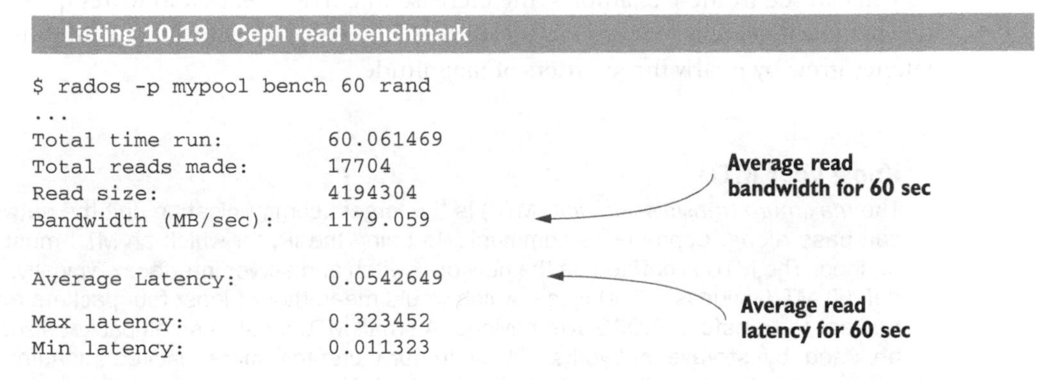

Listing 10.19 Ceph read benchmark

```
$ rados -p mypool bench 60 rand
...
Total time run:        60.061469
Total reads made:      17704
Read size:             4194304
Bandwidth (MB/sec):    1179.059

Average Latency:       0.0542649

Max latency:           0.323452
Min latency:           0.011323
```


In this example, the random read bandwidth is 94% of the theoretical maximum 10 GB bandwidth of 1,250 MB/sec. One would expect a sequential benchmark to achieve even higher numbers.

BENCHMARKING DISK LATENCY

Although bandwidth is a good indicator of network and disk throughput, a better performance indicator for virtual machine workloads is disk latency. In the previous examples, the default concurrency (number of simultaneous reads or writes) was 16. The following two listings show the same benchmarks with a concurrency level of 500.

Listing 10.20 Checking write latency

```
$ rados -p mypool bench 60 write --no-cleanup -t 500
...
Total time run:         60.158474
Total writes made:      16459
Write size:             4194304
Bandwidth (MB/sec):     1094.376

Stddev Bandwidth:       236.015
Max bandwidth (MB/sec): 1200
Min bandwidth (MB/sec): 0
Average Latency:        1.79975
Stddev Latency:         0.18336
Max latency:            2.08297
Min latency:            0.155176
```

Increased bandwidth deviation

Greatly increased latency

Listing 10.21 Checking read latency

```
$ rados -p mypool bench 60 rand -t 500
...
Total time run:          60.846615
Total reads made:        17530
Read size:               4194304
Bandwidth (MB/sec):      1152.406

Average Latency:         1.70021        <-- Greatly increased latency
Max latency:             1.84919
Min latency:             0.852809
```

As you can see in these examples, the increase in current reads and writes quadrupled the maximum latency in both cases. Perhaps more concerning, the minimum read latency grew by nearly three orders of magnitude.

Know your MTU

The *maximum transmission unit* (MTU) is the largest communication unit the network can pass along. Ceph nodes communicate using the IP, for which an MTU must be defined. The MTU is defined on the network switch and server interface. Typically, the default MTU value is 1,500 bytes, which would mean that at least four packets must be sent to transfer a 6,000-byte payload. A small MTU creates small packet sizes to be used by storage networks, which in turn creates more packets, leading to increased network overhead. By increasing the MTU value on storage to a range commonly called a *jumbo frame* (around 9,000 bytes), the payload could be transmitted in a single packet.

The examples in this book are shown with jumbo frames enabled. Using the default MTU of 1,500 bytes for the benchmarks in this section resulted in bandwidth values that were over 40% less than those where jumbo frames were used.

This basic pool-based benchmarking method can be used to determine the underlying performance of your Ceph cluster. Take a look at the online Ceph documentation (http://ceph.com/docs/master/rados/operations/pools/) and experiment with different system-wide and pool settings.

10.5 Summary

- Ceph is a highly scalable cluster storage system based on a common (RADOS) backend storage platform.
- Ceph can be used to provide block-, file-, and object-level storage services using commodity servers.
- The majority of OpenStack deployments make use of Ceph storage.
- Ceph_deploy is a collection of scripts used to deploy Ceph clusters.
- In Ceph, a user-defined grouping of storage is called a pool.

11 Automated HA OpenStack deployment with Fuel

This chapter covers

- Preparing your environment for Fuel
- Installing the Fuel server
- Deploying OpenStack using Fuel

This chapter demonstrates an automated high-availability (HA) deployment of OpenStack using Fuel.

The deployment type is described as *automated* because you'll prepare your hardware for automated deployment and describe your environment, and the automation tool will perform all steps required to deploy OpenStack in your environment, including the deployment of OpenStack components, as covered in chapters 5–8, and Ceph, covered in chapter 10. *High availability* refers to the architectural design, where multiple OpenStack controllers are used.

In chapter 2 you were introduced to the DevStack automation tool. This tool performed automation tasks related to OpenStack deployment, but it was intended as a development tool, not for deploying production environments. A production-focused automation tool must do more than simply configure and install OpenStack; it must also deal with environment preparations, like OS installation and server-side network configurations. The tool demonstrated in this chapter doesn't go this low in the stack, but some automation tools will actually configure network hardware as well. A production-focused automated tool must be auditable, repeatable, and stable, and provide the option of commercial support.

High availability requires the deployment to continue operation within certain limits, even when specific components fail. The deployments described in chapter 2 and part 2 of this book used a single controller. In these types of deployments, if the controller server or one of its dependencies (such as MySQL DB) fails, your OpenStack deployment has failed. Until the single controller is returned to operation, no changes can be made to your infrastructure. In the HA deployment demonstrated in this chapter, the deployment will continue to operate as normal, even if a controller is disabled. Controllers are aware of each others' states, so if a controller failure is detected, services are redirected to another controller.

The OpenStack deployment tool demonstrated in this chapter is called *Fuel*. Fuel was developed and later (2013) open-sourced by Mirantis Inc. (www.mirantis.com). In late 2015 Fuel was formally approved by the OpenStack Foundation to be included as part of their "Big Tent" (governance.openstack.org/reference/projects) project governance model. Several other production OpenStack automation tools are in active development, but I chose Fuel for this demonstration based on its maturity, the stability of the Mirantis OpenStack code, the number of production enterprise deployments, and the availability of commercial support. Although this chapter demonstrates an HA deployment of OpenStack using Fuel, many of the steps would be the same regardless of tool.

First, you'll prepare your environment, then you'll deploy the Fuel tool, and finally you'll use Fuel to deploy your OpenStack environment.

What version of OpenStack are we talking about?

The question in a production environment is not only what version of OpenStack you want to use—you also need to think about the code or package maintainer that you're using.

In the first part of the book, you used DevStack, which pulled OpenStack code directly from the community source (https://github.com/openstack/). In the second part of the book, you used the Ubuntu CloudArchive (https://wiki.ubuntu.com/ServerTeam/CloudArchive#Icehouse) Icehouse packages (the default in Ubuntu 14.04 LTS). In this chapter, you'll not only use Fuel to deploy OpenStack, you'll also use the Mirantis OpenStack version of OpenStack.

Just as various Linux distributions maintain their own kernel and userland packages, the same is true of OpenStack. Determining the appropriate production OpenStack deployment tool involves evaluating both the vendor-specific OpenStack package as well as the capabilities of the deployment tool.

11.1 Preparing your environment

Claims of turnkey automation typically provide a vision of a car-buying experience. You sign on the dotted line, the salesperson tosses you the keys, and, free as a bird, you set out on the open road. Unfortunately, no standard assembly-line deployment of OpenStack exists, so there's no standard deployment support model across the OpenStack domain. For example, a Windows or Linux administrator, on at least a basic level, will understand a Windows or Linux instance installed anywhere. The rules of these systems apply universally. From a deployment standpoint, such universal rules don't exist for OpenStack, the "Cloud Operating System." This, of course, is the intended benefit of this book: understand the framework well enough so that when things fail, or when automation tools evolve, you have some idea of what's going on under the covers.

Preparing an automated deployment environment and configuring it might well take longer than deploying a small environment manually. But for the enterprise, the use of automation tools even in small deployments is very helpful. Helpful, in that the tasks you completed in part 2 of the book can be repeated, support can be purchased, and actions can be tracked and, if necessary, audited.

11.1.1 Network hardware

As you might have noticed in chapters 2 and 6, the Neutron networking component of OpenStack can be confusing, because enterprise systems and network administrators typically have delegated responsibilities. Systems people often don't deal with network virtualization, routers, overlays, and the like, and network administrators don't typically deal with the internal workings of virtual server environments.

Admittedly, there are many moving pieces to manage and configure. This difficulty is amplified if your operational understanding of router and switch configurations is limited. Even if you have a good understanding of routing and switching, you might not have direct access to network hardware to make changes. Or if you have direct access to the network hardware for your OpenStack deployment, you might not have the ability to assign your own addresses and VLANs or to configure upstream network hardware. As you work through this chapter, you'll need to be able to do these things, or to have someone do them for you.

VLANS untagged vs. tagged

The IEEE 802.1Q networking standard provides the ability to designate a virtual network by adding information to the Ethernet frame. Virtual local area networks (VLANs) are considered an OSI L2 function and allow administrators to do things like divide up a switch into separate L2 networks and assign multiple VLANs to a single physical trunk interface. When it's said that a VLAN is *tagged*, the Ethernet frame contains an

(continued)

802.1Q header specifying its VLAN. Likewise, when an Ethernet frame doesn't contain an 802.1Q header, it's said to be *untagged*. It's common for switches to communicate with each other through tagged frames (many VLANs on the same port), whereas servers generally use untagged frames (one VLAN per port).

As you learned in chapter 6, in OpenStack your server acts like a switch, and like a switch it could be using tagged or untagged VLANs. It's important that you understand this concept as you move forward. The examples in this book use tagged VLANs on both the physical switches and physical network interfaces on the server.

CONFIGURING THE DEPLOYMENT NETWORK

You'll accomplish three things in this section:

- Make sure that you can communicate with your automation (Fuel) server.
- Confirm that your automation server can contact all host servers.
- Make sure you have out-of-band access to both your automation and host servers.

The deployment demonstrated in this chapter makes use of two separate physical networks: the automation administration network and the out-of-band (OOB) network. The administration network is used by the automation system to manage host servers on the operating system and OpenStack levels. The OOB network is used to access and configure servers on the hardware level.

Love your OOB network as you love yourself

Working with OpenStack or any large system without an OOB network is like working on an aircraft engine while in the air. The importance of the OOB network for automated deployments can't be overstated. You might be able to run between physically connected consoles in your existing environments or even in a manual OpenStack deployment. But the provisioning processes used in automated deployment all but require remote network access to configure the hardware aspects of your servers. For instance, the configuration of hardware disk controllers or the sequence of boot devices on your server are things you'll want to control not only remotely, but also programmatically.

There's no greater fear in a system or network administrator's heart than the loss of access to the OOB network; this means a trip to the data center.

Just how these networks will be used will be discussed in the following subsections, but for now it's sufficient to know that the example will use two untagged (VLAN) networks on two separate network interfaces for administrative and OOB networks, for each server.

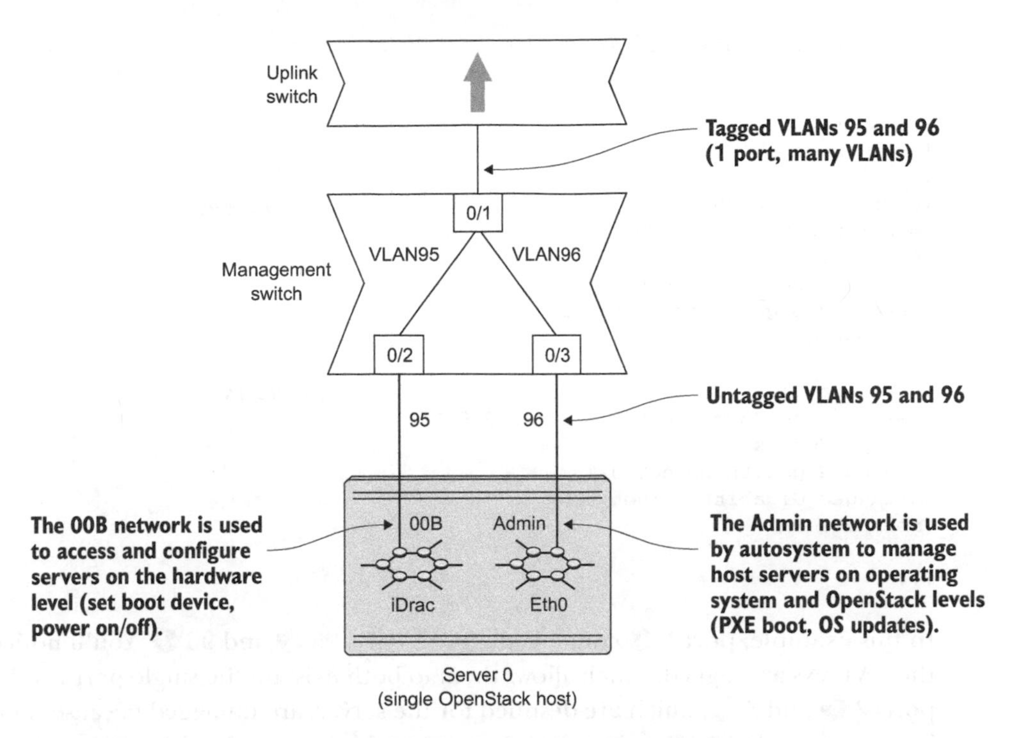

Figure 11.1 Network hardware topology for single-host management

CONFIGURING SWITCH UPLINK PORTS

In order to actually use your management switch, you will need to configure an *uplink* to an existing network. The physical network topology for the demonstration deployment is shown in figure 11.1. It shows the relation between the uplink and management switches and a single OpenStack host.

The switch interface and VLAN configuration for the example OpenStack management switch, a Force10 S60, follows.

Listing 11.1 Out-of-band and administration switch configuration

```
interface GigabitEthernet 0/1                    ❶ Port 1
 description "Uplink Port VLAN 95,96"
 no ip address
 switchport
 no shutdown
!
interface GigabitEthernet 0/2                    ❹ Port 2
 description "OOB Server 0"
 no ip address
 switchport
 no shutdown
!
interface GigabitEthernet 0/3                    ❺ Port 3
```

```
 description "Admin Server 0"
 no ip address
 switchport
 no shutdown
!
...
interface Vlan 95                          ← ❷ VLAN 95
 description "OOB Network 10.33.1.0/24"
 no ip address
 tagged GigabitEthernet 0/1
 untagged GigabitEthernet 0/2
 no shutdown
!
interface Vlan 96                          ← ❸ VLAN 96
 description "Admin Network 10.33.2.0/24"
 no ip address
 tagged GigabitEthernet 0/1
 untagged GigabitEthernet 0/3
 no shutdown
!
...
```

In this example, port 1 ❶ contains the two VLANs 95 ❷ and 96 ❸. You'll notice that these VLANs are tagged, which allows them to both exist on the single port 1. Likewise, ports 2 ❹ and 3 ❺, which are destined for the server, are untagged because there will be only a single VLAN assigned per port. In addition, low-level functions related to automated deployment are often complicated or impossible using tagged VLANs, because the software or hardware used in the deployment process might not support VLAN tagging. It's worth noting that the described networks will not be used directly by OpenStack, but only in the deployment and management of the underlying hardware and OS. These networks will continue to be used for system administrative purposes.

There will be much more network configuration specific to OpenStack Networking (networking that OpenStack controls), but this is sufficient to get started. Next, we'll discuss the hardware preparation.

11.1.2 Server hardware

The network topology you'll set up in this chapter will provide OOB and automation administration for each server, both participating in the OpenStack deployment and the automated deployment (Fuel) server. At this point every server has two physical cables plugged into a management switch. One cable is for the OOB network and the other is for administration, as previously shown in figure 11.1. Here we'll prepare the server hardware to use the two networks.

CONFIGURING THE OOB NETWORK

From a deployment and ongoing management standpoint, the OOB network is critical. The OOB can do the following:

- Manage software-configurable aspects of server and network hardware
- Remotely access a virtual representation of the hardware console

- Remotely mount virtual media used for software installation
- Programmatically access hardware operations (script reboots, boot devices, and so on)

We'll assume at this point that your servers are unconfigured, but that they've been placed in the rack, have physical connections to the previously described networks, and have power. Your next step is to establish OOB connectivity to all your servers. OOB management can be thought of as your lifeline to the physical hardware. This interface is separate from the operating system and often involves a physically separate add-on device.

There are several ways to establish OOB connectivity, some automated and others quite manual. In some cases, the initial configuration tasks might be performed by data center operations, and in other situations this falls to the systems administrator. The size of your deployment and your enterprise operating policy will be factors in determining which process is best for you.

Typically, establishing OOB connectivity is done only once and changes are very infrequent, so unless you have an established method for automating this process, it might be easier to manually configure OOB on each server. Better yet, find an intern and ask them to do it.

Automation: measure once, cut twice

The process of racking, stacking, cabling, and, most importantly, documenting the physical environment is extremely important. You may well have been through painful experiences like hearing, "What do you mean the rack/row was on the same feed/network/unit?"

Through automation, you can easily deploy very complicated configurations across large environments. This effectively creates a configuration needle in a haystack if the underlying infrastructure isn't properly configured. The physical deployment and documentation is the basis for your system, so make sure everything is *exactly* as it is documented before proceeding. Your up-front rigor will be rewarded in the end.

To manually configure the OOB network, you must physically access your server hardware console with monitor and keyboard (or, optionally, use a serial interface). Typically a system configuration screen can be accessed by interrupting the server boot process by pressing designated keys when prompted.

Figure 11.2 shows the system configuration screen related to OOB management. The demonstration system is a Dell server, which contains iDRAC OOB management cards. Although the OOB management cards are vendor-specific, the general configuration and desired results will be the same across vendors. In the figure, you can see where a static address, gateway, and subnet mask have been assigned. Once saved, this information will persist across reboots.

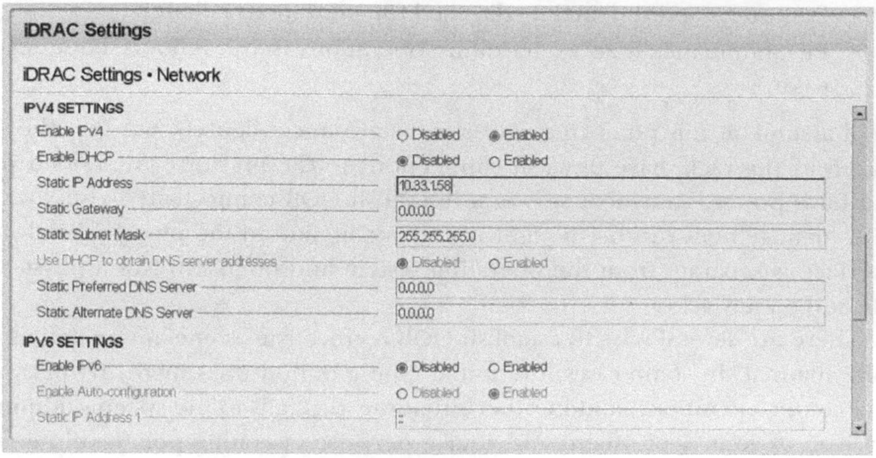

Figure 11.2 OOB management interface configuration on the server

> **DHCP AND OOB MANAGEMENT** Dynamic Host Configuration Protocol (DHCP) can be used to configure the OOB management interface on your server. In larger deployments, this additional degree of automation is usually necessary.

Once the OOB interface on your server has been configured, you'll likely have several ways to access your server, including web interfaces and secure shells (SSH).

ACCESSING AN OOB WEB INTERFACE

Figure 11.3 shows the OOB web interface for the demonstration server.

From the web interface, you can access the virtual console and mount virtual media. The virtual console will display exactly what you'd see on the physical console if you were standing in front of the server.

OOB web interfaces are notorious for requiring specific and typically out-of-date web browsers. The good news is that most of the functions you're likely to be interested in from the OOB perspective can also be configured through an OOB SSH console.

ACCESSING THE OOB MANAGEMENT CONSOLE USING SSH

In addition to the graphical web interface, OOB management systems typically include an SSH or Telnet interface. Although the web-based interface is convenient, it's not easily used from a programmatic perspective. The iDRAC in the demonstration system provides an SSH interface, which allows for the scripted manipulation of OOB-managed hosts. These manipulations range from a simple reboot to complete hardware reconfigurations. For example, for each type of server (Compute, Storage, and the others) you can create a configuration for a specific hardware profile. For example, you can specify BIOS-level settings, RAID configurations, network interface settings, and the like. A role-specific configuration can then be applied through your OOB management interface.

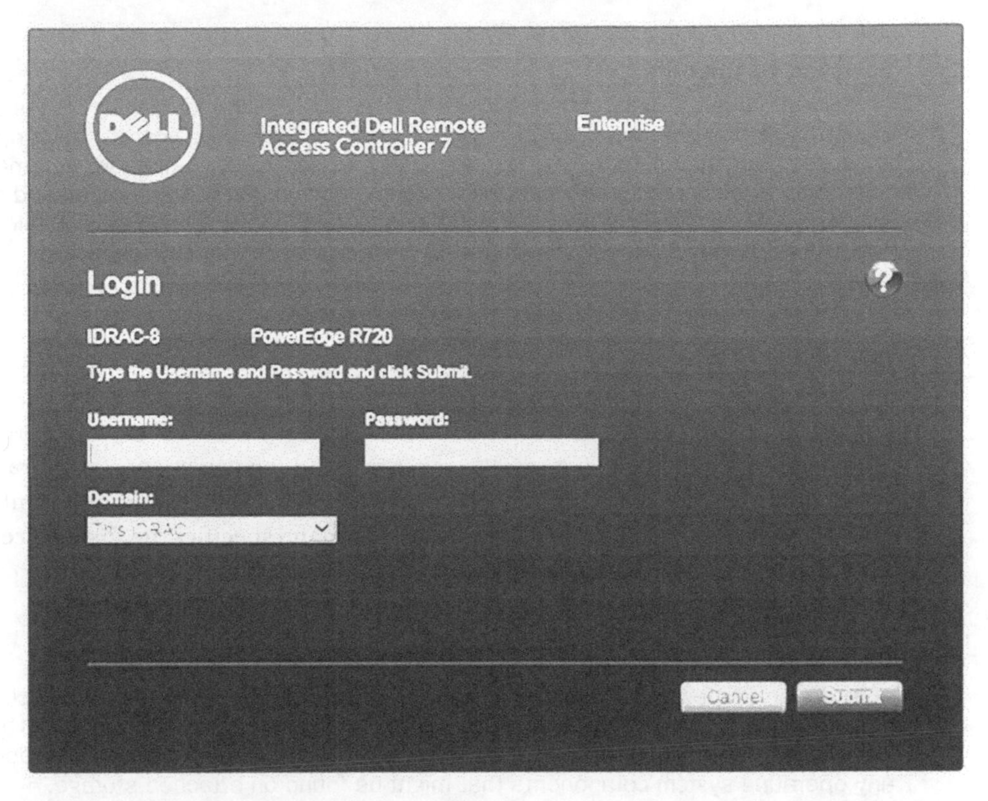

Figure 11.3 Web-based OOB management console

Making use of the sshpass non-interactive SSH password provider software, the following example demonstrates the process of configuring hardware over SSH. Sshpass allows you to script access.

Listing 11.2 Scripting OOB management console actions over SSH

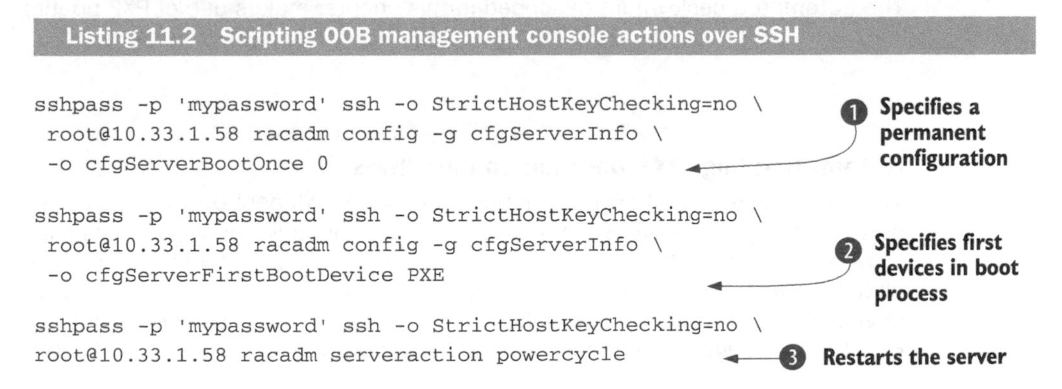

```
sshpass -p 'mypassword' ssh -o StrictHostKeyChecking=no \
 root@10.33.1.58 racadm config -g cfgServerInfo \
 -o cfgServerBootOnce 0

sshpass -p 'mypassword' ssh -o StrictHostKeyChecking=no \
 root@10.33.1.58 racadm config -g cfgServerInfo \
 -o cfgServerFirstBootDevice PXE

sshpass -p 'mypassword' ssh -o StrictHostKeyChecking=no \
root@10.33.1.58 racadm serveraction powercycle
```

Scripted passwords

In general, keeping clear-text passwords in scripts and using them with SSH is considered bad practice. Obviously, low-level credentials can be gained if the script is accessed, but this is also the case with other forms of authentication. Automated SSH login is designed to use public/private key encryption, but an unintended consequence of scripting SSH access could be passwords showing up in the console history and other places where interactive login information is typically redacted.

In this example, three separate commands are sent to the OOB management interface that was configured in the previous figure. The command set for your OOB management interface will vary by vendor. In this iDRAC demonstration, the command `racadm` is used in both the configuration (`-g cfgServerInfo`) and management (`serveraction`) actions. The first command specifies that the following boot configuration is intended as a permanent configuration ❶. The second command specifies the first devices to be used in the boot process ❷. The final command restarts the server ❸.

Configuring first boot device as network

Automation frameworks must have a way to discover new devices, and this discovery is typically done through the network using preboot execution environment (PXE) booting. Servers configured to *PXE boot* will try to boot using the network before accessing any operating system components that might be found on attached storage.

The PXE boot device will transmit a DHCP request, and the DHCP server will return an address assignment along with the location of executable code that can be used to boot the server. Once the DHCP address has been assigned to the server, the boot code will be transmitted over the network, and the server will boot based on this information.

The automated deployment described in this chapter makes use of PXE booting.

Automation bug: PXE boot incompatibilities

You might have several devices that are capable of PXE boot on your server, and the server management software might have a PXE boot option as well. In this chapter's example environment, the PXE boot agent provided by the system-level Unified Extensible Firmware Interface (UEFI) was disabled and the PXE boot agent from the network device itself was used. This was done due to PXE boot incompatibilities between the Fuel PXE boot environment and the PXE agent provided by the server software.

Next, you'll configure your storage hardware.

CONFIGURING SERVER STORAGE

Although private cloud technologies like OpenStack have started to change things in the enterprise, the norm for a typical server is still to use centralized storage in the form of proprietary storage area networks (SANs). There's nothing wrong with this, and as discussed in chapter 9, OpenStack can easily make use of SANs provided by many vendors. But many OpenStack strategies make use of open source server-based storage solutions, like Ceph, described in chapter 10. In fact, based on OpenStack community surveys, server-based open source storage solutions are used in the majority of OpenStack deployments. As a result, most automation frameworks natively support open source storage solutions as part of a fully automated deployment. For these reasons, we'll cover an automated deployment that makes use of the Ceph open source storage software.

> **LOCAL STORAGE ON COMPUTE NODES** Regardless of the role of your server, you'll need to configure its internal storage. Although it's possible to PXE boot servers, often PXE boot is only used during deployment and upgrade phases. Unless it's part of your overall operational strategy, you likely don't want to rely on PXE boot availability for normal operation.

Local disk configuration will be specific to your hardware and the intended purpose of your deployment. Nevertheless, there are some role-based recommendations that can be made based on the environment and automation framework presented in this chapter:

- *Controllers (three nodes)*—The framework used in this chapter (Fuel) places the majority of the administrative and some of the operational load (MySQL, network functions, storage monitor, and so on) on the controllers. For this reason, servers designated as controllers should have fast system volumes (where the OS is installed), and SSD disks if possible. In an HA environment, performance should be valued over redundancy for controllers, because you already have redundancy through duplication of your controllers. An SSD RAID-0 system volume will be used in this chapter's example.
- *Compute (five nodes)*—Your VM storage will be located on separate storage nodes, so once again you only have to worry about the system volume. But because the system volume is required to provide the operating environment for the virtual servers, it should be resilient. Unless you plan on an environment where RAM is expected to be highly over-provisioned, the system volumes don't require SSD performance. An SAS RAID-10 system volume will be used in this example.
- *Storage (three nodes)*—Your VM storage, images, and all storage related to resources provided from OpenStack will be stored here. As previously stated, Ceph will be used to manage these resources, but hardware resources must first be provisioned on the device level. Ceph will work with storage on the device level, as presented by the hosting OS. (Of course, you must have an OS in the first place, so a redundant system volume is needed.) Disks used by Ceph can be separated into

journal and data volumes; you'll want your fastest disks as journal volumes and your largest disks as data volumes. In both cases, Ceph volumes should be configured as JBOD (just a bunch of disks) or RAID-0 if your disk controller doesn't support JBOD. For this chapter's example, an SAS RAID-10 system volume was used, along with 4 SSD RAID-0 journal and 16 SAS RAID-0 data volumes.

Automation bug: disks

In the example environment, the data volumes on the storage nodes were combined in pairs to limit the total number of storage devices on the server to 13. The servers actually contain 24 disk devices, but listing the device and path for all disks creates a long string that caused deployment failures. The string size exceeded an OS-related limit during the automated OS deployment phase. Reducing the number of volumes reduced the configuration string to an acceptable length.

You might wonder, "What about the hardware requirements of the automation server?" Well, it really doesn't matter. Technically, the automation server could be run from a VM or even a laptop. In practice, as will be discussed in the next section, you'll likely want a physical automation server because this server will be both booting (at least initially) your host servers and maintaining their configuration. In addition, you don't want to spread your failure domain if you must use another system to provide a virtual administration node. Performance is less of an issue on admin nodes, but you do want something that can function independently of any other system.

Now let's look at the automation administration network, which will be used to deploy and manage your cluster.

CONFIGURING THE AUTOMATION ADMINISTRATION NETWORK

From deployment and management standpoints, the automation administration network is second only to the OOB network in terms of importance. The automation administration network is how the automation framework communicates with your hosts. This network will be used for the following functions:

- PXE boot network used during install and upgrade
- Administrative traffic between automation (Fuel) server and managed nodes
- OS-level network communication, such as outgoing Network Time Protocol (NTP) and incoming SSH traffic on managed hosts

The good news is that, as with the OOB network, the configuration of this network should be simple. On each server to be used in the deployment, pick an OS-accessible interface (not an OOB hardware interface) and assign these interfaces to the administration network. Do yourself a favor and select the same network interface for all servers—using the same interface for all servers allows you to group all the servers together when you configure the interfaces in the automation framework. In the example environment, the first on-board network interface (`eth0`) will be designated as the automation administration network interface.

Assignment, in this context, means that the designated physical server ports will be connected to switch ports that have been configured with the untagged VLAN 96. These switch ports won't be assigned any other VLANs, and the traffic for VLAN 96 won't contain VLAN tags, as previously described. From the perspective of the endpoint device (server), VLAN 96 doesn't exist, and it's only used by the switch to isolate traffic to ports internally designated on VLAN 96. Figure 11.4 shows how this network is used during the PXE boot process.

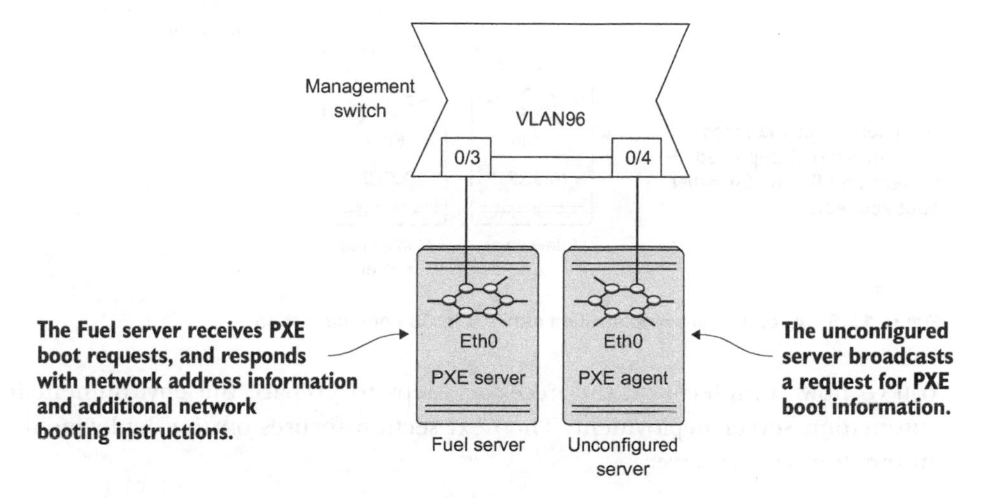

Figure 11.4 Automation administration network in PXE boot

In the figure, you'll see a configured automation (Fuel) server and an unconfigured server on the same untagged network. The Fuel server boots from its local disk and starts listening for DHCP/PXE boot requests on its `eth0` interface. The node labeled "unconfigured server" is a server that has at least been configured with an OOB network (not shown) and been set to PXE boot.

During the PXE boot process, the unconfigured server will broadcast a request for boot information. Because both servers are on the same network, the Fuel server will receive the request and respond to the broadcast with both network address information and additional network booting instructions. The unconfigured server will then proceed with the boot process. Figure 11.5 shows how the network is used once the server has been deployed.

As you can see, the same network that was used in the PXE boot process will be used for OS-level administration once the node has been deployed. In addition, the Fuel server will continue to use this network to both discover new servers and manage the configuration on existing servers.

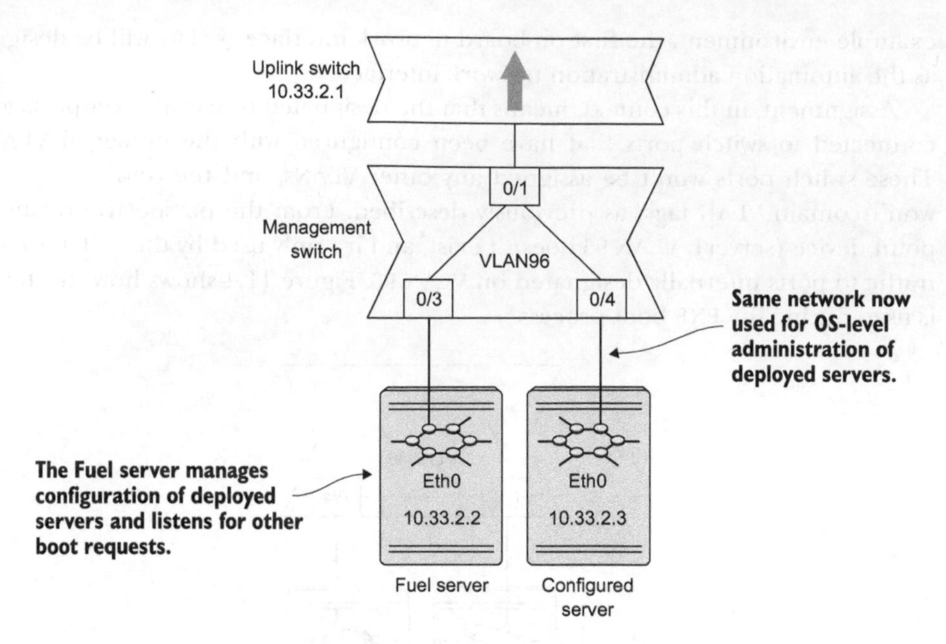

Figure 11.5 Automation administration network in OS communication

You've now completed all the necessary steps to prepare the environment for the automation server deployment. The next section focuses on the deployment of the automation (Fuel) server.

11.2 Deploying Fuel

You are now set up for Fuel, but before you start the Fuel deployment process, make sure that you have OOB connectivity to all servers, including the designated automation server (Fuel). In addition, make sure that all servers use the same interface device name, such as `eth0` used in the previous examples, for their untagged automation administration network assignment.

Next, download the Fuel 7.0 community edition ISO from the Fuel wiki (https://wiki.openstack.org/wiki/Fuel), and start the installation process.

11.2.1 Installing Fuel

At this point you should have OOB network or direct console connectivity to the server designated as your Fuel server, as previously shown in figure 11.1.

Follow these steps to start the Fuel installation process:

1. Using the OOB management capabilities of your deployment server (or manually), mount the Fuel 7.0 ISO on your designated automation server. This process will be specific to your OOB management tool.
2. Reboot your automation server, and based on the vendor-specific instructions for your server, boot from the Fuel 7.0 ISO.

3 As the server starts to boot from the mounted Fuel 7.0 ISO, press the Tab key to interrupt the boot process. The interrupted boot menu is shown in figure 11.6. From the boot menu you have the option of changing initial boot-time settings. You will also have the option to change Fuel settings during the setup phase shown in this section.

If you successfully stopped the boot process, you'll see the screen in figure 11.6.

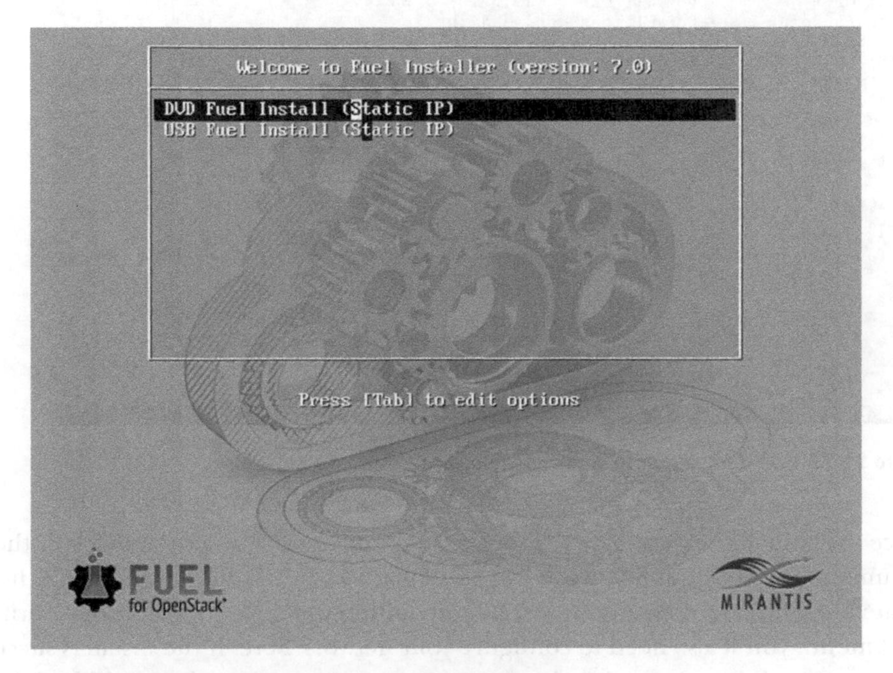

Figure 11.6 Edit settings on the Fuel installer screen

The network settings on your screen won't be the same as the ones shown in the figure. My default Fuel uses an automation administration address range of 10.20.0.0/24, whereas the address range for the demonstration environment is 10.33.2.0/24. In order to use a different range, you must use your cursor and modify the following settings to match your environment:

- `ip`—The address to be used by the Fuel server for both PXE and administration
- `gw`—The network gateway address to be used by both the Fuel server and all servers managed by Fuel
- `dns1`—The name server address to be used by both the Fuel server and all host-managed hosts
- `netmask`—The subnet mask to be used by both the Fuel server and all host-managed hosts
- `hostname`—The host name to be used by the Fuel server

When you've made your changes, press the Enter key to continue the installation.

WARNING If the Fuel installer detects existing partitions on the local disk, you'll be prompted to overwrite the partition. The Fuel installer will overwrite your existing partitions, so make sure you're OK with losing the existing data.

```
Fuel 7.0 setup Use Up/Down/Left/Right to navigate.  F8 exits.
Menu
                     Set Fuel User password.
< Fuel User        > Default user: admin
< Network Setup    > Default password: admin
< PXE Setup        >
< DNS & Hostname   > The password should contain upper and lower-case letters,
< Bootstrap Image  > digits, and characters like !@#$%^&*()_+.
< Root Password    >
< Time Sync        > Fuel password              ******
< Shell Login      > Confirm password           ******_
< Quit Setup       >
                     < Check                                                   >

ASCII characters only
```

Figure 11.7 Fuel data overwrite verification screen

Once the initial Fuel installation process completes, you'll be presented with the Fuel command-line setup utility, as shown in figure 11.7. At a minimum, you'll need to create a password for the root user. If you didn't configure your network during the boot menu, you'll also need to configure your network here. If the install is successful, your server will be rebooted and your automation server console should look like figure 11.8.

Your Fuel server is now installed! In the next section, you'll walk through a basic automated HA deployment using the web interface.

```
########################################
#       Welcome to the Fuel server     #
########################################
Server is running on x86_64 platform

Fuel UI is available on: http://10.33.2.2:8443

Default administrator login:    root
Default administrator password: r00tme

Default Fuel UI login: admin
Default Fuel UI password: admin

Please change root password on first login.

fuel login:
```

Figure 11.8 Fuel post-install console screen

11.3 Web-based basic Fuel OpenStack deployment

The examples in this chapter make use of the Fuel web interface. Fuel also provides a CLI for deploying and managing OpenStack (https://wiki.openstack.org/wiki/Fuel_CLI). Although the Fuel CLI is outside the scope of this book, it's a very powerful tool and is heavily used by serious Fuel users.

Access the Fuel web interface by navigating to http://<fuel server ip>:8443 in your browser. For instance, in the demonstration environment, the address would be http://10.33.2.2:8443. In your browser, you should see the web interface shown in figure 11.9.

Figure 11.9 Log in to the Fuel web interface

Using the default Fuel username and password of `admin/admin`, log in to the Fuel web interface. Once logged in, you'll be taken to the Environments screen as shown in figure 11.10.

Of course, at this point you haven't configured any environments, and more importantly, you haven't discovered any servers to be used for a deployment. You'll discover your servers next.

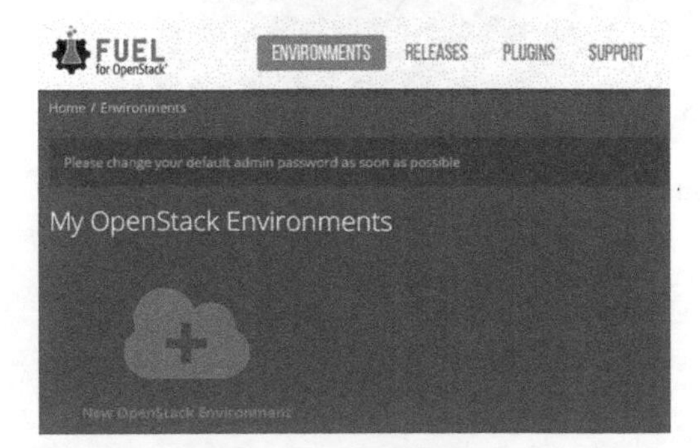

Figure 11.10 Fuel Environments screen

11.3.1 Server discovery

You might be used to management tools that perform server discovery after a service has already been deployed. Often this is done by a network address scan or an embedded agent, or you manually supply a list of host names and addresses. In this case, your servers don't have an address to discover or even an operating system to assign an address.

Server discovery in Fuel is accomplished through a lightweight agent that's placed on each unconfigured server after an initial successful PXE boot. This means that you must first configure all unconfigured servers (excluding the Fuel server) to boot using PXE, as previously discussed. Then each server should be rebooted so Fuel can manage the PXE boot process. The steps for configuring your server hardware for PXE boot and the rebooting process will be specific to your vendor hardware.

Discovery involves the following steps:

1. Set the unconfigured servers to use PXE as the first boot device.
2. Restart the unconfigured servers.
3. The unconfigured servers receive DHCP/PXE information from the Fuel server.
4. The unconfigured servers boot using a management bootstrap image provided by the Fuel server.
5. An agent running under the bootstrap image reports back to the Fuel server once an unconfigured server has booted and a hardware inventory has been collected.
6. The unconfigured server is reported by the Fuel server as an *unallocated server*.

If the discovery process is successful, all of your previously unconfigured servers will now be reported as unallocated servers by the Fuel web interface, as shown in figure 11.11. If a server doesn't show up as discovered, access your virtual OOB console and check the status of the host. You may need to force a cold restart from a particular server's OOB console if the server is in a hung state.

Next, you'll build a new environment, which will be used to deploy OpenStack on your unallocated servers. This environment specifies how you want your servers configured for use with OpenStack.

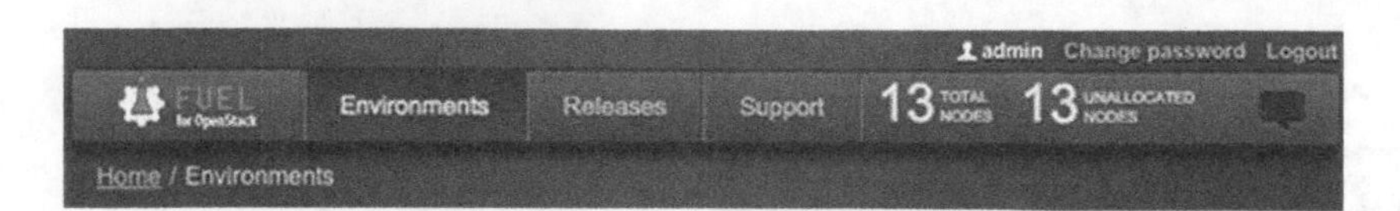

Figure 11.11 Fuel UI with 13 unallocated nodes

11.3.2 Creating a Fuel deployment environment

At this point you're ready to define the environment that Fuel will deploy for you. From the Environments tab, click the New OpenStack Environment icon. You'll be presented with a dialog box and a series of screens like the one shown in figure 11.12.

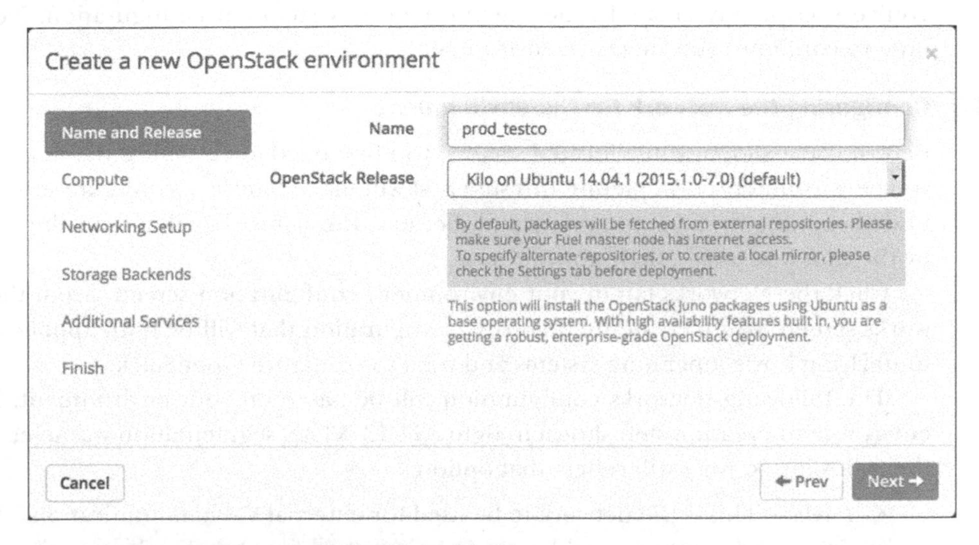

Figure 11.12 Creating a new OpenStack environment using Fuel

You'll be asked to supply the following information:

- *Name and release*—Provide the name of the deployment and the OpenStack release you wish to deploy. The name is purely cosmetic, but the OpenStack release and OS platform selection will determine the underlying OS and release platform for the entire deployment.
- *Deployment mode*—If you have limited hardware, you can select multi-node. In this chapter's demonstration, multi-node with HA will be selected.
- *Compute*—If you're running on hardware, like in the demonstration environment, select KVM. If you're running everything virtually, you should select QEMU.
- *Networking setup*—There are several choices for network type. Likely you'll want to use Neutron for networking, which narrows it down to using GRE or VLAN for network segregation. This chapter's example environment will use VLAN segregation, which is a common choice in production environments for performance reasons.
- *Storage backends*—You can select either Linux LVM as the storage backend or Ceph, both of which will be configured for you. In the example environment, Ceph will be selected for both Cinder and Glance storage.

- *Additional services*—You can optionally install additional services provided by the OpenStack framework.
- *Finish*—When you've completed your configuration, click the Create button to build your environment configuration.

You've successfully created a new deployment environment configuration. Now it's time to configure your network environment.

11.3.3 Configuring the network for the environment

Before you assign any unallocated servers, you first need to configure the network for your environment. Completing this step first will make interface configuration during node assignment easier to understand, because the networks will already be defined by the time you assign them to your hosts.

Click the Networks tab in your environment configuration screen. From the Networks screen, you'll create the network configuration that will be both applied to the underlying hosts' operating systems and used to configure OpenStack.

The following networks configuration will be based on your environment. In the environment creation step shown in figure 11.13, VLAN segmentation was selected, so the following settings will reflect that option:

- *Public*—This is the network to be used for external VM communications. The IP Range is the range of addresses to be reserved for OpenStack operations, such as external router interfaces. The CIDR is the full subnet used for all external (OpenStack and floating) addresses. The Gateway setting specifies the network gateway for the subnet. In the example environment, this network will use the tagged VLAN 97, as indicated by the checked box.
- *Management*—This is the network used by OpenStack nodes to communicate on the API level. This should be considered an internal network for OpenStack components.
- *Storage*—This network will carry the storage traffic from the Ceph nodes to the compute and Glance nodes.
- *Neutron L2 Configuration*—This is the internal or private VM network. Set a range of VLANs to be used for VM-to-VM communication between compute nodes.
- *Neutron L3 Configuration*—This CIDR and gateway will be used internally when creating internal OpenStack networks. Floating IP Ranges specifies the range of addresses reserved from the public network to be available to VMs as floating external addresses.

When your configuration is complete, choose Save Settings.

In the next section, you'll allocate nodes to your environment. Once hosts are assigned, you'll return to this screen and verify your network configuration.

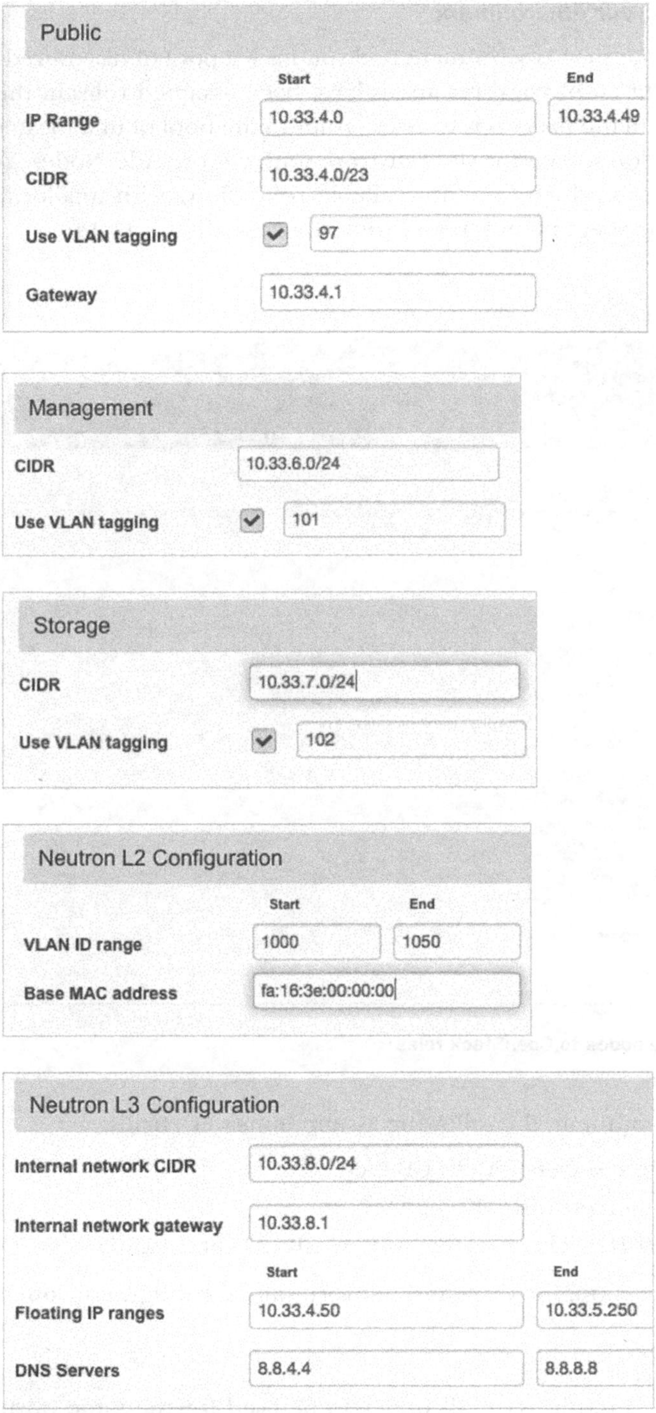

Public

Start End

IP Range 10.33.4.0 10.33.4.49

CIDR 10.33.4.0/23

Use VLAN tagging 97

Gateway 10.33.4.1

Management

CIDR 10.33.6.0/24

Use VLAN tagging 101

Storage

CIDR 10.33.7.0/24

Use VLAN tagging 102

Neutron L2 Configuration

Start End

VLAN ID range 1000 1050

Base MAC address fa:16:3e:00:00:00

Neutron L3 Configuration

Internal network CIDR 10.33.8.0/24

Internal network gateway 10.33.8.1

Start End

Floating IP ranges 10.33.4.50 10.33.5.250

DNS Servers 8.8.4.4 8.8.8.8

Figure 11.13 Fuel environment network configuration

11.3.4 Allocating hosts to your environment

You've configured your new environment and you have a pool of discovered, yet unassigned, resources. But no physical resources have been assigned roles in the environment. The next step in the process is to assign roles to the pool of unallocated servers.

In the configuration screen for your environment, choose Add Nodes. You will be taken to the Nodes tab, which lists the available roles along with unallocated server candidates. The node assignment screen should look like figure 11.14.

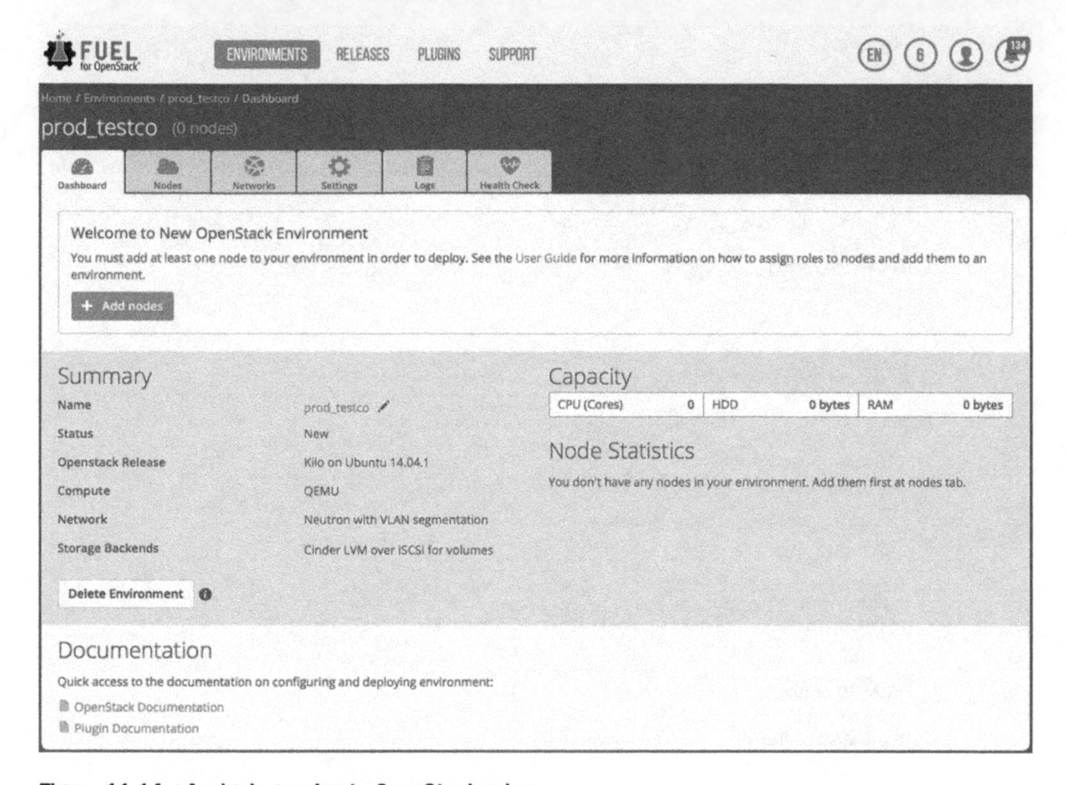

Figure 11.14 Assigning nodes to OpenStack roles

In the example environment, the following assignments were made:

- *Controller*—Three servers with 64 GB RAM
- *Compute*—Five servers with 512 GB RAM
- *Storage, Ceph (OSD)*—Three servers with 48 GB RAM and 16.5 TB disk space

In the following discussion, the disks and network will be configured from the allocation screen.

CONFIGURING INTERFACES

If you've assigned all interfaces on all nodes to be used for the same purpose, then this step is easy; otherwise, you'll have to repeat the interface configuration process for every group of hosts that have unique interface configurations. If interfaces vary based on server or role, simply repeat the configuration for each node.

From the Nodes tab in your environment, toggle the Select All check box and then choose Configure Interfaces. The node interface dialog box should appear, as shown in figure 11.15.

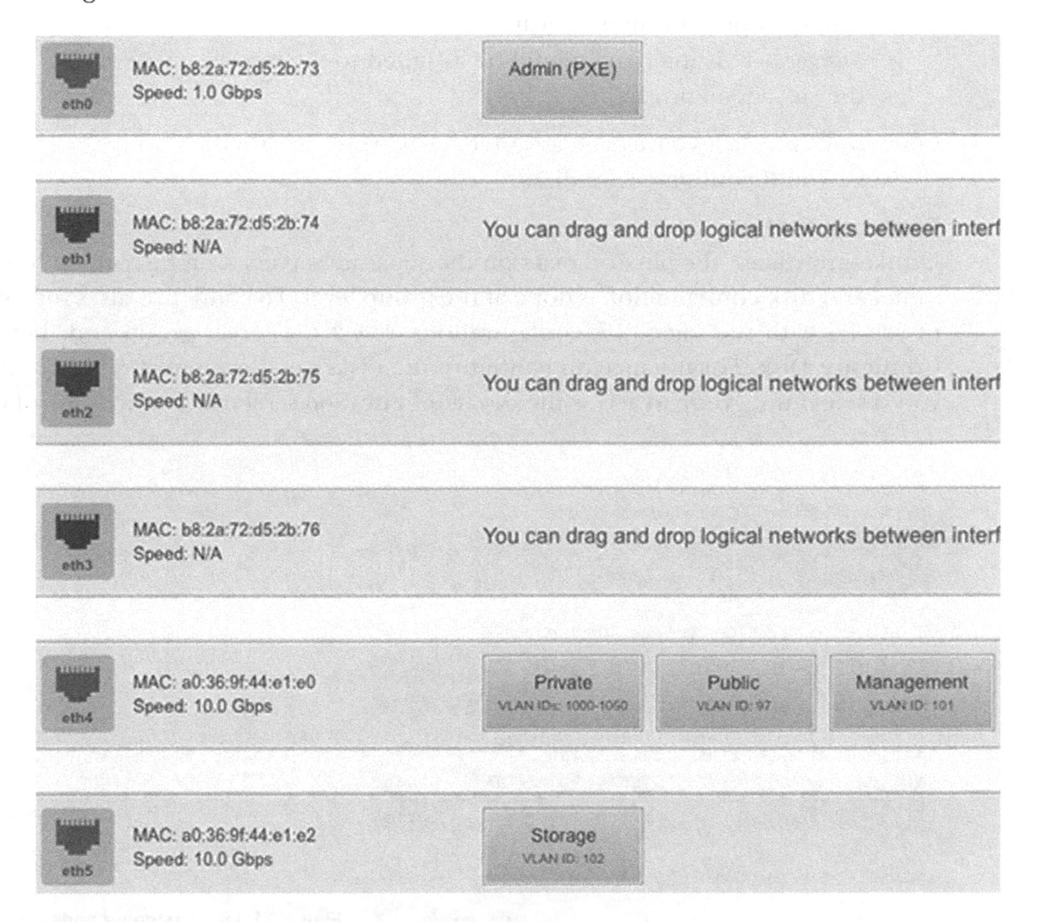

Figure 11.15 Assigning node interface configuration

The networks you configured in the previous section, like the public network shown in figure 11.13, must be assigned to physical interfaces on your servers. The interfaces shown in the screen will be specific to your deployment, but the following network assignments must be made regardless of deployment configuration:

- *Admin (PXE)*—This should be assigned to the automation administration interface. In the example, this will remain on interface `eth0`.
- *Private*—This should be assigned to the interface that will carry VM-to-VM traffic internal to the OpenStack deployment. In this example, VLAN isolation was

selected, so the interface `eth4` is expected to have tagged access to VLANs for transmitting this traffic between nodes.

- *Public*—This should be assigned to the interface that will carry traffic external to the OpenStack environment.
- *Management*—This should be assigned to the interface that will be used for intra-component communication.
- *Storage*—This interface should be assigned to the interface that is connected to the storage network.

Once your interfaces have been assigned, click Apply to save the settings.

Next, you'll configure your disks.

CONFIGURING DISKS

Unlike interfaces, the physical disks on the separate servers aren't expected to be the same, and disk configuration is done at the group level. To configure disks for a group of servers with the same disk configurations, select the server group and then click Configure Disk. For instance, to configure the disks on a group of Ceph OSD nodes, you'd select the group to access the disk configuration screen shown in figure 11.16.

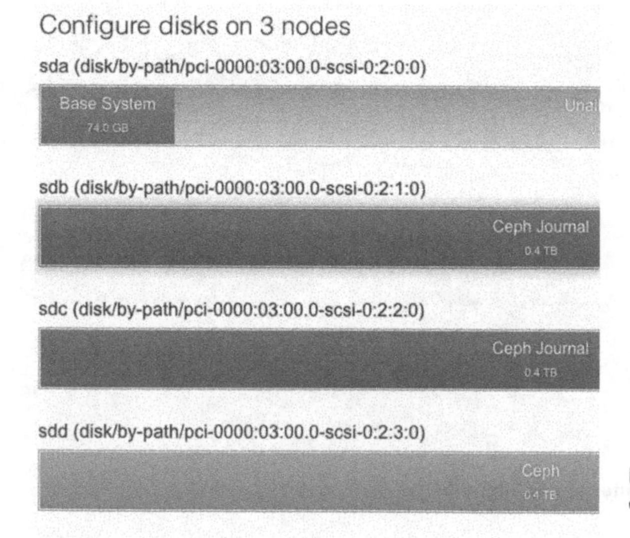

Figure 11.16 Assigning node disk configuration

In this figure, you can see that some disks have been assigned as Ceph journal devices and others have been assigned as Ceph data devices. The configuration will be assigned for all nodes selected in the group.

Once you've configured the disks on all of your nodes, you'll need to make the final configuration settings and verify your network.

11.3.5 Final settings and verification

Click the Settings tab in your environment configuration screen. Under this tab, you'll find your existing configuration, which will be based on the answers you provided during environment creation. From this screen, you can more finely tune your deployment.

At a minimum, I recommend that you change any passwords related to your deployment. In addition, if you're using Ceph for your storage, you might consider assigning Ceph as the back end for both Nova and the Swift API. Once you've made all your changes, click Save Settings.

Once again, click the Networks tab in your environment configuration screen. Scroll down to the bottom of the screen and choose Verify Networks. If all your network settings are correct, you'll see "Verification succeeded," as shown in figure 11.17.

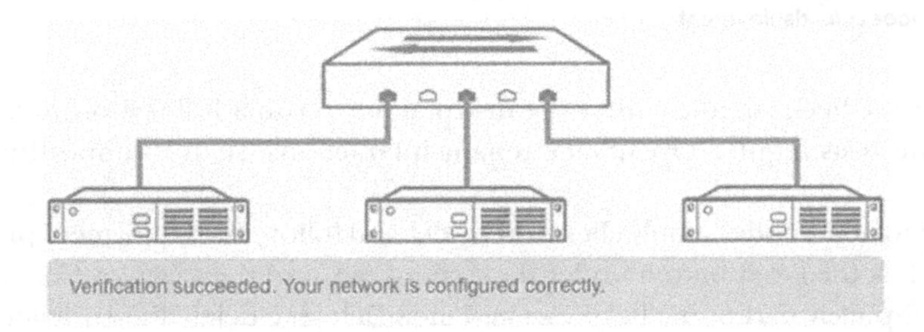
Verification succeeded. Your network is configured correctly.

Figure 11.17 Verified network configuration

> **Help! My network isn't documented!**
>
> Things can admittedly get complicated during this process. Something as simple as determining which server is plugged into which switch can be complicated. If you get completely lost, go back to the basics.
>
> You have no idea where the server is plugged in? You can either start chasing cables in the data center, or you could use the Link Layer Discovery Protocol (LLDP). Consider booting the servers in question using a live CD that contains support for LLDP. Consult your switch documentation for how to enable LLDP on your network hardware. Running LLDP on the switch and server will allow both devices to report link-level locations.
>
> You know where the server is attached, but VLAN verification is failing? Once again, boot a server and confirm connectivity from an untagged VLAN. This can be done by assigning an IP address on the switch and the server, and then using `Ping` to confirm communication between the devices. Once communication is confirmed, repeat the process using tagged VLANs.

STOP! VERIFY OR ELSE Do not proceed until your network configuration can be verified. You'll be guaranteed to have major problems if your network configuration isn't validated on all participating nodes.

11.3.6 Deploying changes

Do you see the blue button labeled Deploy Changes? If you're confident of your hardware configuration and your environment settings, go ahead and click it! This will start the deployment process, which includes OS installation and OpenStack deployment. The deployment progress will be indicated by the green progress bar at the top right of the screen. If you're interested in deployment details, you can click the Logs tab or the paper icon next to the individual server's deployment progress bar.

When the process completes successfully, you'll be provided with your deployment's Horizon web address, as shown in figure 11.18.

Figure 11.18 Successful deployment

Click the Health Check tab and follow the instructions to run a full test on your environment. Some tests require OpenStack tenant interactions, such as importing specific images.

If you need to add a node, simply click Add Nodes and follow the deployment process. To redeploy, click the Reset button under the Actions tab, and then Deploy Changes. If you want to completely start over, click Reset and then Delete to delete the environment.

Enjoy your new HA OpenStack environment.

11.4 Summary

- Fuel can be used in the automated high-availability (HA) deployment of OpenStack.
- Fuel uses the Mirantis version of OpenStack.
- Commercial support is offered by Mirantis for both Fuel and their distribution of OpenStack.
- Fuel provides PXE boot services, allowing your servers to boot from the Fuel server.
- An out-of-band (OOB) network is critical to large-scale automated deployments.

12 Cloud orchestration using OpenStack

This chapter covers

- Application orchestration using OpenStack Heat
- Application orchestration using Ubuntu Juju

One definition of *orchestrate* is to arrange or manipulate, especially by means of clever or thorough planning or maneuvering. You are likely very familiar with the first part of this definition in relation to computing. You must arrange layers of underlying hardware and software dependencies in order to deploy applications. This chapter, and in some respect this entire book, is about the *clever* part of orchestration. Specifically, this chapter covers application orchestration tools that make use of OpenStack resources. Some of the tools we'll explore are official OpenStack projects, and others are related projects.

Even within official OpenStack orchestration tools, dependency hierarchies exist. For instance, the Murano project (not covered in this chapter), which provides users with an application catalog, depends on the Heat project to deploy infrastructure and application components. There are also standalone tools, like

Ubuntu's Juju, that interface directly with core OpenStack APIs to deploy infrastructure dependencies, which Juju then uses to deploy applications.

This chapter starts with the official OpenStack Heat project, which operates between the infrastructure and application levels. Next, we'll look at the standalone Ubuntu tool Juju.

12.1 OpenStack Heat

The OpenStack Heat project is considered the foundation of orchestration in OpenStack. In many ways, Heat does for applications what OpenStack infrastructure components (Nova, Cinder, and the like) do for vendor hardware and software—it simplifies the integration. Large automated compute clusters existed long before OpenStack, but the cluster-management systems were either home-grown or vendor-specific. OpenStack provided a common interface for managing infrastructure resources.

But in the scope of application deployment, infrastructure is only part of the process. Even if you had a system that could provision limitless virtual machines (VMs) instantly, you'd still need additional tools to manage applications on VMs. In addition, you'd want infrastructure and application layers to actively adapt to changes on either level. For instance, if an application performance threshold is breached, you might want additional infrastructure to be added without human interaction. Likewise, if infrastructure resources become limited, you might want the least important applications to gracefully release resources.

Consider for a moment the infrastructure resources that make up a VM. At a minimum, a VM is composed of CPU, RAM, and disk resources. In OpenStack and similar environments, defined formats exist for describing both individual resources and how resources are related to form VMs. Now, suppose you could describe all the steps required in the manual deployment of an application. *Templates* are the textual description of resource dependencies and application-level install instructions.

12.1.1 Heat templates

OpenStack Heat translates templates into applications, making use of OpenStack-provided infrastructure. The process of generating an application *stack* from a template is called *stacking*. Of course, you need a template to make use of Heat capabilities. Heat designers no doubt wanted the project to be useful to the OpenStack community as soon as possible, and to this end they adopted the existing Amazon Web Services (AWS) CloudFormation template format, which is designated in Heat as the Heat CloudFormation-compatible format (CFN). AWS CloudFormation was released in April 2011, several years before Heat, and many CFN templates are available: https://aws.amazon.com/cloudformation/aws-cloudformation-templates/.

The anatomy of the CFN template is shown in the following listing.

Listing 12.1 AWS CloudFormation template format

```
{
  "AWSTemplateFormatVersion" : "version date",

  "Description" : "JSON string",

  "Parameters" : {

  },

  "Mappings" : {

  },

  "Conditions" : {

  },

  "Resources" : {

  },

  "Outputs" : {

  }
}
```

Declares stack input types and values

Assigns key/value pairs for reference in resource and output steps

Specifies logical conditions for the creation of the stack

Declares resource dependencies and application installation procedures

Declares output data to be provided on stack-processing completion

Although supporting the CFN format was valuable in making a large number of existing templates useful for Heat, it was still, after all, a format designed for AWS. Heat project members determined that an OpenStack-specific format was needed, and the Heat OpenStack Template (HOT) was created. The CFN template is structured in JavaScript Object Notation (JSON), whereas HOT templates are structured in YAML format. The HOT specification has been considered the standard template version for Heat since the Icehouse release (April 2014).

Take a look at the following abridged HOT template, which is based on a WordPress deployment template found in the official OpenStack Heat documentation.

Listing 12.2 Example Heat OpenStack Template (HOT)

```
heat_template_version: 2013-05-23

description: >
  Heat WordPress template to support F20, using only Heat OpenStack-native
  resource types, and without the requirement for heat-cfntools in the image.
  WordPress is web software you can use to create a beautiful website or blog.
  This template installs a single-instance WordPress deployment using a local
  MySQL database to store the data.

parameters:

  image_id:
    type: string
    description: >
      Name or ID of the image to use for the WordPress server.
```

Describes the template

Declares input types and values

```
      Recommended values are fedora-20.i386 or fedora-20.x86_64;
      get them from http://cloud.fedoraproject.org/fedora-20.i386.qcow2
      or http://cloud.fedoraproject.org/fedora-20.x86_64.qcow2 .
    default: fedora-20.x86_64

resources:                                   <-- Declares resource dependencies and application installation procedures
  wordpress_instance:
    type: OS::Nova::Server
    properties:
      image: { get_param: image_id }
      ...
      user_data:
        str_replace:
          template: |
            #!/bin/bash -v

            yum -y install mariadb mariadb-server httpd wordpress
            ...
outputs:                                     <-- Declares output data to be provided on stack-processing completion
  WebsiteURL:
    description: URL for WordPress wiki
    ...
```

Both the CFN and HOT templates share foundational components, but they are separate template languages. These template languages can be thought of as programming languages. While initially you might not think of an orchestration template as a programming language, consider the fundamental attributes of a language. A computer language is a formal language used to communicate instructions to a computational system. If a template is a formal language, then Heat serves as an interpreter of that language. The intermediate (process step) output of template interpretation is the sum total of all instructions required to deploy applications on OpenStack-provided infrastructure. The final output of interpretation is a deployed system implemented by template language instructions, with resulting application outputs defined in the template *outputs* section. The individual applications that comprise the Heat project are listed in table 12.1.

Table 12.1 Heat applications

Name	Description
heat	CLI tool that communicates with Heat APIs
heat-api	OpenStack-native REST API
heat-api-cfn	AWS-style query API (AWS CloudFormation API compatibility)
heat-engine	Engine that takes input from APIs and interprets template languages

Next, we'll walk through the creation of a stack using Heat.

12.1.2 A Heat demonstration

In this section, you'll see the deployment of a simple application stack using Heat command-line tools. But Heat does more than simply deploy applications. Heat can be used in conjunction with OpenStack Ceilometer (the central measurement service) to dynamically scale resources based on policies described in templates. A full description of autoscaling with Heat is beyond the scope of this book, but you can find all the details in the official OpenStack Heat documentation: http://docs.openstack.org/developer/heat/.

Back in chapter 2, you used DevStack to deploy OpenStack. The examples here will be deployed in a DevStack environment, but any working Heat environment could be used. If you already have a working Heat environment, skip ahead to the subsection "Confirming stack dependencies."

ENABLING HEAT IN DEVSTACK

If you're using the DevStack environment you deployed in chapter 2, you must make some configuration changes to your local.conf script to enable Heat. Access the command shell in your DevStack environment and add the following lines to /opt/devstack/local.conf.

Listing 12.3 Enable Heat in your DevStack local.conf

```
# Enable Heat (orchestration) Service
enable_service heat h-api h-api-cfn h-api-cw h-eng
HEAT_BRANCH=stable/juno
```

Your new configuration specifies that you want all Heat services enabled, as well as identifies the code branch and release you want to use. In your configuration, make sure the release name in `HEAT_BRANCH` matches the release name of the rest of the components in your configuration.

Once the Heat configurations are in place in your local.conf file, repeat the stacking (`stack.sh`) and unstacking (`unstack.sh`) process from chapter 2 (section _2310_35359_190446), and set your environmental variables as shown in the following listing.

Listing 12.4 Set environmental variables

```
$ source openrc
```

Run this command from the /opt/devstack directory.

As you might recall from chapter 3, the `openrc` script provided by DevStack sets variables in your shell that allow you to interact with OpenStack services.

You should now have a working Heat environment and console access to your environment. Next, you need to confirm that all dependencies are in place for your first stack.

CONFIRMING STACK DEPENDENCIES

You need to take a few additional steps to verify that your environment is ready for stacking. First, you want to make sure that you have command-line access to OpenStack components, including the `heat` application mentioned in table 12.1. A full command reference for the `heat` application can be found in the official OpenStack documentation: http://docs.openstack.org/cli-reference/content/heatclient_commands.html. Run the `heat` command shown in the following listing to confirm basic Heat operation.

Listing 12.5 List Heat stacks

```
$ heat stack-list
+----+------------+--------------+---------------+
| id | stack_name | stack_status | creation_time |
+----+------------+--------------+---------------+
+----+------------+--------------+---------------+
```

As expected, no stacks were listed in this example, but this does confirm that appropriate variables have been set and Heat has been installed.

IS IT HEAT OR THE ENVIRONMENT? If you experienced an error during the previous step, look closely at the error. Does the error look like a Heat-specific error or more like a lack of credentials? To confirm that your variables have been set, try to access a known working service like Nova: `nova list`. If you can access Nova, then your problem is likely with Heat; if you can't access Nova, your problem likely has to do with your environmental variables.

Now that you've confirmed access to OpenStack components, you need to see what images are available in the existing system. You might recall from earlier chapters that the Glance service is responsible for images. Using Glance, list all images as follows.

Listing 12.6 List Glance images

```
$ glance image-list
+--------------------------------------+-------------+..+
| ID      | Name                       | Disk Format |..
+--------------------------------------+-------------+..+
| b5...d9 | Fedora-x86_64-20-20140618-sda | qcow2    |..
+--------------------------------------+-------------+..+
```

If you enabled Heat in a DevStack environment, you'll notice that a new image was added in the DevStack stacking process. The new Fedora image will be used in this Heat stacking example. If you aren't using DevStack and don't have a Fedora image listed, please add one as demonstrated in chapter 5, under the heading "Image management" in section 5.2.2.

IMAGES THAT WORK WITH HEAT Any image could be specified in a Heat template, but some templates make use of Heat CloudFormation tools that must be pre-installed on the image. Fedora F20 images (http://cloud.fedoraproject.org/fedora-20.x86_64.qcow2) contain `heat-cfntools` and are a common choice for Heat images.

The final step in your preparation is creating an SSH key pair that can be injected into the host during the stacking process. As with the image, any key pair can be specified in a template. For this demonstration, you'll create a new key pair named `heat_key` to be used for Heat instances, as shown in the next listing. Make sure you save the `heat_key` in case you need to access your instances directly.

Listing 12.7 Generate Heat SSH key pair

```
$ nova keypair-add heat_key > heat_key.priv
$ chmod 600 heat_key.priv
```

You're now ready to create a Heat stack using the image, flavor, and key pair for your environment. It's time to complete the stack process.

LAUNCHING A HEAT STACK

At this point you have everything you need, with the exception of the template. The good news is that there are many existing templates in the OpenStack Git repository (https://github.com/openstack/heat-templates), and many more can be found on the AWS CloudFormation Templates site (http://aws.amazon.com/cloudformation/aws-cloudformation-templates/). The last step in the process is to select a template and define template parameters. You might recall from earlier sections in this chapter that template parameters are simply key/value pairs that are used to describe specific attributes of a specific stack.

For this we'll use a Heat template that's used to deploy WordPress, an open source content management system.

> **What's the big deal?**
>
> You might well have experienced the pain of deploying highly dependent software packages like WordPress from the ground up. If you haven't, you need to understand the many layers of interdependent complexities in these systems to know just how amazing application orchestration is.
>
> Before package management systems like apt and yum were common, open source tools had to be compiled from source code. Often the source code of one package had many dependencies from other packages. Not only did you have to compile many separate packages from source, but specific libraries often had to be explicitly referenced as part of the linking (compilation) process. Anyone who has compiled the Apache web server from scratch with support for PHP, MySQL, LDAP, SSL, and so on, will know this pain. Binary packages aren't immune to dependency problems either, which gave rise to the colloquial term "dependency hell."
>
> It's amazing that tools like Heat can take a template and a few parameters and provide in minutes what it might take a person weeks to do manually.

The following listing deploys a WordPress stack with the name mystack. Modify any parameters in the example that might differ on your system.

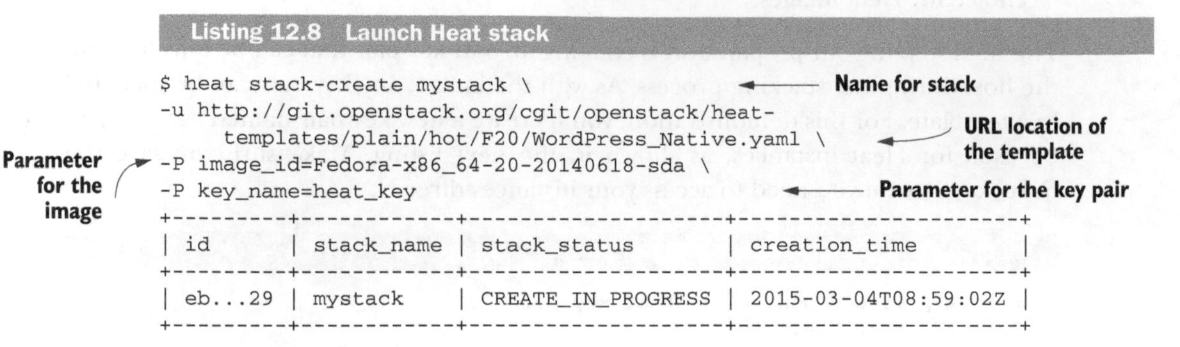

Listing 12.8 Launch Heat stack

```
$ heat stack-create mystack \
-u http://git.openstack.org/cgit/openstack/heat-
     templates/plain/hot/F20/WordPress_Native.yaml \
-P image_id=Fedora-x86_64-20-20140618-sda \
-P key_name=heat_key
+---------+------------+--------------------+----------------------+
| id      | stack_name | stack_status       | creation_time        |
+---------+------------+--------------------+----------------------+
| eb...29 | mystack    | CREATE_IN_PROGRESS | 2015-03-04T08:59:02Z |
+---------+------------+--------------------+----------------------+
```


If the command executed successfully, you'll see the initial status of your mystack stack listed as CREATE_IN_PROGRESS.

In order to check the status of your stacks, run the command heat stack-list as shown here.

Listing 12.9 Listing Heat stack status

```
$ heat stack-list
+---------+------------+-----------------+----------------------+
| id      | stack_name | stack_status    | creation_time        |
+---------+------------+-----------------+----------------------+
| eb...29 | mystack    | CREATE_COMPLETE | 2015-03-04T08:59:02Z |
+---------+------------+-----------------+----------------------+
```

When your stack has completed, its status will change to CREATE_COMPLETE.

If you experience issues or simply want to check on events related to your stack, run the following command.

Listing 12.10 List mystack events

```
$ heat event-list mystack
+--------------------+...+----------------+--------------------+..+
| resource_name      |...| status_reason  |       status       |..
+--------------------+...+----------------+--------------------+..+
| wordpress_instance |...| state changed  | CREATE_COMPLETE    |..
| wordpress_instance |...| state changed  | CREATE_IN_PROGRESS |..
+--------------------+...+----------------+--------------------+..+
```

You can see in this example that only two events exist for this simple deployment. In an autoscaling environment, however, you might see many events related to the cooperative interaction between infrastructure and the applications in the stack.

Now that the stack has completed, you'll want to take a closer look at the details of the stack with the heat stack-show command, as follows.

Listing 12.11 Show mystack details

```
$ heat stack-show mystack
+-----------------+----------------------------------------------------+..+
| Property        | Value                                              |..
+-----------------+----------------------------------------------------+..+
| capabilities    | []
..
| outputs         | [
|                 |  "output_value": "http://10.0.0.4/wordpress",
|                 |  "description": "URL for WordPress wiki",
|                 |  "output_key": "WebsiteURL"
...
|                 | ]
| parameters      | {
|                 |   "OS::stack_id": "eb...29",
|                 |   "OS::stack_name": "mystack",
|                 |   "image_id": "Fedora-x86_64-20-20140618-sda",
|                 |   "db_password": "******",
|                 |   "instance_type": "m1.small",
|                 |   "db_name": "wordpress",
|                 |   "db_username": "******",
|                 |   "db_root_password": "******",
|                 |   "key_name": "heat_key"
...
```

As you can see in this listing, details of `mystack` are shown, including the `outputs` values. In this case, the `output_value` provides a reference to the WordPress site created by the stacking process. Of course, the output values could be anything defined by the template.

Finally, to remove all data and release all infrastructure related to your stack, you can use the `stack-delete` process shown in the following listing.

Listing 12.12 Delete mystack

```
$ heat stack-delete mystack
+---------+------------+--------------------+---------------------+
| id      | stack_name | stack_status       | creation_time       |
+---------+------------+--------------------+---------------------+
| eb...29 | mystack    | DELETE_IN_PROGRESS | 2015-03-04T08:59:02Z |
+---------+------------+--------------------+---------------------+
```

In this section, you've learned about OpenStack Heat and completed the `mystack` example in your environment.

We've now looked at the OpenStack Heat orchestration project, which is developed and maintained as part of the OpenStack framework. However, there are other automation tools that can make use of OpenStack resources, but aren't official OpenStack projects. One such OpenStack-related project is Ubuntu Juju. In the next section, you'll learn how Juju is used with OpenStack in application orchestration.

12.2 Ubuntu Juju

The Ubuntu Juju project is all about orchestration, such as the system-level deployment of OpenStack or application-level deployment of WordPress. Juju can be used for bare-metal deployments, but that's beyond the scope of this book. Here we'll focus on application-level deployments using OpenStack resources.

You can think of Juju as an agent-based orchestration system. You'll configure a Juju client on your personal computer to make use of an OpenStack instance for infrastructure resources. Juju deploys a bootstrap agent to a tenant in your OpenStack deployment, which it then uses to deploy additional applications and dependencies via its own orchestration engine.

> **OpenStack Heat vs. Ubuntu Juju**
>
> Both OpenStack Heat and Ubuntu Juju have their own strengths and weaknesses depending on your use case. Heat is more closely integrated with OpenStack, but Juju allows you to use other cloud framework resources (such as Amazon, HP, and OpenStack). In terms of the resulting output (namely, the automated deployment of applications on cloud resources), the two tools accomplish the same tasks. It's up to you to determine the appropriate tool for the job.

Before you get started, you'll need to make sure you have the rights to create instances using Nova and to store objects using OpenStack Swift.

12.2.1 Preparing OpenStack for Juju

It's possible to use Juju without setting any OpenStack variables in your shell by simply providing information about your OpenStack environment in your Juju configuration. But this process can be tedious, so I'll try to reduce the number of manual configurations you must make. Luckily, the OpenStack Dashboard can be used to generate a script that sets appropriate shell variables.

> **Where does Juju live?**
>
> Unlike almost everything else in this book (excluding Ceph and Fuel), the Juju installation doesn't have to take place on an OpenStack node. In fact, Juju installers exist for Linux, Mac OS X, and even Windows. The operations performed in section 12.2.1 will prepare your OpenStack environment for use with Juju. There are no server-side Juju components to install on OpenStack nodes. This distinction was outlined in chapter 4, which explained the differences between official core OpenStack projects and related projects. Juju falls into the *related* category and even supports cloud frameworks other than OpenStack.

Log in to your OpenStack Dashboard with the userid you wish to use with Juju, and select the tenant you want to contain Juju resources. Under the Projects drop-down,

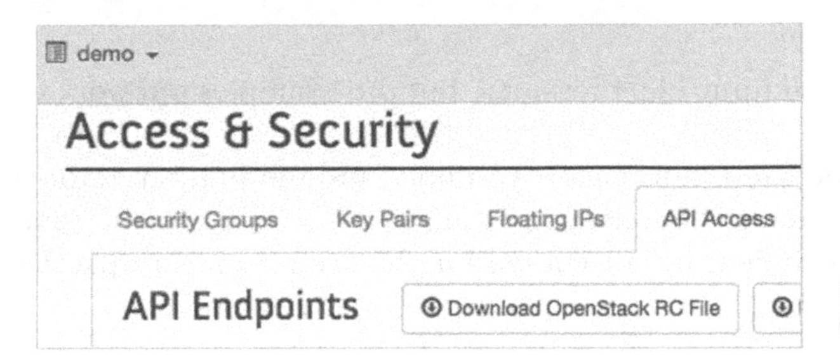

Figure 12.1 Download OpenStack RC

click on Access & Security and navigate to the API Access tab. From this tab, click on Download OpenStack RC File as shown in figure 12.1.

This generates a script that sets identity variables based on your selected tenant and current user. You'll be prompted to download the script, which will use the naming convention *[tenant name]-openrc.sh.*

Copy the file to your OpenStack environment and then process the script as follows.

Listing 12.13 Running the openrc.sh script

```
$ source demo-openrc.sh
Please enter your OpenStack Password:
```

It's a good idea at this point to access your OpenStack service to make sure the variables were set correctly. For example, make sure the commands like `glance image-list` execute without errors.

The next step is to make sure you have an image that can be used with Juju. Determining what images are supported will be covered shortly; for now it's sufficient to know that Ubuntu 12.04 is a popular image used with Juju. If you have an Ubuntu 12.04 image in your environment, make note of the image ID using the `glance list-image` command. If you don't have an Ubuntu 12.04 image, register one as shown in the following listing, and make a note of the image ID, which will be used in the next section.

Listing 12.14 Register an image for Juju

```
$ glance image-create --name="Ubuntu 12.04" \
--is-public=true --disk-format=qcow2 \
--container-format=bare \
--location http://cloud-images.ubuntu.com/precise/current/precise-server-
     cloudimg-amd64-disk1.img
+------------------+--------------------------------------+
| Property         | Value                                |
+------------------+--------------------------------------+
...
| id               | ce7616a6-b383-4704-be3a-00b46c2de81d |   <-- Make note of image ID
...
| name             | Ubuntu 12.04                         |
...
+------------------+--------------------------------------+
```

You're now ready to install Juju.

12.2.2 Installing Juju

You'll install Juju on an Ubuntu 14.04 instance, but the examples will work on other Juju-supported platforms.

In order to install the latest Juju release, you must add the Juju repository to your package management system. The following listing shows how to add the Juju repository, update the package index, and install the Juju binaries from your package management system.

Listing 12.15 Install Juju binaries

```
$ sudo add-apt-repository ppa:juju/stable
$ sudo apt-get update
$ sudo apt-get install juju-core
```

If the previous commands complete without errors, you'll now have the Juju binaries installed on your system.

> **WHICH REPO SHOULD BE USED?** The official Juju documentation suggests that you install Juju from the Juju-specific repository, ppa:juju/stable, and fall back to the Ubuntu universe repository if you experience problems. The command `sudo apt-get install juju` will install Juju version 1.20.11-0ubuntu0.14.04.1 if you're using Ubuntu version 14.04.1. Over the course of writing this chapter, I found that the Juju packages that are part of the normal Ubuntu universe repository worked better with OpenStack.

Next, generate a Juju configuration file as shown in the following listing. The file will be generated in the ~/.juju directory.

Listing 12.16 Generate a Juju configuration file

```
$ juju init
A boilerplate environment configuration file has been written
to /home/sysop/.juju/environments.yaml.
Edit the file to configure your juju environment and run bootstrap.
```

Using your preferred text editor, take a look at the environments.yaml file you just generated. You'll see example configurations for several framework types including `ec2`, `openstack`, `manual`, `maas`, `joyent`, and `azure`.

As shown in the next listing, modify your Juju configuration file (environments.yaml) by adding the new `myopenstack` environment and setting it as the default environment.

Listing 12.17 Modify your Juju configuration

```
default: myopenstack                  <-- Set your configuration to default

environments:
    myopenstack:
        type: openstack
        use-floating-ip: false        <-- If you're using floating-ips, set to true
```

```
use-default-secgroup: true
network: "private"                  ← Set to your desired tenant network
region: "RegionOne"                 ← Set to your OpenStack region
auth-mode: userpass
default-series: precise
```

By modifying the environments.yaml file, you've provided Juju with basic information about your OpenStack deployment. But Juju requires additional deployment-specific information to function.

Juju *charms* were introduced back in chapter 4—they're collections of installation scripts that define how services and applications integrate into virtual infrastructure. Charms are used by Juju in the same way HOT and CNF templates are used by OpenStack Heat. Charms reference image types as requirements, not specific images. For example, a charm might require an Ubuntu 12.04 operating system, but it wouldn't require a specific instance of that image. So how does the Juju system know where to find an Ubuntu 12.04 (or any other) image? You must define your existing OpenStack Glance images for Juju to make use of them.

In the following listing, the image metadata to be used by Juju is generated for an existing Ubuntu 12.04 image.

Listing 12.18 Generate image metadata

```
$ juju metadata generate-image \
 -i 0d6d8f6d-870c-4c58-aa96-ccc0e65df206 \      ← Image ID (from Glance) of the Ubuntu 12.04 image
 -r RegionOne \
 -u http://192.168.1.178:5000/v2.0/ \           ← Keystone auth URL
 -d /home/sysop \                               ← Root path for metadata
 -s precise \                                   ← Version of the OS distribution
 -a amd64                                       ← Architecture of the OS distribution
image metadata files have been written to:
/home/sysop/images/streams/v1.
For Juju to use this metadata, the files need to be put into the
image metadata search path. There are 2 options:

1. Use the --metadata-source parameter when bootstrapping:
   juju bootstrap --metadata-source /home/cody

2. Use image-metadata-url in $JUJU_HOME/environments.yaml
Configure a http server to serve the contents of
/home/sysop
and set the value of image-metadata-url accordingly.
```

The image metadata generated in the previous listing will be stored in the [root path]/images directory. This metadata will be used by Juju to translate charm requirements to resource requests on your OpenStack deployment.

Juju has now been configured with your system and image information, but once an appropriate image has been launched for a specific charm, additional application installation procedures must take place on that instance. Juju uses an agent-based model, which places Juju tools (agents) on images, and then uses the agents to complete the charm-deployment process. It's possible to use various Juju tool versions on the same OpenStack deployment across tenants, just as various OS images are used.

To allow tool flexibility across the deployment, you must also generate tool metadata, just as you did with the image metadata. The following listing shows how you can generate tool metadata.

Listing 12.19 Generate tool metadata

```
$ juju metadata generate-tools -d /home/sysop
Finding tools in /home/sysop
```

The tool metadata generated here will be stored in the [root path]/tools directory. This metadata contains information related to the version of the tool used to generate the metadata, which information will be used by Juju to determine what tools it should use for a specific deployment.

Now that you've configured Juju with system, image, and tool information, you're ready to bootstrap your tenant. Bootstrapping is the process used by Juju to place a control (bootstrap) instance in a tenant. The bootstrap instance communicates with the Juju client and your OpenStack deployment to coordinate application orchestration.

The bootstrap allows application nodes to remain isolated on internal OpenStack networks until they're explicitly exposed. For instance, if you deploy a load-balanced WordPress environment (more than one web server), only the load balancers need to be exposed to the outside world, and the web and database servers can remain isolated. Of course, you could do this manually, but often the access required to manually manage these nodes makes isolation impractical.

The following listing shows the OpenStack bootstrap process.

Listing 12.20 Bootstrapping an OpenStack tenant

```
$ juju bootstrap \
 --metadata-source /home/sysop \      <-- Root path for image and tool directories

 --upload-tools -v                    <-- Uploads tools to deployment

WARNING ignoring environments.yaml:
using bootstrap config in file
 "/home/sysop/.juju/environments/myopenstack.jenv"
Bootstrapping environment "myopenstack"
Starting new instance for initial state server
Launching instance
 - 25e3207b-05e5-428c-a390-4fb7d6849d6d
Installing Juju agent on bootstrap instance
Waiting for address
Attempting to connect to 10.0.0.2:22
Logging to /var/log/cloud-init-output.log on remote host
Installing add-apt-repository
Adding apt repository: ...
Running apt-get update
Running apt-get upgrade
```

```
Installing package: git
Installing package: curl
Installing package: cpu-checker
Installing package: bridge-utils
Installing package: rsyslog-gnutls
Fetching tools: ...
Bootstrapping Juju machine agent
Starting Juju machine agent (jujud-machine-0)
```

Once the bootstrap successfully completes, you'll be ready to use Juju in your tenant. To check on the state of your Juju environment at any time, run the following command.

Listing 12.21 Check Juju status

```
$ juju status
environment: myopenstack
machines:
  "0":
    agent-state: started
    agent-version: 1.20.11.1
    dns-name: 10.0.0.2
    instance-id: 6a61c0db-9a32-4bba-b74d-44e09692210c
    instance-state: ACTIVE
    series: precise
    hardware: arch=amd64 cpu-cores=1 mem=2048M root-disk=20480M
    state-server-member-status: has-vote
services: {}
```

Now it's time to deploy WordPress on OpenStack using a Juju charm.

12.2.3 *Deploying the charms CLI*

Using the following listing as a reference, deploy WordPress using Juju.

Listing 12.22 Deploy WordPress using Juju

```
$ juju deploy wordpress
Added charm "cs:precise/wordpress-27" to the environment.
```

When you issue the deployment command, your request will be sent to your bootstrap instance. You can check your deployment status using the following command.

Listing 12.23 Check WordPress deployment

```
$ juju status
environment: myopenstack
machines:
  "0":
  ..
  "1":                                  <-- New instance for WordPress
    agent-state: pending
    dns-name: 10.0.0.3
    instance-id: ed50e7af-0426-4574-a0e0-587abc5c03ce    <-- OpenStack instance ID
```

```
      instance-state: ACTIVE
      series: precise
      hardware: arch=amd64 cpu-cores=1 mem=2048M root-disk=20480M
services:
  wordpress:                                   # New WordPress service
    charm: cs:precise/wordpress-27
    exposed: false                             # Service is not exposed
    relations:
      loadbalancer:
      - wordpress
    units:
      wordpress/0:
        agent-state: pending                   # Indication of service state
        machine: "1"
        public-address: 10.0.0.3
```

From the OpenStack perspective, you can watch the bootstrap node deploying instances by using the following Nova command.

Listing 12.24 Nova list instances

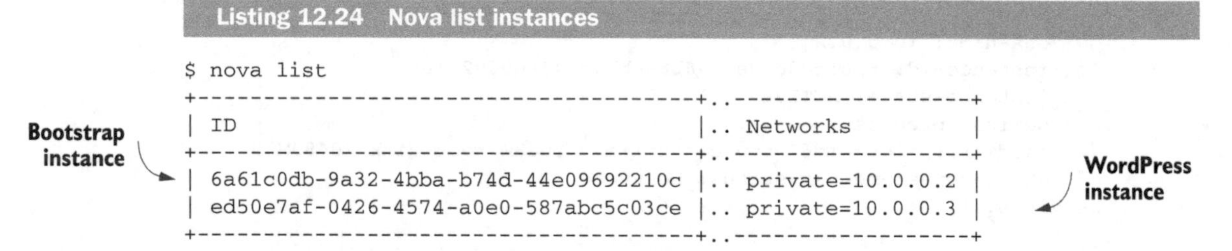

```
$ nova list
+--------------------------------------+..------------------+
| ID                                   |.. Networks         |
+--------------------------------------+..------------------+
| 6a61c0db-9a32-4bba-b74d-44e09692210c |.. private=10.0.0.2 |
| ed50e7af-0426-4574-a0e0-587abc5c03ce |.. private=10.0.0.3 |
+--------------------------------------+..------------------+
```


Your WordPress service is now in the process of being provisioned, but the WordPress charm has a dependency on MySQL. Just as you deployed the WordPress service, you must now deploy MySQL and relate it to the WordPress instance.

Listing 12.25 Deploy remaining dependencies

```
$ juju deploy mysql
...
$ juju add-relation mysql wordpress
...
```

> **JUJU DEPENDENCIES** Juju doesn't currently provide automated dependency resolution. You'll have to read the charm documentation to determine the dependencies. Documentation for charms can be found here: https://jujucharms.com/.

Once all the resources have been provisioned from OpenStack, agents on each instance will install WordPress and the dependent components. Components on each node will be configured as specified by the Juju WordPress charm. For instance, web servers will be configured to consume resources from database servers, and load balancers will be configured to balance traffic on the web servers.

When your WordPress service's `agent-state` changes from `pending` to `started`, as indicated by the `status` command in listing 12.23, your service is ready. But before you can access your service, you must first expose it.

Listing 12.26 Expose the WordPress service

```
$ juju expose wordpress
```

Your WordPress deployment will now be accessible on the public address indicated by the `status` command in listing 12.23.

You've now completed a Juju deployment of WordPress using charms. If you want to gain console access to instances deployed using Juju, determine the machine ID with the `status` command in listing 12.23, and then use the following command. This example demonstrates how to gain console (SSH) access to machine 1.

Listing 12.27 SSH to Juju instance

```
$ juju ssh 1          <-- Machine ID
```

Next, you'll deploy the Juju GUI, which can be used to graphically deploy Juju charms in your environment.

12.2.4 Deploying the Juju GUI

Once you've bootstrapped your OpenStack tenant, you can deploy the Juju GUI. We'll first walk through the deployment of the GUI, and then through the deployment of WordPress using the GUI.

Once again, check the status of your Juju environment as follows.

Listing 12.28 Check Juju environment status

```
$ juju status
environment: myopenstack
machines:
  "0":
    agent-state: started
    agent-version: 1.20.11.1
    dns-name: 10.33.4.54
    instance-id: de0fbd71-a223-4be4-862b-8f1cb6472640
    instance-state: ACTIVE
    series: precise
    hardware: arch=amd64 cpu-cores=1 mem=1024M root-disk=25600M
    state-server-member-status: has-vote
services: {}
```

You can see in the preceding listing that the only node in the `myopenstack` Juju environment is the bootstrap node, machine 0.

As shown in the following listing, deploy and expose the Juju GUI charm to your environment.

Listing 12.29 Deploying and exposing the Juju GUI

```
$ juju deploy juju-gui
Added charm "cs:precise/juju-gui-109" to the environment.
$ juju expose juju-gui
```

Now that the Juju GUI has been deployed and exposed, check the status of the deployment.

Listing 12.30 Check Juju environment status

```
$ juju status
environment: myopenstack
...
services:
  juju-gui:
    charm: cs:precise/juju-gui-109
    exposed: true                  <-- Service is exposed
    units:
      juju-gui/0:
        agent-state: started
        agent-version: 1.20.11.1
        machine: "1"                <-- Running on machine 1
        open-ports:
        - 80/tcp                   <-- Using ports 80/443
        - 443/tcp
        public-address: 10.33.4.53     <-- Using IP 10.33.4.53
```

In the previous example, you can see that the service agent has started (`agent-state: started`), that the service is now exposed (`exposed: true`), and that the public address of the service is 10.33.4.53 (`public-address: 10.33.4.53`).

Using a web browser, try to access the Juju GUI using the public address listed for your environment. Once you access the Juju GUI site, you should be presented with a login screen, as shown in figure 12.2.

Figure 12.2 Juju GUI login

During the bootstrap process, an admin secret was generated for your environment, and this is the password for your Juju GUI login. You can retrieve your admin secret using the following command.

Listing 12.31 Retrieve admin secret

```
$ more ~/.juju/environments/myopenstack.jenv | grep admin-secret
  admin-secret: bc03a7948a117561eb5111437888b5f9
```

Using the `admin` username and your admin secret as the password (`bc03a7948a117561eb5111437888b5f9` in the example), log in to the Juju GUI.

Once you log in, you'll be presented with a home screen that provides a graphical interface for your bootstrapped environment. Using the search bar at the top left of the home screen, search for a WordPress charm, as shown in figure 12.3.

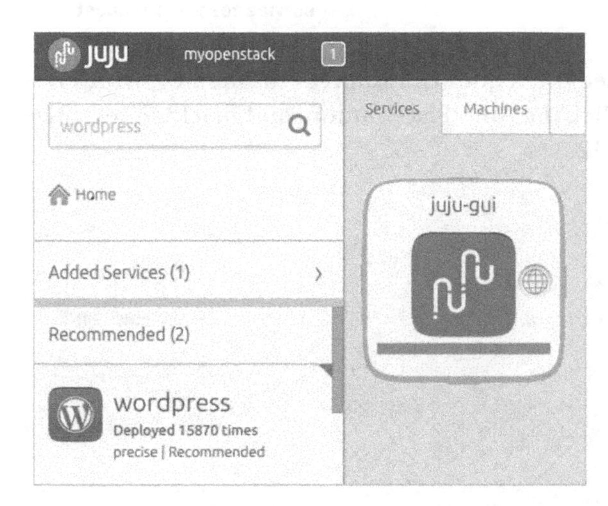

Figure 12.3 Juju GUI homepage

When you click on the name of a charm, you'll be taken to the charm panel. Click on a WordPress charm, which will take you to a charm panel like the one in figure 12.4.

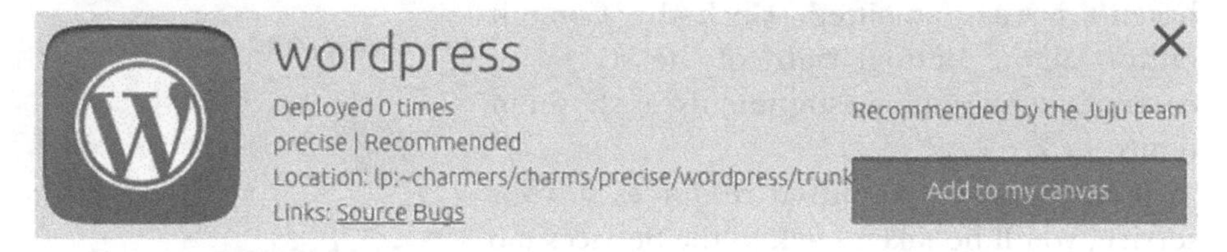

Figure 12.4 Adding WordPress charm to canvas

Once in the WordPress charm panel, click Add to My Canvas. The WordPress charm will be added as a service to your canvas, but no machine resources will be assigned. Click the Machines tab on the home screen to see the machine resource assignments, as shown in figure 12.5.

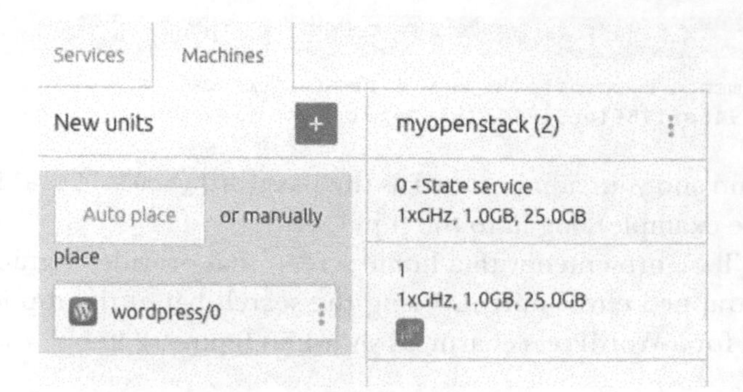

Figure 12.5 View unassigned service resource request

In the Machines tab, you'll see the requested resources for the new WordPress service in the New Units column. Click the Auto Place button, and machine resources will be assigned, as shown in figure 12.6.

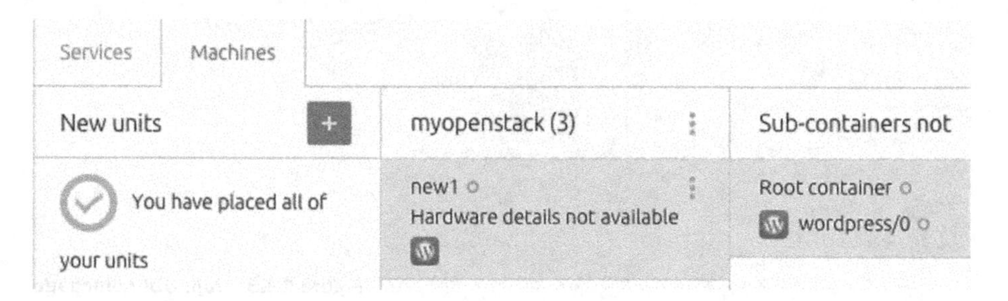

Figure 12.6 View assigned service resource assignments

You have now assigned machine resources to the new WordPress service, but these resources haven't been committed. Click the Commit button at the bottom right of the screen to commit your resource assignments, as shown in figure 12.7.

Once you've committed resources to your service, you'll be taken back to the Services tab on the home screen.

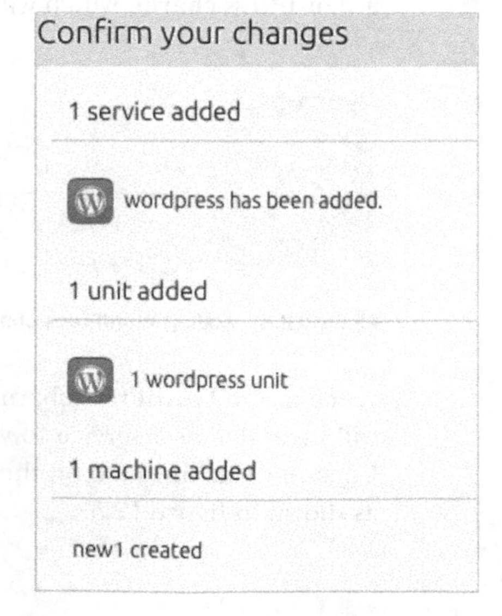

Figure 12.7 Confirm service provisioning

At this point Juju will be busy building your WordPress service, but not all external dependencies and relationships will have been built. The WordPress service is dependent on a MySQL service.

Repeat the process you followed to install the WordPress service from the wordpress charm to install the MySQL service from the mysql charm. Once both WordPress and MySQL services are running (once they're green), build a relationship between WordPress and MySQL on the canvas screen, and commit the change. Once the services have been started and the relationships built, your canvas should look like figure 12.8.

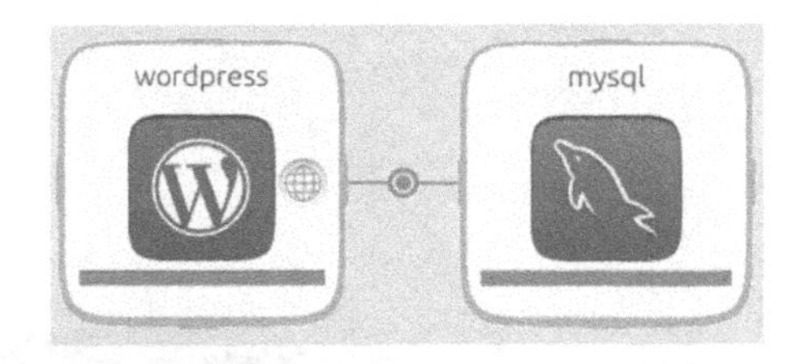

Figure 12.8 Service relationship between WordPress and MySQL

You're now ready to use your services, but not only are the services not exposed, you also don't know what addresses are being used for your services. Click on the WordPress node on your canvas, which will bring up the WordPress service panel. From the WordPress panel, you can expose your service by toggling the Expose slider and confirming the expose action.

Once the service is exposed, click on the running instance of the WordPress service (wordpress/0), which will load the service inspector panel, shown in figure 12.9. In the service inspector panel, you can see the public address and ports used by the service.

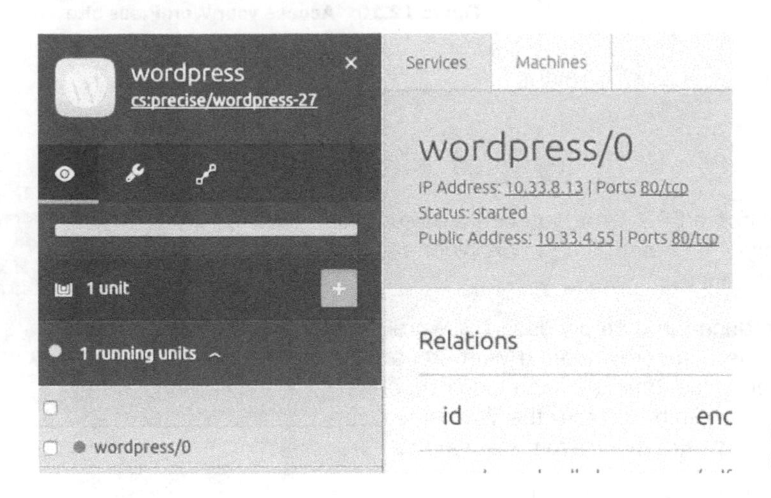

Figure 12.9 View the service inspector panel

Using a web browser, attempt to access your new WordPress site using your public address, as shown in figure 12.10.

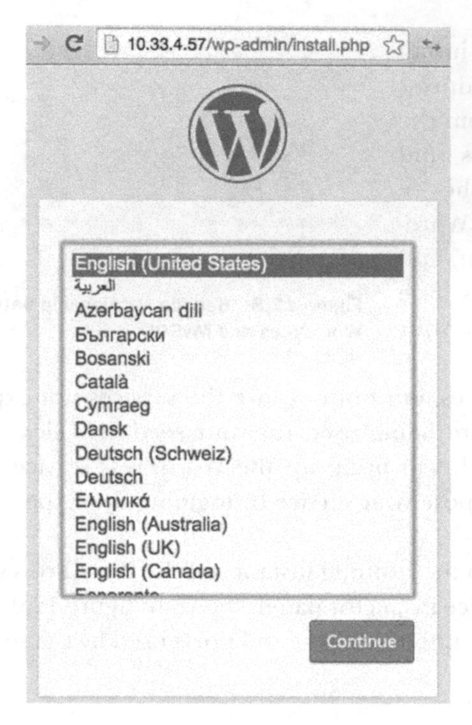

Figure 12.10 Access your WordPress site

> **503 Error?**
>
> If you experience a 503 error, you could be running out of memory—the default charm provisions a very small VM.
>
> One way to reduce the memory usage of WordPress is to reduce the number of loaded WordPress plugins by changing the tuning of the service, before resources are committed. From the WordPress service panel, change the tuning from *standard* to *bare*.
>
> Alternatively you could modify the deployment to make use of larger instances.

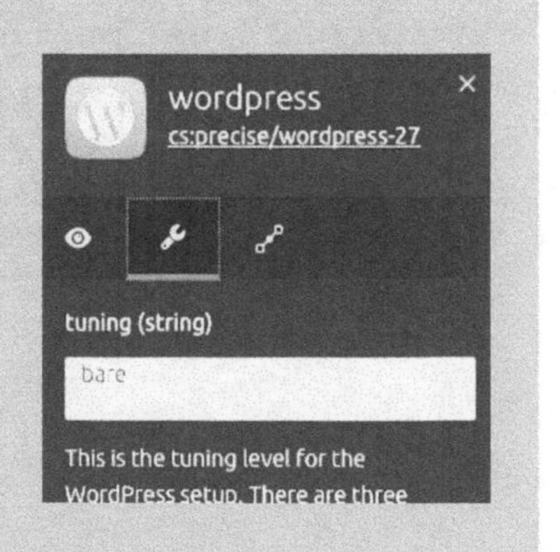

You have now used the Juju GUI to deploy WordPress and its related dependencies. You can repeat this process using charms to deploy many applications using both public and private clouds.

12.3 Summary

- The OpenStack Heat project can be used to automatically deploy applications on OpenStack clusters.
- Heat is an official OpenStack project.
- Heat can make use of both the Amazon Web Services (AWS) CloudFormation template format and its own HOT format.
- Ubuntu Juju can be used to automatically deploy applications on Amazon- and OpenStack-based public and private clouds.
- Juju is a OpenStack-related project.
- Juju is an agent-based orchestration tool.

appendix *Installing Linux*

This appendix covers the basic install of Ubuntu Linux version 14.04 LTS on a single physical server. Even if you've worked with Linux in the past, you might want to review this tutorial, if only to understand the underlying system configuration used for the examples in the book.

> **UBUNTU LINUX VERSION 14.04 LTS** In connection with version 14.04, the acronym *LTS* means Long Term Support. There will be support for the 14.04 branch of Ubuntu until at least April 2019.

Although Linux is used as an underlying operating system throughout this book, this is not a book on Linux. This appendix is included to provide simple install instructions for those without Ubuntu Linux experience. These instructions will walk you through every step in the installation process.

The install process is trivial if you know how to answer the questions asked by the installer. This appendix provides common answers for each configuration step in the process. If at any time you get turned around in the process, simply start over. An entire install can be completed in 15–20 minutes once you know what you're doing. If you run into hardware-specific problems, or you want to learn more about the process, take a look at the Ubuntu community page: https://help.ubuntu.com/community/Installation.

A.1 Getting started

You'll need a few items to get started. First, you'll need some physical hardware. This could be a full-blown server or simply an old desktop or laptop.

For the Linux installation demonstrated in this appendix, I use a server with four wired Ethernet network cards. If you have a wired Network Interface Card (NIC) in your hardware, the network device names and quantity of adapters on your server might or might not differ from the examples. Where appropriate, make modifications to match your environment.

Finally, you'll need to download the latest stable copy of Ubuntu 14.04 LTS ISO, which can be found here: http://releases.ubuntu.com/14.04/. The instructions will be based on using the command line, so the server-specific install ISO works just fine. If you'd prefer to use the desktop version, it will work as well, but it won't be needed for any examples in this book.

Make sure you download the proper ISO for your hardware architecture. In general, choose the x86 version for older (32-bit) hardware (5+ years) and the x86-64 version for newer (32- or 64-bit) hardware. When in doubt, look up your specific CPU to see if it's a 32- or 64-bit architecture.

This appendix covers these primary installation steps:

- *Initial configuration*—Setting language and location
- *Network configuration*—Connecting the hardware to the network
- *User configuration*—Creating new users for the operating system
- *Disks and partitions*—Building a disk configuration for the operating system
- *Base system configuration*—Software installation and initial service configuration

Let's get started.

A.2 Initial configuration

In this section, you'll provide information about language and location.

As shown in figure A.1, you pick your language and press Enter. This book is in English, but you're free to pick any language. Note that this is the language for your Ubuntu install; additional software, like OpenStack, will have its own language settings.

Language			
Amharic	Gaeilge	Malayalam	Thai
Arabic	Galego	Marathi	Tagalog
Asturianu	Gujarati	Nepali	Türkçe
Беларуская	עברית	Nederlands	Uyghur
Български	Hindi	Norsk bokmål	Українська
Bengali	Hrvatski	Norsk nynorsk	Tiếng Việt
Bosanski	Magyar	Punjabi (Gurmukhi)	中文(简体)
Català	Bahasa Indonesia	Polski	中文(繁體)
Čeština	Íslenska	Português do Brasil	
Dansk	Italiano	Português	
Deutsch	日本語	Română	
Dzongkha	ქართული	Русский	
Ελληνικά	Қазақ	Sámegillii	
English	Khmer	සිංහල	

Figure A.1 Initial screen, selecting language

As prompted in figure A.2, you can test the install disk or the memory of your hardware, if you've had problems in the past or you're the cautious type. Otherwise, continue with the installation by selecting Install Ubuntu Server with your keyboard arrows and pressing Enter.

Figure A.2 Select install Ubuntu server

After selecting Install Ubuntu Server, you'll once again be asked to pick your language, as shown in figure A.3. Select your language and move forward.

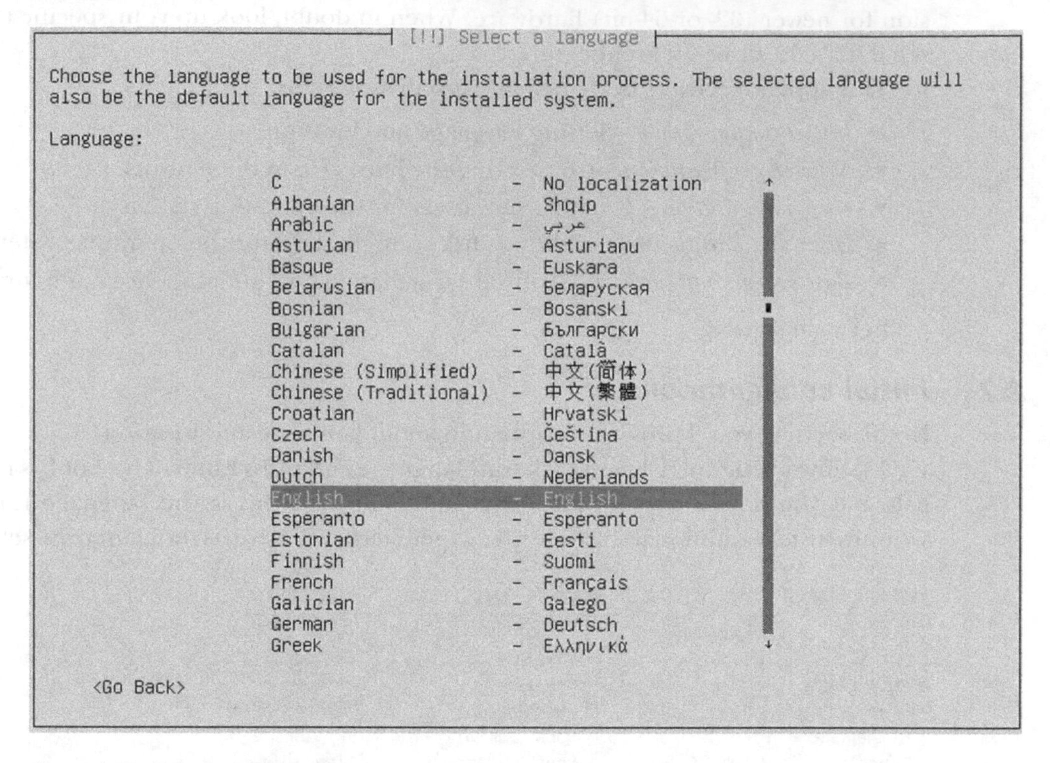

Figure A.3 Select a language

Next, select your location, as shown in figure A.4. This sets the time zone along with other locality settings.

If you want to enable keyboard detection, select Yes in the next screen (figure A.5). I've never bothered with layout detection and simply answer additional questions.

After you answer the questions about language and location, additional install packages will be loaded from the ISO, as shown in figure A.6. These components will be used in the next install steps. No user interaction is required during this step.

In the next section, you'll perform basic network configuration.

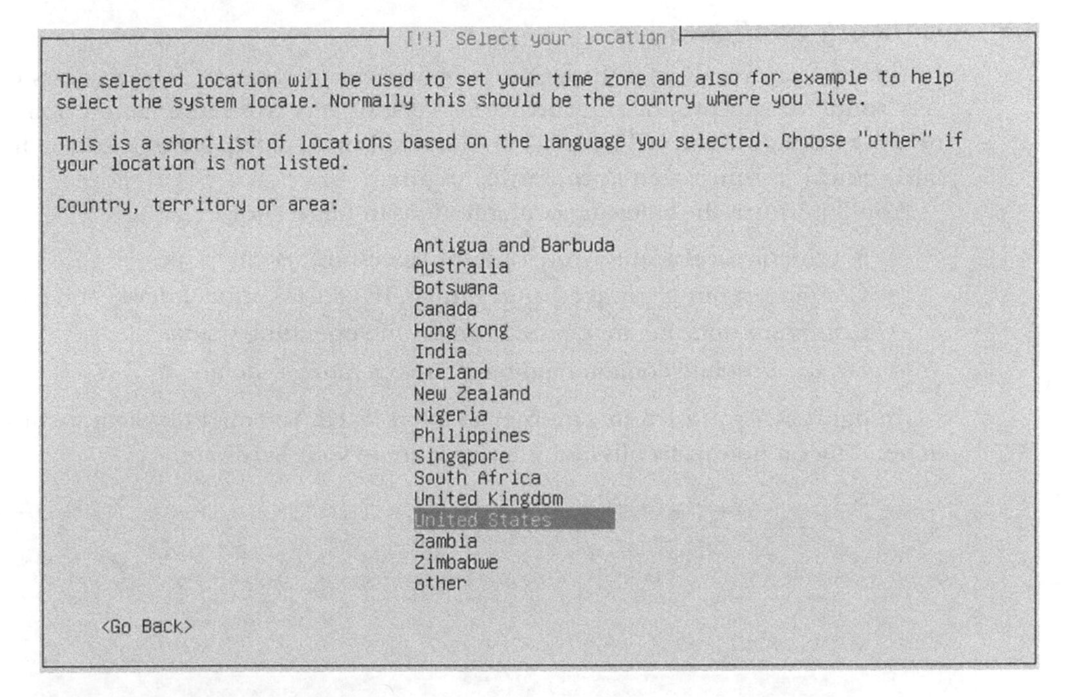

Figure A.4 Select your location

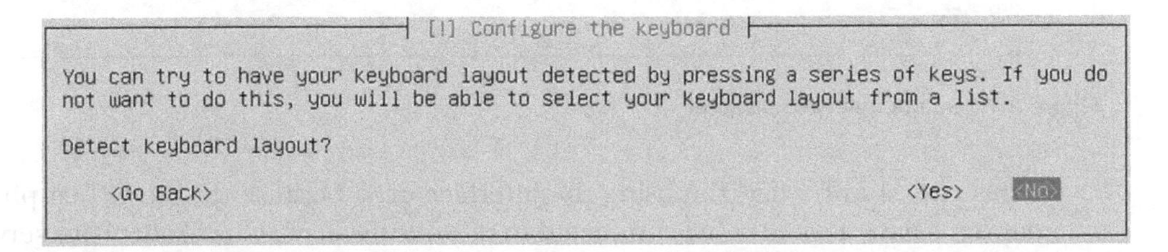

Figure A.5 Configure the keyboard

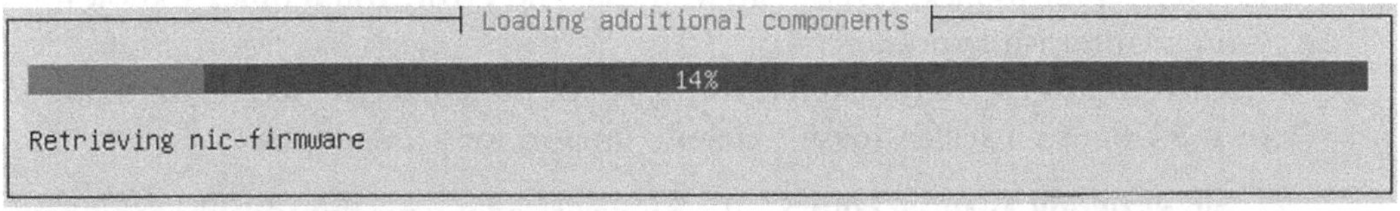

Loading additional components

14%

Retrieving nic-firmware

Figure A.6 Loading additional components

A.3 Network configuration

In this section, you'll walk through the network configuration. This process can be very simple if you provide the correct answers, or very frustrating if you don't. This network configuration will be used to download up-to-date packages, so you'll know fairly quickly if things aren't properly configured.

You'll perform the following configurations in this section:

- Set the physical adapter you want the operating system to use
- Configure your physical adapter with an IP address, subnet mask, and gateway
- Configure domain name resolution for the operating system
- Set the host and domain names for the operating system

In figure A.7 you see four Ethernet adapters listed. You might have more or fewer, depending on how many physical interfaces are in your hardware.

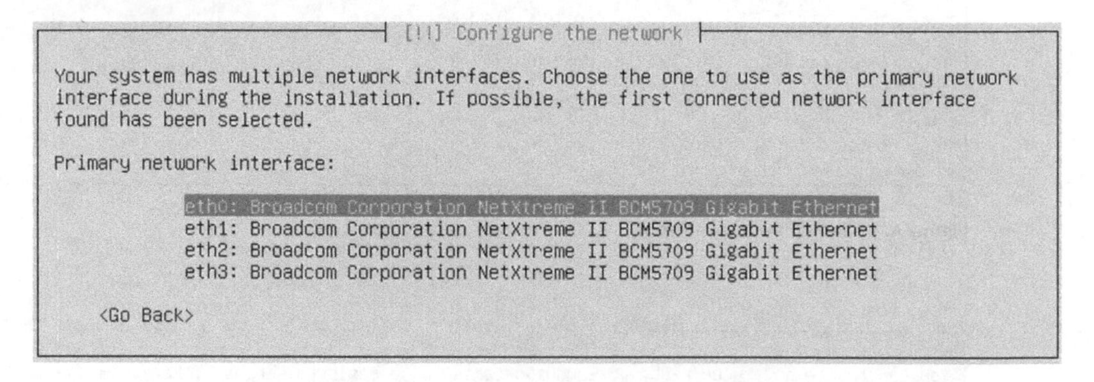

Figure A.7 Select a network adapter

In this case, I know that I'm using the interface `eth0` because this is my first physical adapter. I have a wired cable connected to this interface, which will allow this server to communicate on my network. Advanced Linux administrators have ways of identifying specific adapters, but this is outside the scope of this appendix. With any luck, you'll only have one adapter, or you'll know which adapter to choose. If you don't know which adapter to pick, you can go with trial and error. You'll quickly know if the network adapter isn't working.

If you don't see any adapters listed, it's possible that you have a hardware failure, or the Ubuntu installer doesn't include support for your adapter.

> **NO NETWORK ADAPTER FOUND** If no network adapter is found, either you have a bad adapter or your adapter is not supported. The best thing to do is consult the Ubuntu hardware and community support pages (https://wiki.ubuntu.com/HardwareSupport and https://help.ubuntu.com/community/Installation). It can often be easier to find another piece of hardware than to go ahead with unsupported hardware.

At this point the installer will send out a Dynamic Host Configuration Protocol (DHPC) request. Based on the response, or lack thereof, one of two things can happen. If the response fails, meaning that your network is not configured for DHCP, you'll have to manually configure the network. But if your network is configured for DHCP, you can skip ahead.

Take a look at figure A.8. If your screen looks like the image to the left, where you're asked to enter an IP address, you'll need to continue with the manual configuration in section A.3.1. But if your screen looks like the image on the right, where you're asked for a host name, you can skip ahead to section A.3.2.

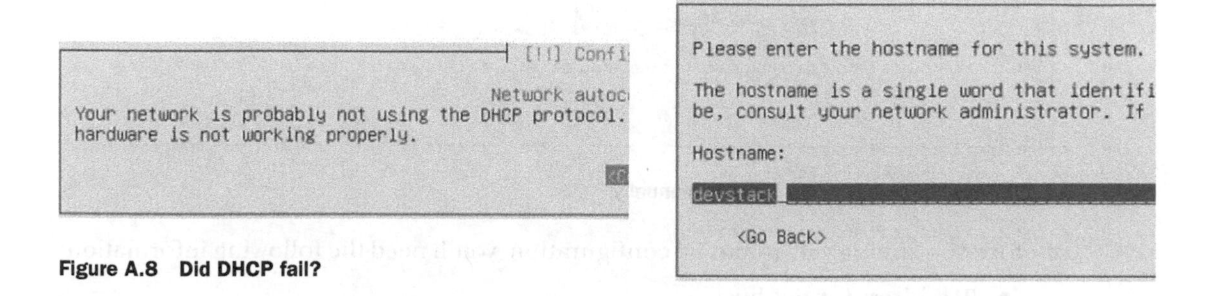

Figure A.8 Did DHCP fail?

In either event, the install process is sequential and you shouldn't have to backtrack.

A.3.1 Manually configuring the adapter

If DHCP configuration failed, your screen should look like figure A.9. This could have occurred because you selected the incorrect adapter or because DHCP isn't configured on your network.

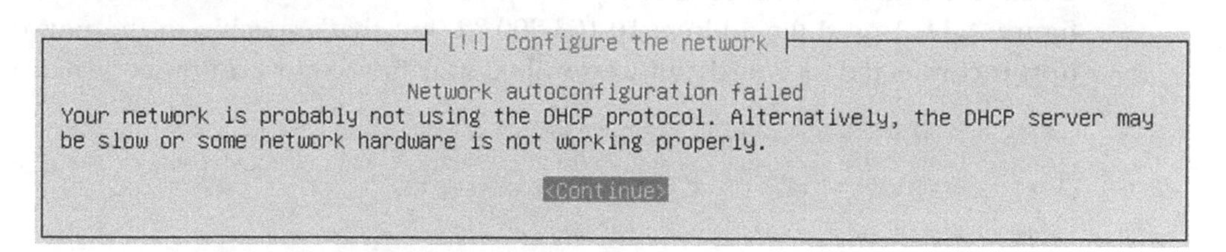

Figure A.9 Continue with a manual configuration

If you were expecting DHCP to work, you can retry the configuration with another adapter, as shown in figure A.10. On the other hand, if you've manually selected the correct adapter and you're not using DHCP, you can proceed with a manual network configuration.

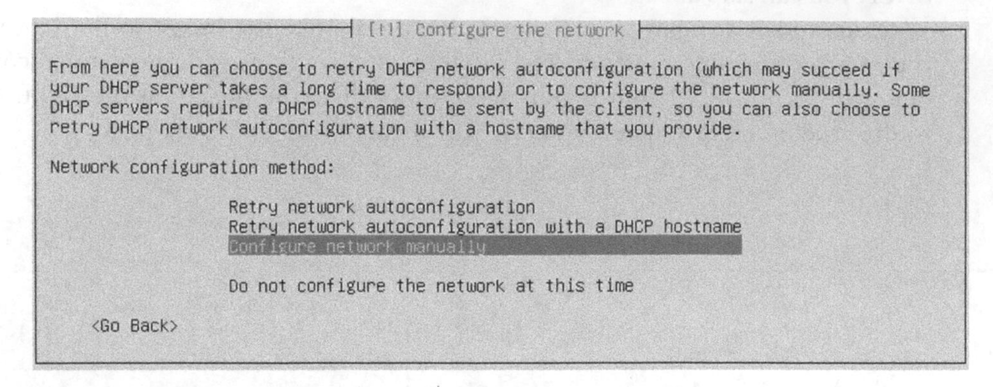

Figure A.10 Configure the network manually

Before continuing with a manual configuration, you'll need the following information:

- IP address for this host
- Subnet mask for the host address
- Router gateway address for the host network subnet
- Domain Name System (DNS) server addresses for name resolution

If you don't have this information, speak to someone familiar with your network to get these addresses.

CONFIGURING THE HOST IP ADDRESS

To start the manual configuration, you need to enter the IP address for *your* host, as in figure A.11. I used the address 10.163.200.32, but that's specific to the network this host is connected to. Your IP address will be specific to your host for your network.

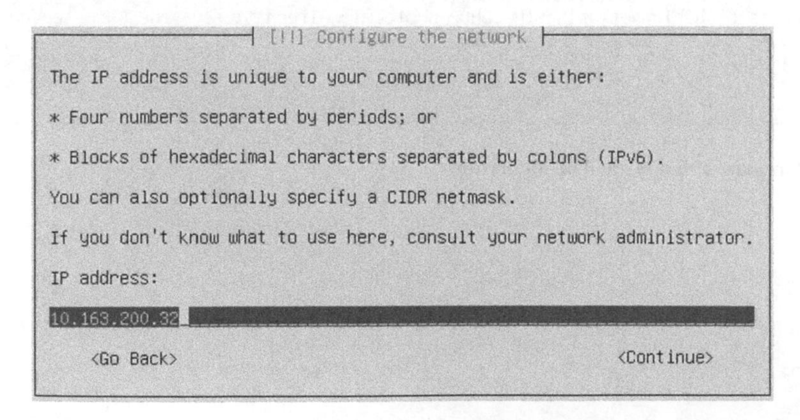

Figure A.11 Configure the IPv4 address for the interface

CONFIGURING THE SUBNET MASK

Next, you need to enter your subnet mask, as shown in figure A.12. You don't need to know what this mask does, but it's necessary in the network configuration.

```
[!!] Configure the network

The netmask is used to determine which machines are local to your network.  Consult your
network administrator if you do not know the value.  The netmask should be entered as
four numbers separated by periods.

Netmask:

255.255.255.0

<Go Back>                                                              <Continue>
```

Figure A.12 Configure the subnet mask

Put simply, the IP address is like a street address. The street address is unique, but the street might contain many houses. The subnet address is like a specification for the street, telling you where the street starts and ends. Hosts sharing the same IP address range and subnet are said to be in the same broadcast domain, so you can think of this as being on the same street.

CONFIGURING THE NETWORK GATEWAY ADDRESS

If properly configured, your host can now properly communicate with a host on its same network (broadcast domain). But you'll want to communicate with networks outside your domain, like those on the internet, even before this installation is complete. For this, you must specify a gateway address, as shown in figure A.13.

The gateway address is an address in the same broadcast domain as your host that will be used to route traffic from your host to other routing domains on other networks.

```
[!!] Configure the network

The gateway is an IP address (four numbers separated by periods) that indicates the
gateway router, also known as the default router.  All traffic that goes outside your LAN
(for instance, to the Internet) is sent through this router.  In rare circumstances, you
may have no router; in that case, you can leave this blank.  If you don't know the proper
answer to this question, consult your network administrator.

Gateway:

10.163.200.1

<Go Back>                                                              <Continue>
```

Figure A.13 Configure the gateway

CONFIGURING THE DNS SERVER

Finally, you want to be able to resolve host addresses to names. When accessing Google, you don't want to say, "Let's go to 74.125.225.148"; you want to say, "Let's go to www.google.com." In this example, *www* is the host name and *google.com* is the domain name. In order to perform this name resolution, you need to provide the address of a DNS server.

Unlike the network address, DNS addresses are not necessarily specific to a location or a network. In figure A.14, Google public DNS servers are shown. If in doubt about the DNS address, speak with your network administrator.

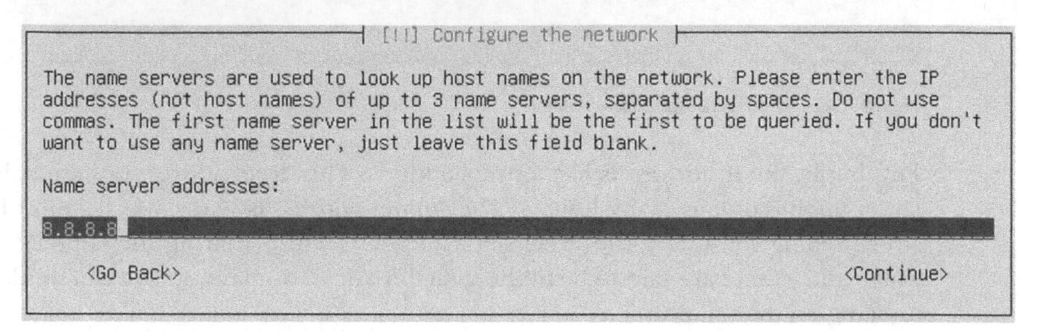

Figure A.14 Configure the DNS server

At this point, your core network configuration is complete. In the next section, you'll configure host and domain names.

A.3.2 Configuring host and domain names

There are two components to a computer name: the host name, which identifies a specific host or service, and the domain name, which specifies a higher-order name often related to an organization. Without getting into too much detail, suppose you wanted to call your server *devstack* and your domain name was *example.com*. The fully qualified domain name (FQDN) of that your host would be *devstack.example.com*.

You'll want to set your host and domain names during the installation process, and there are two ways to do this. The first way is the way you'd configure something in production: use one FQDN for each host. This means that a record relating the host and domain names to the IP address must exist in some DNS server. Fortunately, if you're loading Linux on your laptop at home, a real FQDN isn't a requirement, especially if you're using Linux for a single-node deployment of OpenStack. Basically, if you're in an environment where you can use a real FQDN, you should do so. If an FQDN isn't available to you, simply invent host and domain names.

Enter your host name as shown in figure A.15.

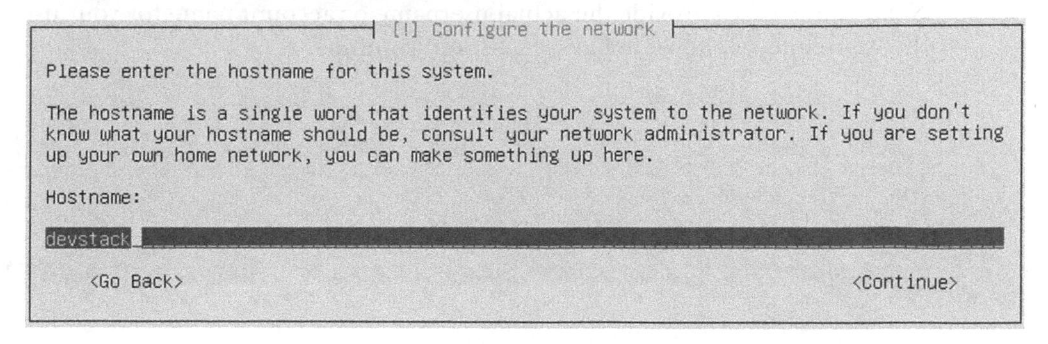

Figure A.15 Configure the host name

Enter your domain name as shown in figure A.16.

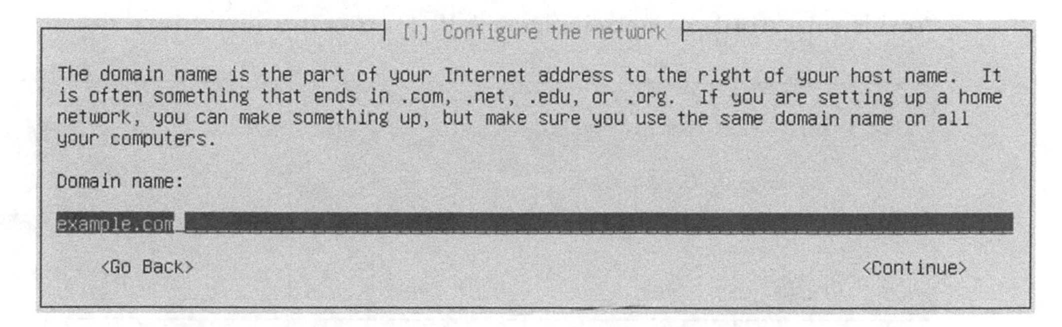

Figure A.16 Configure the domain name

At this point the network configuration is complete.

A.4 User configuration

The user configuration is short and simple. Basically, you provide information related to creating a user for your installation.

First, enter the real name of your user, as shown in figure A.17, and continue.

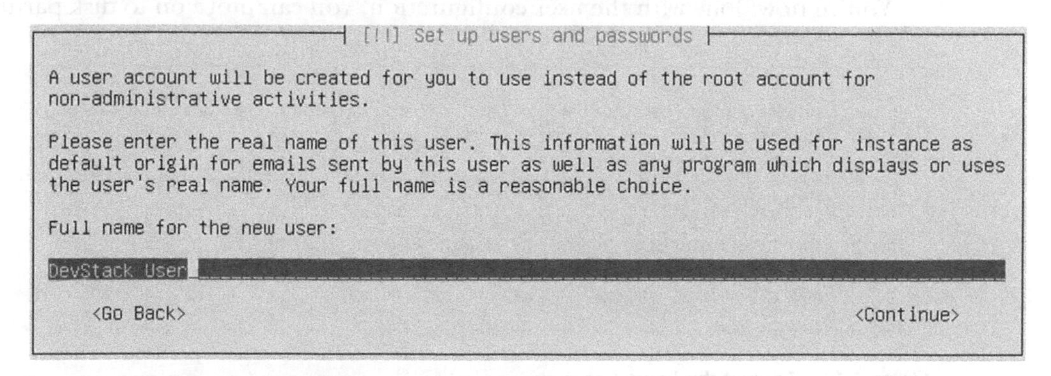

Figure A.17 Set up full name for user

Next, you need to provide the actual username or account name for your user. Enter the username, as shown in figure A.18, and continue.

```
[!!] Set up users and passwords

Select a username for the new account. Your first name is a reasonable choice. The
username should start with a lower-case letter, which can be followed by any combination
of numbers and more lower-case letters.

Username for your account:

devstack

<Go Back>                                                              <Continue>
```

Figure A.18 Set up username for your account

Provide a password, as shown in figure A.19, for your new user.

```
[!!] Set up users and passwords

Please enter the same user password again to verify you have typed it correctly.

Re-enter password to verify:

**********

<Go Back>                                                            <Continue>
```

Figure A.19 Set up password

The next screen allows you to encrypt the home directory (figure A.20). This is arguably more useful for multi-user or desktop deployments. Although the OpenStack servers will be accessed by many people, these people won't have accounts directly on the operating system. I generally don't encrypt the user directory in these types of systems, but that's up to you.

You're now done with the user configuration. You can move on to disk partitions.

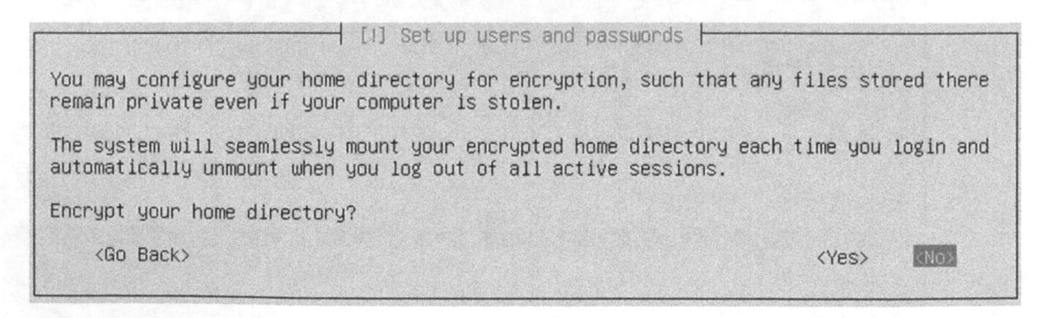

[!] Set up users and passwords
You may configure your home directory for encryption, such that any files stored there remain private even if your computer is stolen.
The system will seamlessly mount your encrypted home directory each time you login and automatically unmount when you log out of all active sessions.
Encrypt your home directory?
<Go Back>
<Yes>
<No>

Figure A.20 Encrypt the home directory

A.5 Disks and partitions

Configuring disks and partitions on servers can be one of the most important configuration steps, because these actions can be difficult to reverse. It's beyond the scope of this appendix to describe the best practices for configuring storage devices or Linux filesystem partitioning. What we will do is walk through a basic manual disk and partition configuration (see figure A.21).

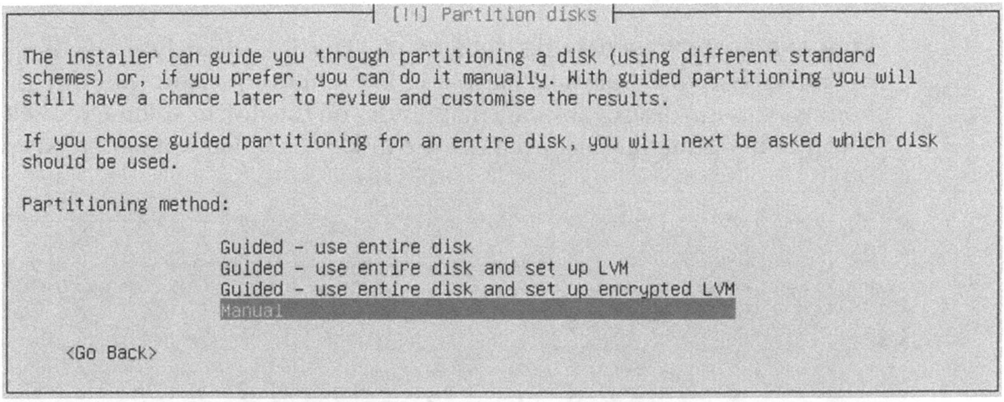

```
[!!] Partition disks

The installer can guide you through partitioning a disk (using different standard
schemes) or, if you prefer, you can do it manually. With guided partitioning you will
still have a chance later to review and customise the results.

If you choose guided partitioning for an entire disk, you will next be asked which disk
should be used.

Partitioning method:

          Guided - use entire disk
          Guided - use entire disk and set up LVM
          Guided - use entire disk and set up encrypted LVM
          Manual

  <Go Back>
```

Figure A.21 Manually partition disks

Alternatively, you can use one of the Guided setup options, which will provide default values for filesystems and mount points (locations in the directory structure). In Linux administration, it's important to understand mount points and filesystems, but that's beyond the scope of this appendix.

With OpenStack, we'll be most concerned with the devices and servers providing the OpenStack Storage resources, not with the storage related to specific servers.

NO VOLUME OR DISK FOUND As with the network adapter, if no storage device is found, either you have a bad adapter or your adapter isn't supported. The best thing to do is to consult the Ubuntu hardware and community support pages (https://wiki.ubuntu.com/HardwareSupport and https://help.ubuntu.com/community/Installation). It can often be easier to find another piece of hardware than to go ahead with unsupported hardware.

We're going to look at the simplest manual configuration possible, but it won't be the configuration you'd use for a production server. If you're interested in learning more about partitions, you can take a look at the Ubuntu 14.04 installation guide: https://help.ubuntu.com/14.04/installation-guide/amd64/.

In our simple configuration, we'll create two partitions. The first partition will be the swap and the second will be the root volume. It isn't important that you know the functions of these volumes, but you should know that they're the minimum volumes needed for an installation.

A.5.1 Configuring the block device (hard drive)

In figure A.22, a single volume is shown. This volume comprises several physical disks, but it's presented to the operating system as a single volume by the storage adapter. This is a common thing for servers to do. If you're using a desktop or laptop, you might have several disks or a single physical disk. Select the disk you want to use, and move on to partitioning. Beware, this is your last chance. If you write changes to the disk, there's no going back.

> **STOP AND READ: YOUR DATA DEPENDS ON IT** This is a tutorial for installing an operating system on physical hardware. If you move forward with disk partitioning, you *will* destroy any data that resides on the disk or volume you select. If you're in doubt, physically remove the disk or data that you want to keep.

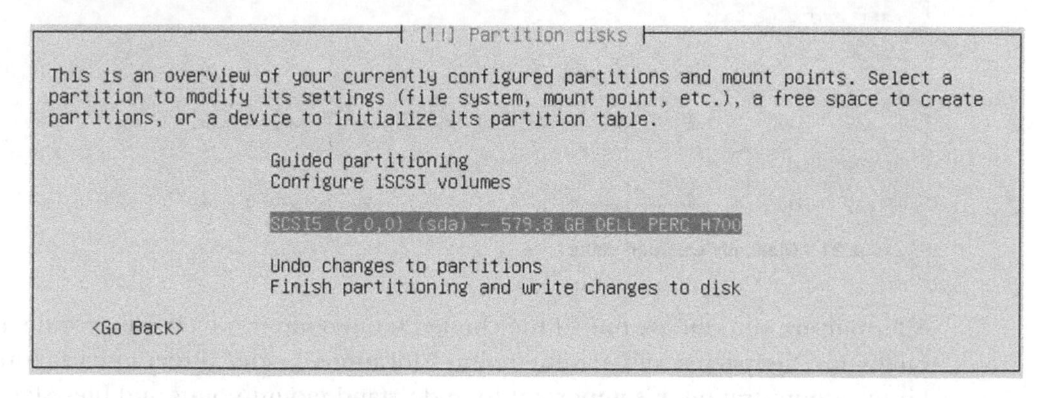
[!!] Partition disks
This is an overview of your currently configured partitions and mount points. Select a partition to modify its settings (file system, mount point, etc.), a free space to create partitions, or a device to initialize its partition table.
Guided partitioning
Configure iSCSI volumes
SCSI5 (2,0,0) (sda) - 579.8 GB DELL PERC H700
Undo changes to partitions
Finish partitioning and write changes to disk
<Go Back>

Figure A.22 Select disk to partition

You're all in now. Create a new empty partition on the device, as shown in figure A.23.

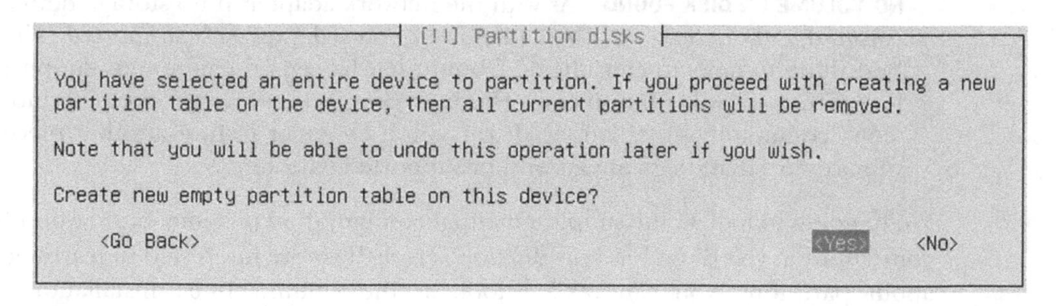
[!!] Partition disks
You have selected an entire device to partition. If you proceed with creating a new partition table on the device, then all current partitions will be removed.
Note that you will be able to undo this operation later if you wish.
Create new empty partition table on this device?
<Go Back>
<Yes>
<No>

Figure A.23 Select entire disk

You've now created a new partition table on your device. As you can see in figure A.24, all the space on my volume is designated as FREE SPACE.

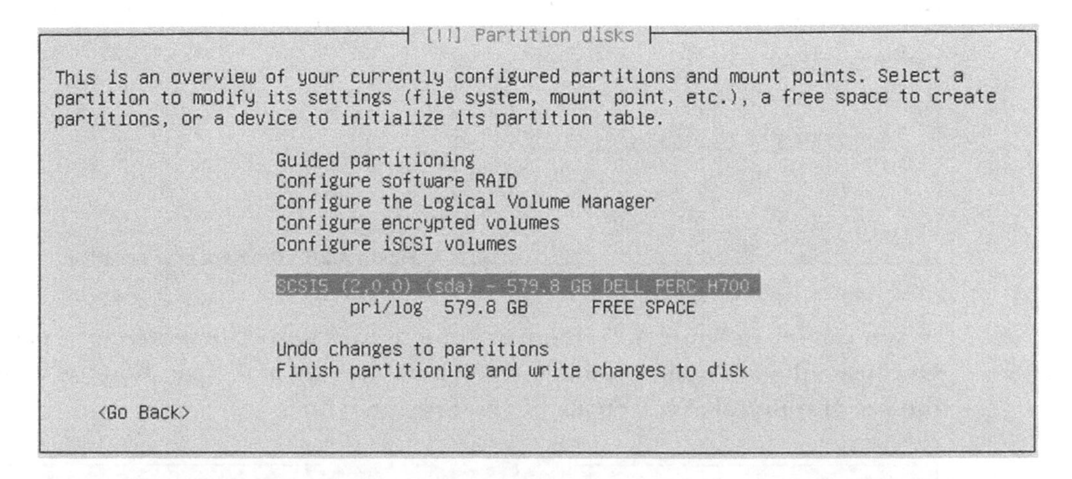
[!!] Partition disks
This is an overview of your currently configured partitions and mount points. Select a partition to modify its settings (file system, mount point, etc.), a free space to create partitions, or a device to initialize its partition table.
Guided partitioning
Configure software RAID
Configure the Logical Volume Manager
Configure encrypted volumes
Configure iSCSI volumes
SCSI5 (2,0,0) (sda) - 579.8 GB DELL PERC H700
pri/log 579.8 GB FREE SPACE
Undo changes to partitions
Finish partitioning and write changes to disk
<Go Back>

Figure A.24 Partition menu

Select your FREE SPACE, as shown in figure A.25, and press Enter.

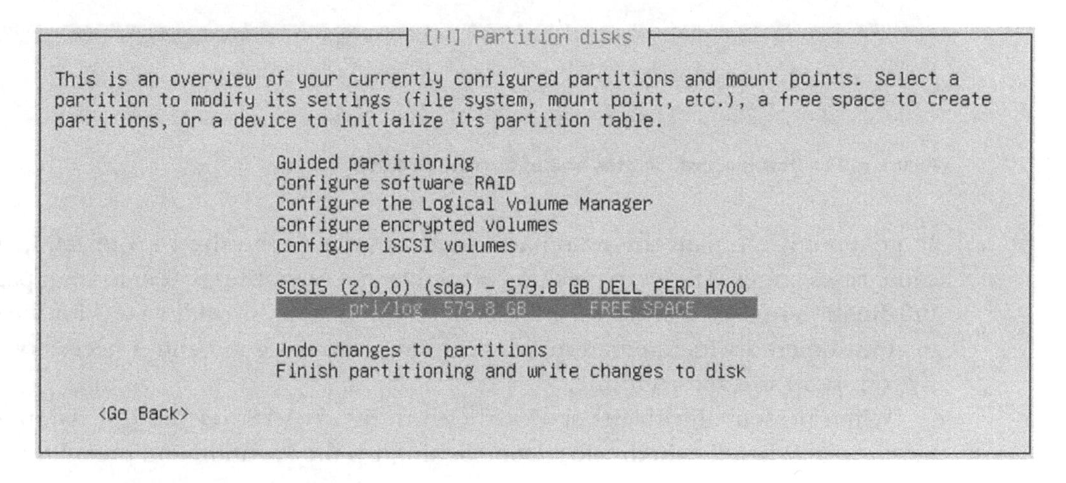
[!!] Partition disks
This is an overview of your currently configured partitions and mount points. Select a partition to modify its settings (file system, mount point, etc.), a free space to create partitions, or a device to initialize its partition table.
Guided partitioning
Configure software RAID
Configure the Logical Volume Manager
Configure encrypted volumes
Configure iSCSI volumes
SCSI5 (2,0,0) (sda) - 579.8 GB DELL PERC H700
pri/log 579.8 GB FREE SPACE
Undo changes to partitions
Finish partitioning and write changes to disk
<Go Back>

Figure A.25 Select FREE SPACE

Now it's time to configure your swap and root partitions from the free space available on your disk.

A.5.2 Configuring root and swap partitions and mount points

You'll want to create a new partition when presented with the menu shown in figure A.26.

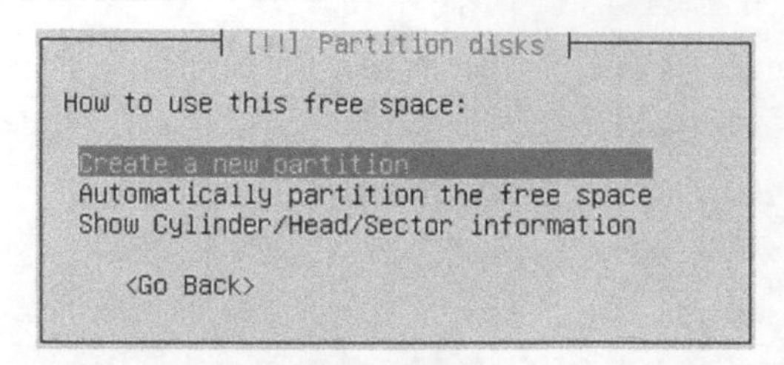

Figure A.26 Create a new partition

As you can see in figure A.27, there are various ways to specify the size of a partition. Of course, the size of the partition depends on the type of the partition you want and the size of your disk. We'll create the *swap* partition first.

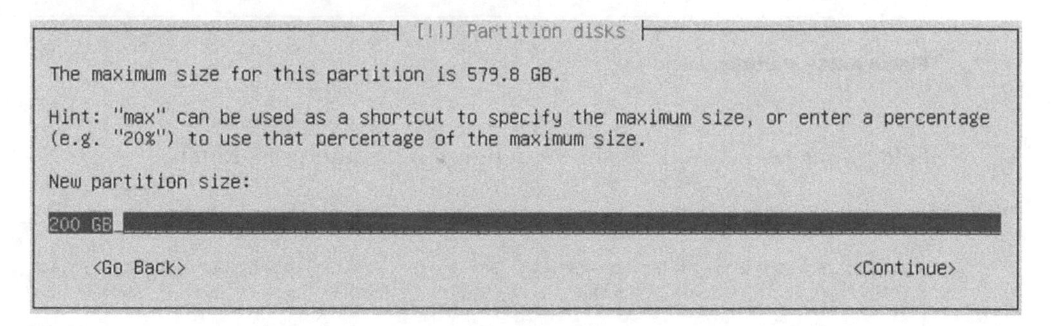

Figure A.27 Partition disk: set the size of the new partition

In general, you'll want the swap partition to be at least the size of your RAM, and in some cases larger. The swap partition is used by the operating system to swap pages of information in RAM to frames on the swap partition. Swap provides a way for the OS to deal with memory fragmentation. In short, swap space is good and is necessary. Specify your swap size and continue.

When presented with the screen shown in figure A.28, set the Use As option to Swap Area. When finished, select Done Setting Up the Partition and press Enter.

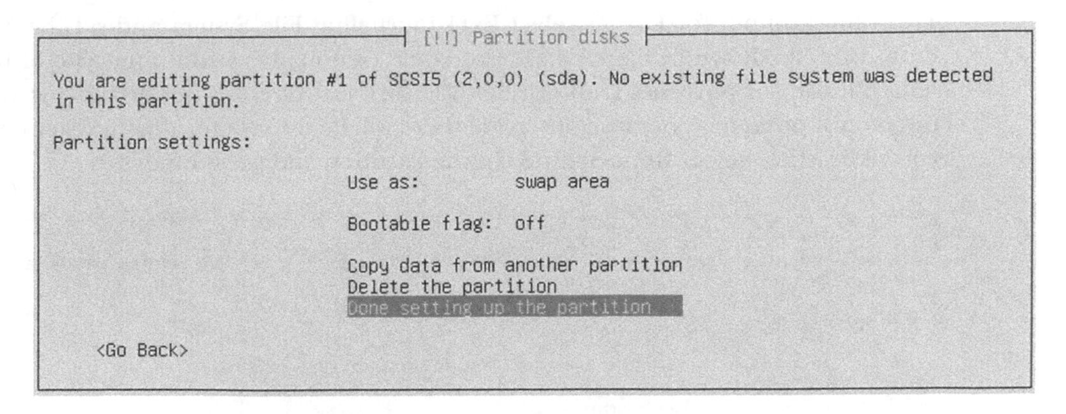

Figure A.28 Set partition as "swap area"

You can see in figure A.29 that the swap partition, along with `FREE SPACE`, is displayed. You'll now repeat the last several steps to create the root partition. Select `FREE SPACE` and press Enter.

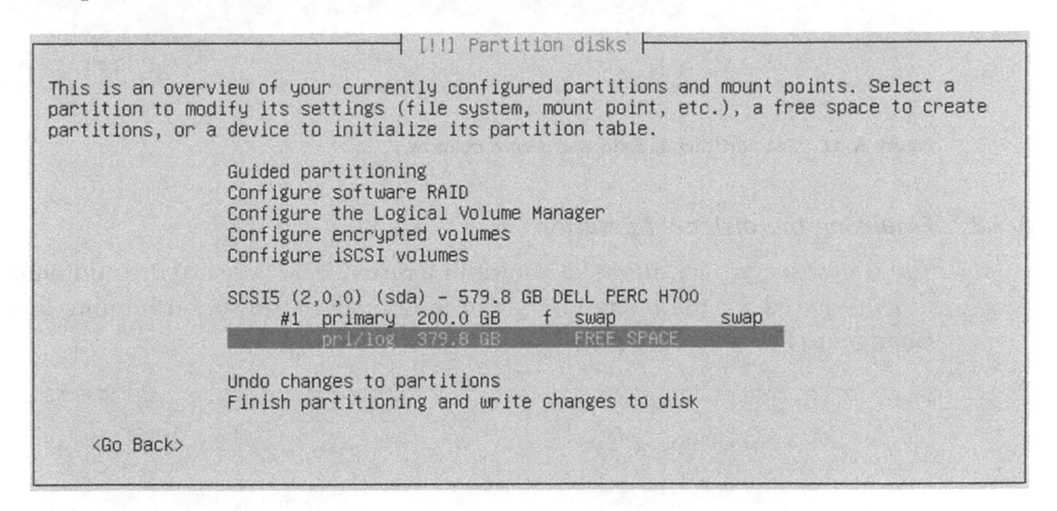

Figure A.29 Select `FREE SPACE`

Once again, you'll want to select Create a New Partition when presented with the menu shown in figure A.30.

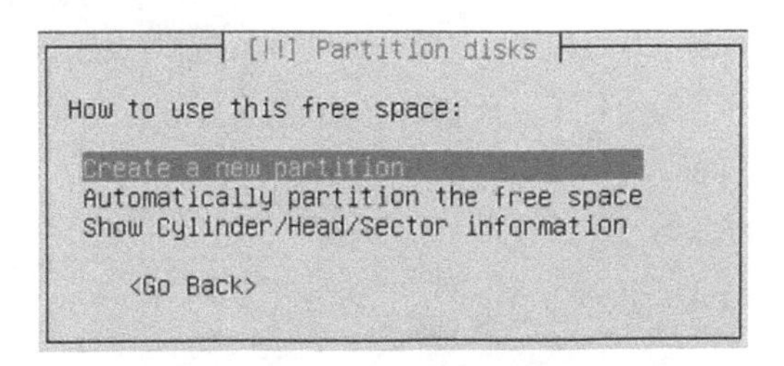

Figure A.30 Create a new partition

This time around, for Use As select Ext4 Journaling File System and set the Mount Point to /, as shown in figure A.31. You could optionally use another type of filesystem, but for our purposes Ext3 or Ext4 will work just fine. It's important that you set the mount point to /, because otherwise there will be no root for the filesystem. When you're finished, select Done Setting Up the Partition and press Enter.

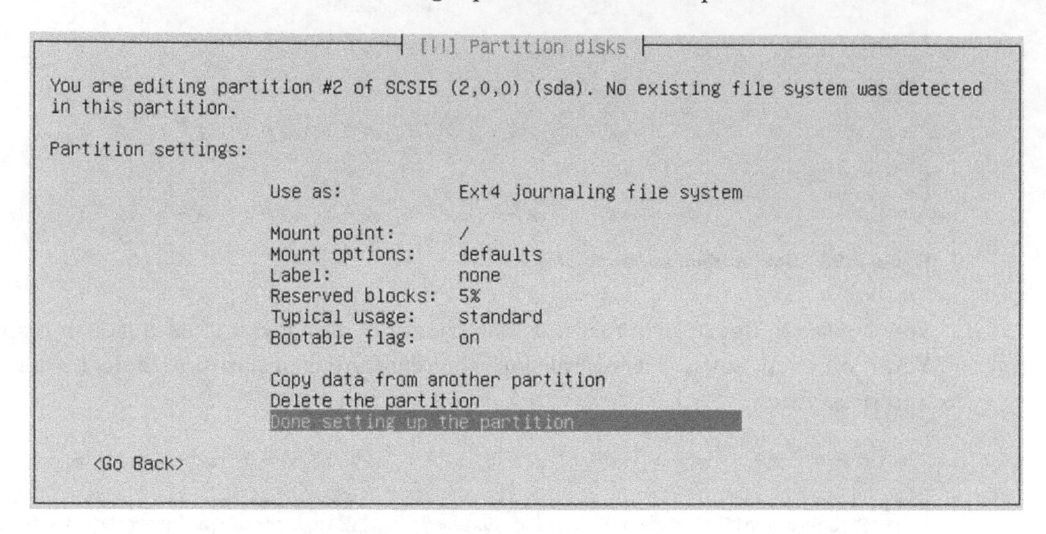

Figure A.31 Set partition as Ext4 and mount point as /

A.5.3 *Finalizing the disk configuration*

You'll now see two partitions, as shown in figure A.32. These are the minimum partitions required for the operating system install. Select Finish Partitioning and Write Changes to Disk and then press Enter.

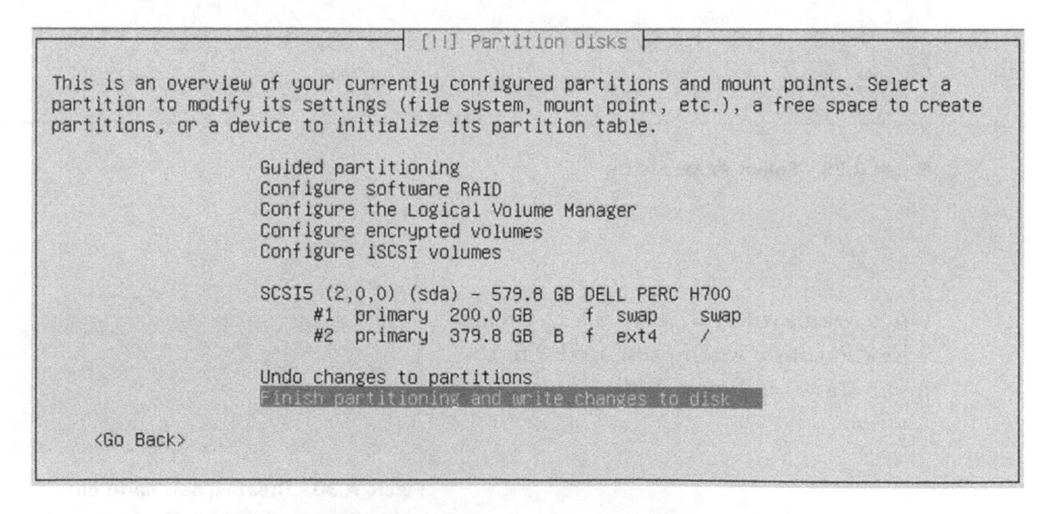

Figure A.32 Finish partition configuration

You'll be provided with a report of what your volume will look like after the partitions are written, as shown in figure A.33. If you're confident that you aren't destroying anything important, go ahead and write the changes to disk.

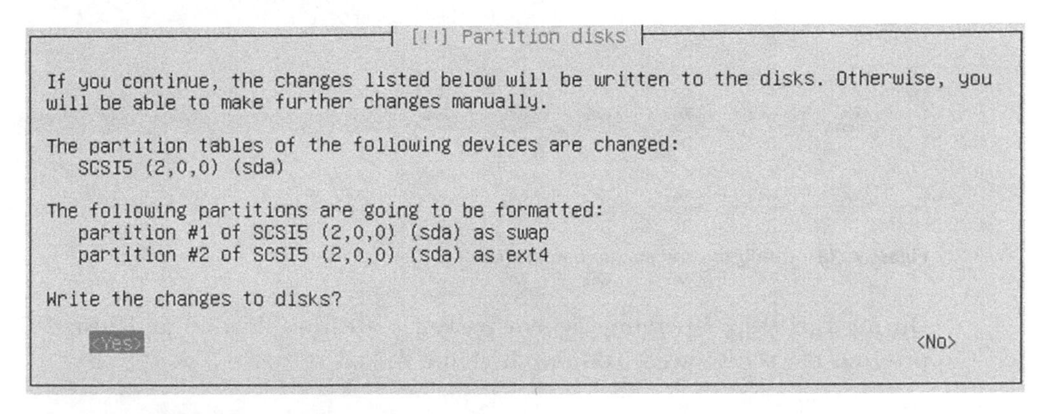

[!!] Partition disks
If you continue, the changes listed below will be written to the disks. Otherwise, you will be able to make further changes manually.
The partition tables of the following devices are changed:
SCSI5 (2,0,0) (sda)
The following partitions are going to be formatted:
partition #1 of SCSI5 (2,0,0) (sda) as swap
partition #2 of SCSI5 (2,0,0) (sda) as ext4
Write the changes to disks?
<Yes>
<No>

Figure A.33 Finish disk configuration

Your language, network, user, and disk information have now been configured. You're ready to proceed to the final installation steps.

A.6 *Base system configuration*

In this final section, there's very little configuration left. Most of your time will be spent waiting on the system to install packages and deploy your configuration.

Once the partition changes are written to disk, the installer will start the base package installation, as shown in figure A.34.

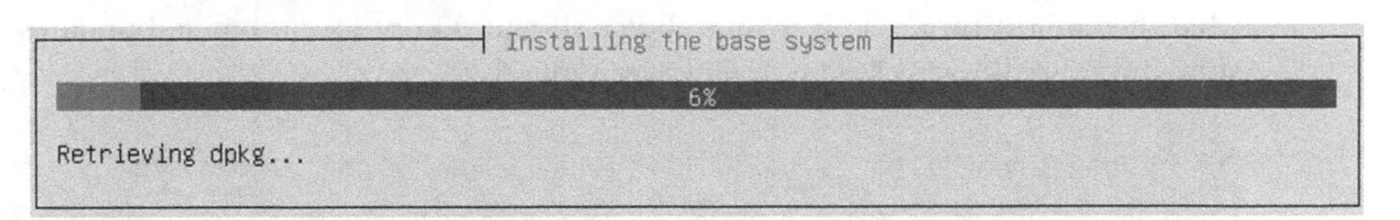

Figure A.34 Installing base packages

It's unlikely that you'll have to enter proxy information, but if you do, figure A.35 shows how you can configure a proxy for the package manager.

If you didn't provide proxy information in the previous step and are unsure if it's needed, ask someone in your security or networking team if proxies are needed. If you're setting this up on a home network, it's highly unlikely that a proxy server is used, and it could cause additional connectivity problems.

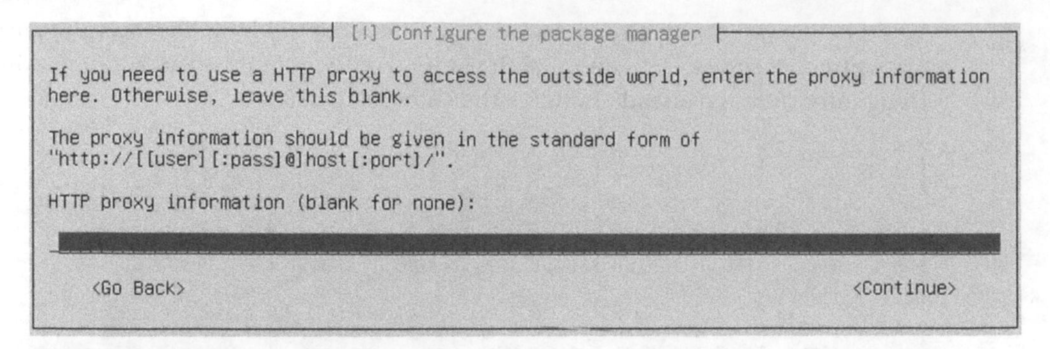

Figure A.35 Configure the package manager

During this step, operating system packages are downloaded and installed, and a progress bar is displayed, as shown in figure A.36.

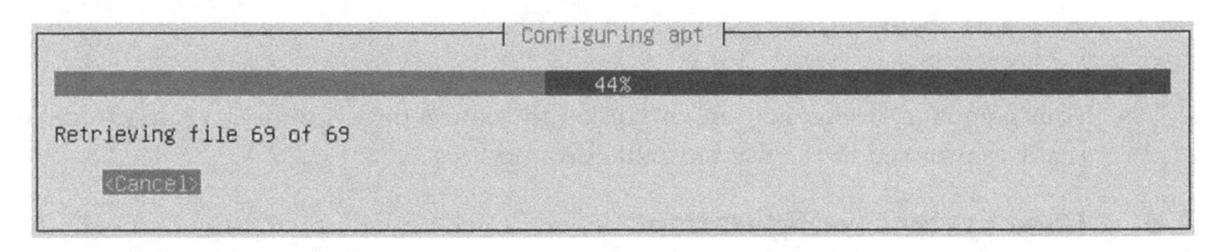

Figure A.36 Retrieving packages

The next screen lets you configure how updates will be applied (see figure A.37). In general, I don't install security updates automatically on servers, because I don't know what they might break. This is your choice. If you do opt for automated updates and something breaks, check if an update was applied.

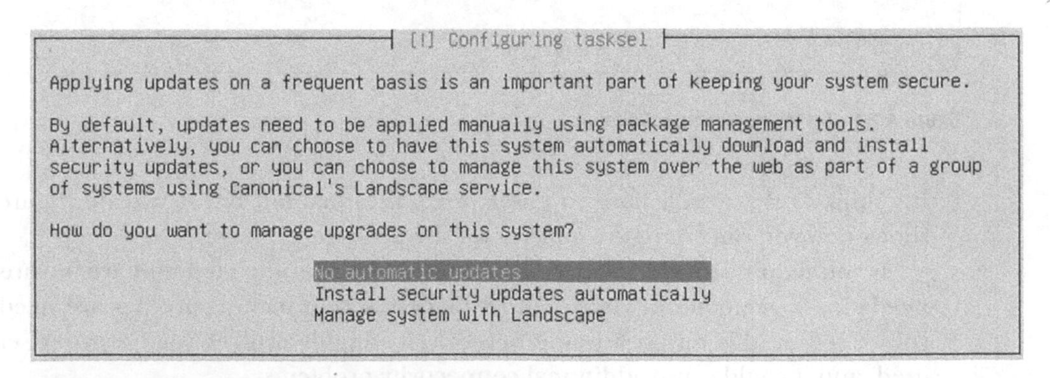

Figure A.37 Configure update automation

There's really only one service that you'll want to install from the menu, shown in figure A.38, and that's OpenSSH. OpenSSH is used for a host of functions, but mainly for accessing this operating system from another computer. Once it's installed, you can use OpenSSH, commonly known as `ssh`, to access this host using the name or IP address and username you previously specified.

```
[!] Software selection

At the moment, only the core of the system is installed. To tune the system to your
needs, you can choose to install one or more of the following predefined collections of
software.

Choose software to install:

        [*] OpenSSH server
        [ ] DNS server
        [ ] LAMP server
        [ ] Mail server
        [ ] PostgreSQL database
        [ ] Print server
        [ ] Samba file server
        [ ] Tomcat Java server
        [ ] Virtual Machine host
        [ ] Manual package selection

              <Continue>
```

Figure A.38 Configure initial services

The additional services you selected, such as OpenSSH, will be installed (see figure A.39).

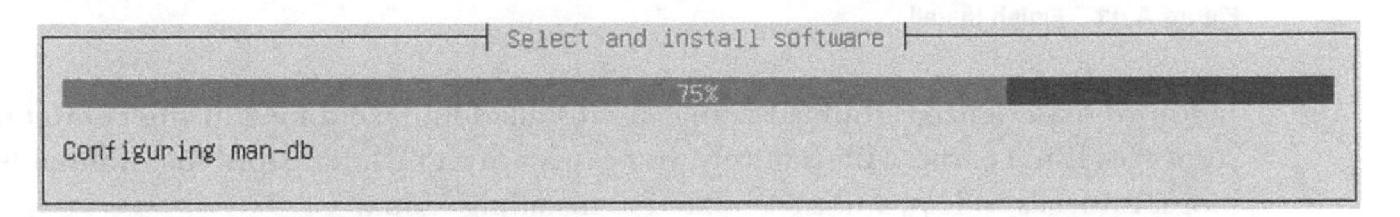

Figure A.39 Base system: installing packages

In the next screen (figure A.40), install the GRUB boot loader. If you're using this host for OpenStack, there's no reason not to.

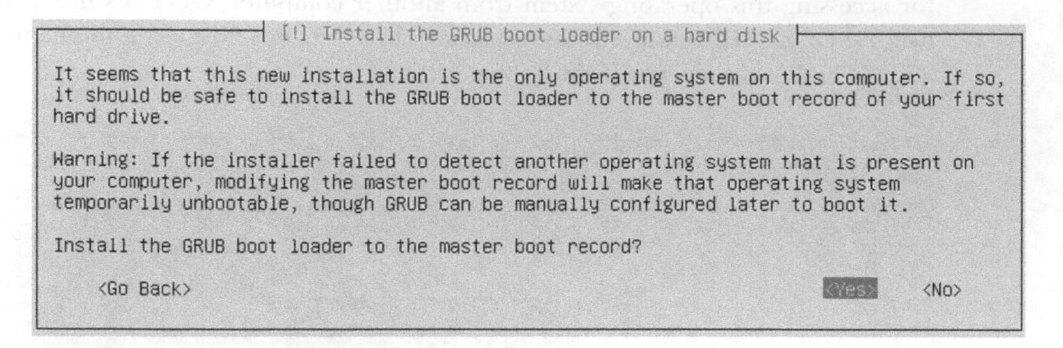

Figure A.40 Install boot loader

OK. That's it. With any luck, you'll be able to restart your host as shown in figure A.41 and remotely access it via ssh.

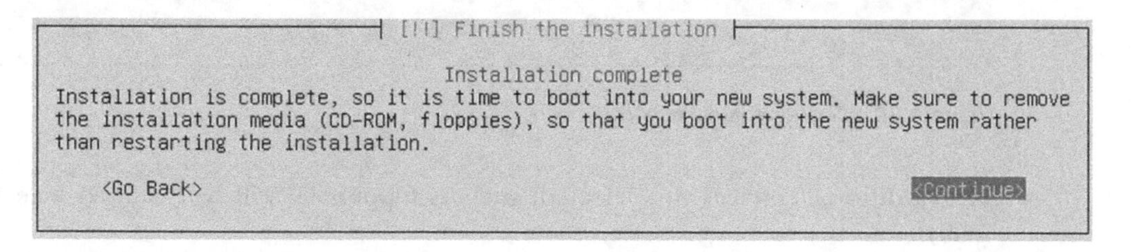

Figure A.41 Finish install

If you've experienced trouble during the installation, take a look at the community resources listed earlier. Alternatively, you can learn a great deal from the thousands of blogs, tutorials, videos, and sites related to installing Linux.

index

A

abstraction and OpenStack 10–11
Access & Security screen, Dashboard 38–41
AD (Active Directory) 125
admin node, Ceph 260
admin user 62
adminurl endpoint 130–131, 137, 145, 150, 155
Advanced Message Queuing Protocol. *See* AMQP
Advanced Packaging Tool. *See* APT
agents, DHCP 71
AKI (Amazon kernel image) format 42, 141
Amazon ramdisk image format. *See* ARI
Amazon Web Services. *See* AWS
AMI (Amazon machine image) format 42, 141
AMQP (Advanced Message Queuing Protocol) 121
APIs (application programming interfaces)
  commands usage 58–59
  OpenStack management options 4
APT (Advanced Packaging Tool) 26
architecture considerations
  private clouds
    control over performance and quality 252
    data gravity 252–253
    hybrid cloud usage 253
    networking for 255–257
    OpenStack deployment tools 254–255
    public clouds vs. 251
    storage in 257–258
  replacing existing virtual server platforms
    deployment considerations 245
    network topologies 246–247
    overview 242–244
    server types 250
    storage considerations 247–249
ARI (Amazon ramdisk image) format 42, 141
automated deployment 277, 283
Availability Zone 48
AWS (Amazon Web Services) 5, 304

B

bare format 141
bare-metal servers 250
benchmarking Ceph clusters
  disk latency 275–276
  reading benchmarks 275
  writing benchmarks 274–275
Block Storage 18, 86
  Cinder
    configuring 209–210
    creating volumes using CLI 211–213
    creating volumes using Dashboard 213–215
    installing 208
    overview 206–208
    restarting services and verifying installation 210
  configuring Keystone user for 144–145
  creating data store 143–144
  creating service and endpoints 145–146
  defined 47
  installing 146–147
  LVM
    creating volumes 204
    device identification and 203
    installing 202
    overview 202
    physical volume operations 204–205
    volume-group operations 205–206
  overview 15–16, 142–143
  prerequisites
    network interfaces configuration 199–201

Block Storage, prerequisites *(continued)*
network interfaces review 198–199
updating packages 201
required for booting instance 44
boot devices 286, 289, 294
branches, Git 28, 31

C

cache-enabled volumes 202
Ceilometer project 18, 86
CentOS images 41
central processing units. *See* CPUs
Ceph node deployment
benchmarking clusters
disk latency 275–276
reading benchmarks 275
writing benchmarks 274–275
Ceph pools 273–274
clusters
configuring 265–266
deploying configuration 267–268
deploying software to nodes 266
creating user for 261–262
installing 264–265
installing different version 266
node types 260–261
node-to-node authentication 262–264
OSD resources
creating OSDs 271–272
overview 268–269
preparing devices 269–271
uninstalling 266
certifications, Microsoft 12
CFN (CloudFormation) format 304
charms, Juju 89, 315
chpasswd command 261
CIDR (Classless Inter-Domain Routing) 69–70, 186
Cinder project 18, 21, 86
configuring 209–210
creating volumes using CLI 211–213
creating volumes using Dashboard 213–215
driver used by 257
installing 208
overview 206–208
restarting services and verifying installation 210
system support for 249
Classless Inter-Domain Routing. *See* CIDR
CLI (command-line interface)
commands usage 56–58
creating Cinder volumes using 211–213
–debug argument 59
for Fuel 300
for Ubuntu Juju 317–319
OpenStack management options 4
clock synchronization 261
cloud computing
OpenStack and 9–10
private clouds
control over performance and quality 252
data gravity 252–253
hybrid cloud usage 253
networking for 255–257
OpenStack deployment tools 254–255
public clouds vs. 251
storage in 257–258
terminology used in 17
cloud orchestration
OpenStack Heat
confirming dependencies 308–309
enabling in DevStack 307
Heat templates 304–306
launching Heat stack 309–311
overview 304
Ubuntu Juju
deploying charms CLI 317–319
deploying GUI 319–324
installing 314–317
overview 312
preparing OpenStack for 312–313
CloudFormation format. *See* CFN format
cloud-init 180
clusters, Ceph
benchmarking
disk latency 275–276
reading benchmarks 275
writing benchmarks 274–275
configuring 265–266
deploying configuration 267–268
deploying software to nodes 266
watching activity in 271
command-line interface. *See* CLI
companion VM 23–24
components, OpenStack
communication
dashboard authentication communication 86–87
interaction with Juju 88–90
overview 85–86
resource provisioning 87–88
resource query and request 87
distributed computing model
component communication at VM-level 93–95
general discussion 91
OpenStack distributed component model 92
VM provisioning interaction 93
overview 85

compute node deployment 86
  configuring Keystone user for 154–155
  creating data store 153–154
  creating service and endpoints 155
  defined 18, 115
  hypervisor installation
    installing software 228
    kernel module installation 228
    KVM overview 227–228
    removing default network bridge 229
    verifying host support 226–227
    verifying KVM-accelerated QEMU environment 229
  installing 156–158
  Neutron
    configuring 230
    configuring ML2 plug-in 231
    installing 230
  Nova
    configuring 232–233
    creating instance using CLI 234–237
    installing 231–232
    verifying in Dashboard 233–234
  overview 152–153
  prerequisites
    network interfaces configuration 220–222
    network interfaces review 219–220
    Open vSwitch configuration 225–226
    Open vSwitch installation 223–225
    server-to-router configuration 222–223
    updating packages 222
containers, Linux 14, 250
controller node deployment
  defined 115
  deploying Block Storage service
    configuring Keystone user for 144–145
    creating data store 143–144
    creating service and endpoints 145–146
    installing 146–147
    overview 142–143
  deploying Compute service
    configuring Keystone user for 154–155
    creating data store 153–154
    creating service and endpoints 155
    installing 156–158
    overview 152–153
  deploying Dashboard Service
    accessing 159
    debugging 160
    installing 158–159
  deploying Identity Service
    assigning roles 134
    configuring data store 126–128
    creating roles 133–134
    creating service and endpoints 129–132
    creating tenants 132
    creating users 133
    initializing database 128
    initializing variables 128–129
    installing 126
    listing roles 134–135
    overview 125–126
  deploying Image Service
    configuring Keystone user for 136–137
    creating data store 135–136
    creating service and endpoints 137–138
    image formats 140–142
    installing 138–140
    overview 135
  deploying Networking service
    configuring Keystone user for 149–150
    creating data store 148–149
    creating service and endpoints 150–151
    installing 151–152
    overview 147–148
  deploying shared services 124–125
  environment preparations 116
  installing dependencies
    MySQL 122–124
    MySQL console access 124
    RabbitMQ 121–122
  network interface configuration
    configuring interfaces in Ubuntu 117–120
    determining existing interfaces 117
  updating packages 120–121
core projects 18, 85
cpu-checker utility 226
CPUs (central processing units) 8–9, 327
CRC (cyclic redundancy check) 173, 224
credentials 40

D

Dashboard 18, 86
  Access & Security screen 38–41
  accessing 159
  authentication communication 86–87
  creating Cinder volumes using 213–215
  debugging 160
  Images & Snapshots screen 41–44
  installing 158–159
  Instances screen
    Access & Security tab 49
    Details tab 48–49
    Networking tab 49–51
    overview 47–48
  overview 36–37
  Overview screen 38
  viewing Neutron networks in 193–194
  Volumes screen 44–47
data gravity 252–253

Database Service project 86
DDOS (distributed denial of service) attacks 169
dependencies 309, 318
deployment
  Block Storage
    configuring Keystone user for 144–145
    creating data store 143–144
    creating service and endpoints 145–146
    installing 146–147
    network interfaces configuration 199–201
    network interfaces review 198–199
    overview 142–143
    updating packages 201
  Cinder
    configuring 209–210
    creating volumes using CLI 211–213
    creating volumes using Dashboard 213–215
    installing 208
    overview 206–208
    restarting services and verifying installation 210
  Compute service
    configuring Keystone user for 154–155
    creating data store 153–154
    creating service and endpoints 155
    installing 156–158
    overview 152–153
  considerations for replacing existing virtual server platforms 245
  Dashboard service
    accessing 159
    debugging 160
    installing 158–159
  DevStack
    configuring options 30–31
    creating server 25
    general discussion 23–25
    preparing for installation 28–29
    running build script 31–32
    server environment setup 26–27
    testing stack 32–35
  Identity Service
    assigning roles 134
    configuring data store 126–128
    creating roles 133–134
    creating service and endpoints 129–132
    creating tenants 132
    creating users 133
    initializing database 128
    initializing variables 128–129
    installing 126
    listing roles 134–135
    overview 125–126
  Image Service
    configuring Keystone user for 136–137
    creating data store 135–136
    creating service and endpoints 137–138
    image formats 140–142
    installing 138–140
    overview 135
  LVM
    creating volumes 204
    device identification and 203
    installing 202
    overview 202
    physical volume operations 204–205
    volume-group operations 205–206
  multi-node 114–115
  network node
    environment configuration 164
    installing Linux Bridge and VLAN utilities 168–169
    network interfaces configuration 165–167
    network interfaces review 164–165
    Open vSwitch configuration 174–176
    Open vSwitch installation 171–174
    server-to-router configuration 169–171
    updating packages 167–168
  Networking service
    configuring Keystone user for 149–150
    creating data store 148–149
    creating service and endpoints 150–151
    installing 151–152
    overview 147–148
  Neutron components
    configuring 178
    DHCP agent configuration 180
    installing 177
    L3 agent configuration 179–180
    Linux network namespaces and 191–193
    Metadata agent configuration 180–181
    ML2 plug-in configuration 178–179
    overview 177
    restarting services and verifying installation 181–182
    viewing in Dashboard 193–194
  Neutron networks
    external networks 188–189
    external subnets 189–191
    internal networks 183–184
    internal subnets 184–186
    Network console 183
    overview 182
    routers 186–188
DevStack
  deploying
    configuring options 30–31
    creating server 25
    general discussion 23–25
    preparing for installation 28–29

DevStack, deploying *(continued)*
running build script 31–32
server environment setup 26–27
testing stack 32–35
enabling OpenStack Heat in 307
overview 22–23, 35–36
rebooting 35–36
DHCP (Dynamic Host Configuration Protocol) 66, 71, 284, 331
DHCP agent
configuring 180
defined 177
purpose of 188
discover command 129
disks, Linux
overview 342–343
partitioning 337–339
root and swap partitions 340–342
distributed computing model
component communication at VM-level 93–95
general discussion 91
OpenStack distributed component model 92
VM provisioning interaction 93
distributed denial of service attacks. *See* DDOS attacks
Distributed Virtual Routing. *See* DVR
DKMS (Dynamic Kernel Module Support) 174
DNS (Domain Name System) 332, 334
Docker 14
driver_version statistic 100, 207
dubs service 229
DVR (Distributed Virtual Routing) 95, 255
Dynamic Host Configuration Protocol. *See* DHCP
Dynamic Kernel Module Support. *See* DKMS

E

EHA (Ethernet Hardware Address) 171
endpoints 126
environment variables, setting 56–57, 182
ephemeral storage 248
Ethernet Hardware Address. *See* EHA
external networks
creating 73–74
in Neutron 188–189
external subnets
creating 75–78
in Neutron 189–191

F

FC (Fiber Channel) 97
FCoE (Fiber Channel over Ethernet) 97
fdisk command 270
file storage 47
flat networks 52, 65, 182, 246–247
flavors, OpenStack 49
floating IPs
assigning to instance 53
defined 40
permitting network traffic to 53–54
removing IP before removing external network 76
FQDN (fully qualified domain name) 334
free_capacity_gb statistic 100, 207
Fuel
allocating hosts 298
configuring network for environment 296
configuring node disks 300
configuring node interfaces 298–300
creating deployment environment 295–296
deploying changes 302
installing 290–292
overview 293
server discovery 294
verifying installation 301
fully orchestrated clouds 10
fully qualified domain name. *See* FQDN

G

gating projects 85
Generic Routing Encapsulation tunnel. *See* GRE tunnel
Git
defined 27
using specific branch 28
Glance project 18, 86
GRANT command 135
graphical user interface. *See* GUI
GRE (Generic Routing Encapsulation) tunnel 176, 256
gre module 173, 224
GUI (graphical user interface) 158

H

HA (high availability)
Fuel installation 290–292
Fuel web interface
allocating hosts 298
configuring network for environment 296
configuring node disks 300
configuring node interfaces 298–300
creating deployment environment 295–296
deploying changes 302
overview 293
server discovery 294
verifying installation 301

HA (high availability) *(continued)*
network hardware preparations
deployment network 280
overview 279–280
switch uplink ports 281–282
overview 277–278
server hardware preparations
accessing OOB management console using SSH 284–286
accessing OOB web interface 284
configuring automation administration network 288–290
configuring OOB network 282–284
configuring server storage 287–288
hardware
Nova compatibility 246
preparations for HA
accessing OOB management console using SSH 284–286
accessing OOB web interface 284
configuring automation administration network 288–290
configuring OOB network 282–284
configuring server storage 287–288
deployment network 280
overview 279–280
switch uplink ports 281–282
Heat OpenStack Template specification. *See* HOT specification
Heat project 18, 86
confirming dependencies 308–309
enabling in DevStack 307
Heat templates 304–306
launching Heat stack 309–311
overview 304
Ubuntu Juju vs. 312
high availability. *See* HA
high-performance computing. *See* HPC
Horizon project 18, 86
HOT (Heat OpenStack Template) specification 305–306
HPC (high-performance computing) 254
hub-and-spoke distribution 91
hybrid clouds 10, 253
hypervisors
installing software 228
kernel module installation 228
KVM overview 227–228
OpenStack and 11–14
removing default network bridge 229
verifying host support 226–227
verifying KVM-accelerated QEMU environment 229

I

IaaS (infrastructure as a service) 9, 17
Identity Service, deploying 18, 86
assigning roles 134
configuring data store 126–128
creating roles 133–134
creating service and endpoints 129–132
creating tenants 132
creating users 133
initializing database 128
initializing variables 128–129
installing 126
listing roles 134–135
overview 125–126
ifconfig command 117
Image Service, deploying 18, 86
configuring Keystone user for 136–137
creating data store 135–136
creating service and endpoints 137–138
image formats 140–142
installing 138–140
overview 135
image-list command 57
Images & Snapshots screen, Dashboard 41–44
images, VM 41
incubated projects 85
infinite free space value 100
infrastructure as a service. *See* IaaS
Instance Console
assigning floating IP to instance 53
logging into VM 51–52
permitting network traffic to floating IP 53–54
instances
creating using CLI 234–237
VMs vs. 38
Instances screen, Dashboard
Access & Security tab 49
Details tab 48–49
Networking tab 49–51
overview 47–48
internal interfaces 118, 164, 198, 219
internal networks
creating 67–68
in Neutron 183–184
internal subnets
creating 68–70
in Neutron 184–186
internalurl endpoint 130–131, 137, 145, 150, 155
International Standards Organization. *See* ISO
Internet Small Computer System Interface. *See* iSCSI
IP (Internet Protocol) 176
ip netns command 192
IPv4 addresses 40–41

IPv6 addresses 41
iSCSI (Internet Small Computer System Interface) 97
ISO (International Standards Organization) 41–42, 141

J

JSON (JavaScript Object Notation) 305
Juju project. *See* Ubuntu Juju
jumbo frame 276

K

key-pair authentication 262–263
Keystone project 18, 86
KVM (Kernel-based Virtual Machine)
  installing software 228
  kernel module installation 228
  overview 227–228
  removing default network bridge 229
  support for 12–13
  verifying KVM-accelerated QEMU environment 229

L

L3 agent
  configuring 179–180
  defined 177
L3 services 106, 255
LDAP (Lightweight Directory Access Protocol) 125
libcrc32c module 173, 224
library projects 85
Libvirt 227
Lightweight Directory Access Protocol. *See* LDAP
Link Layer Discovery Protocol. *See* LLDP
Linux
  containers 14
  disk configuration
    overview 342–343
    partitioning 337–339
    root and swap partitions 340–342
  distributions supported 25
  installing packages 343–346
  network configuration
    DNS server 334
    host and domain names 334–335
    host IP address 332
    network gateway address 333
    overview 330–332
    subnet mask 333
  network namespaces in 191–193
  preparing for installation 326–328
  user configuration 335–336
Linux Bridge 168–169
Linux-IO Target 209
list-image command 313
LLDP (Link Layer Discovery Protocol) 301
Logical Volume Manager. *See* LVM
logical volumes 204
losetup command 36
lsmod command 224
LTS (Long Term Support) 116, 326
lvdisplay command 212
LVM (Logical Volume Manager) 100
  creating volumes 204
  device identification and 203
  installing 202
  overview 202
  physical volume operations 204–205
  volume-group operations 205–206

M

MAC (Media Access Control) 171
maximum transmission unit. *See* MTU
MDS (metadata server) node 260
mechanism drivers 107
Media Access Control. *See* MAC
mesh distribution 91
Metadata agent
  configuring 180–181
  defined 177
Metadata Server node. *See* MDS node
micro-op decoding 9
Microsoft certification 12
Microsoft Hyper-V 250
ML2 (Modular Layer 2) 107, 151
  configuring plug-in 178–179
  defined 177
modprobe command 224
MON nodes 267
mount points 342
MTU (maximum transmission unit) 276
multi-tenancy 10, 35
Murano project 303
MySQL
  console access 124
  installing 122–124
  performance considerations 123

N

namespace isolation 179
NAT translation 52
Nebula tool 18

nested virtualization 25
Network File System. *See* NFS
Network Interface Cards. *See* NICs
network node deployment 18, 86
  configuring Keystone user for 149–150
  creating data store 148–149
  creating Neutron networks
    external networks 188–189
    external subnets 189–191
    internal networks 183–184
    internal subnets 184–186
    Network console 183
    overview 182
    routers 186–188
  creating service and endpoints 150–151
  defined 115
  installing 151–152
  Neutron components
    configuring 178
    DHCP agent configuration 180
    installing 177
    L3 agent configuration 179–180
    Linux network namespaces and 191–193
    Metadata agent configuration 180–181
    ML2 plug-in configuration 178–179
    overview 177
    restarting services and verifying installation 181–182
    viewing in Dashboard 193–194
  overview 147–148
  prerequisites
    environment configuration 164
    installing Linux Bridge and VLAN utilities 168–169
    network interfaces configuration 165–167
    network interfaces review 164–165
    Open vSwitch configuration 174–176
    Open vSwitch installation 171–174
    server-to-router configuration 169–171
    updating packages 167–168
Network Time Protocol. *See* NTP
networks
  considerations for replacing existing virtual server platforms 246–247
  hardware preparations for HA
    deployment network 280
    overview 279–280
    switch uplink ports 281–282
  Linux configuration
    DNS server 334
    host and domain names 334–335
    host IP address 332
    network gateway address 333
    overview 330–332
    subnet mask 333
  OpenStack and 14–15
  SDN 107
  tenant
    connecting router to public network 72–73
    creating external networks 73–74
    creating external subnets 75–78
    creating internal networks 67–68
    creating internal subnets 68–70
    creating routers 70–72
    Neutron console 67
    overview 65–67
  vendor technologies
    examples of 107–108
    how networking is used by VMs 102–105
    OpenStack support 105–107
    overview 101–102
Neutron 18, 21, 86
  configuring 178, 230
  configuring ML2 plug-in 231
  DHCP agent configuration 180
  installing 177, 230
  L3 agent configuration 179–180
  Linux network namespaces and 191–193
  Metadata agent configuration 180–181
  ML2 plug-in configuration 178–179
  networks in
    external networks 188–189
    external subnets 189–191
    internal networks 183–184
    internal subnets 184–186
    Network console 183
    overview 182
    routers 186–188
  overview 177
  restarting services and verifying installation 181–182
  viewing in Dashboard 193–194
Neutron console 67
NFS (Network File System) 97
NICs (Network Interface Cards) 171, 326
–no-cleanup flag 274
node interfaces 118, 164, 198, 219
Nova 21, 86
  configuring 232–233
  creating instance using CLI 234–237
  defined 18
  installing 231–232
  verifying in Dashboard 233–234
NTP (Network Time Protocol) 261, 288

O

object storage 18, 86
  defined 47
  overview 16–17

Object Storage Device node. *See* OSD node
OOB (out-of-band) networks 280, 282–286
Open Government Directive 18
Open Networking Foundation 107
Open System Interconnection. *See* OSI
Open vSwitch. *See* OVS
openrc script 307
OpenSSH 345
OpenStack
  abstraction and 10–11
  Architecture Design Guide 242
  cloud computing and 9–10
  cloud terminology used with 17
  component communication
    dashboard authentication communication 86–87
    interaction with Juju 88–90
    overview 85–86
    resource provisioning 87–88
    resource query and request 87
  components of 18, 85
  distributed computing model
    component communication at VM-level 93–95
    general discussion 91
    OpenStack distributed component model 92
    VM provisioning interaction 93
  history of 18–19
  hypervisors and 11–14
  Linux versions and 120
  network services and 14–15
  overview 4–9
  storage and
    block storage 15–16
    object storage 16–17
    overview 15
OpenStack Heat
  confirming dependencies 308–309
  enabling in DevStack 307
  Heat templates 304–306
  launching Heat stack 309–311
  overview 304
  Ubuntu Juju vs. 312
openvswitch module 173, 224
Orchestration project 18, 86
orchestration, defined 303
OS (operating system) 94, 168
OSD (Object Storage Device) node
  creating 271–272
  defined 260
  overview 268–269
  preparing devices 269–271
OSI (Open System Interconnection) 66
out-of-band networks. *See* OOB networks
OVA format 141
overlay networks 172–173
Overview screen, Dashboard 38
OVF format 141
OVS (Open vSwitch) 14, 172
  configuring 174–176, 225–226
  installing 171–174, 223–225
  usage statistics for 256

P

P2V (Physical to Virtual) 242
PaaS (platform as a service) 17
package management systems 309
password security 286
PCI (Peripheral Component Interconnect Express) 97
performance
  MySQL considerations 123
  separating storage and network traffic 200
persistent storage 248
Physical to Virtual. *See* P2V
physical volumes 204
PI (policy inaccessible) 39
platform as a service. *See* PaaS
pools, Ceph 273–274
ports
  checking for listening on 37
  creating by adding router to subnet 187
Preboot Execution Environment. *See* PXE
private clouds 9
  control over performance and quality 252
  data gravity 252–253
  hybrid cloud usage 253
  networking for 255–257
  OpenStack deployment tools 254–255
  projects, defined 23
  public clouds vs. 251
  storage in 257–258
public clouds 9, 251
public interfaces 118
publicurl endpoint 130–131, 137, 145, 150, 155
pvcreate command 204
pvdisplay command 204–205
pvscan command 204, 206
PXE (Preboot Execution Environment) 286, 289, 294
python-cinderclient package 211
python-mysqldb package 210

Q

QCOW (QEMU Copy On Write) format 42, 141
QEMU (Quick Emulator) 227
qemu-utils package 215

quotas
  for tenant users 80–82
  for tenants 79–80
  general discussion 78–79, 82–83

R

RabbitMQ 121–122
rados tool 274
RAW format 42, 140
RBAC (role-based access control) 125
rebooting DevStack 35–36
related projects 85
releases, OpenStack 19
reserved_percentage statistic 100, 207
reverse-path filter 169, 223
Rocket 14
role-based access control. *See* RBAC
roles
  assigning 64–65, 134
  creating 133–134
  listing 63, 134–135
root partition 340–342
root user 26, 28
router-gateway-set command 72
routers
  connecting to public network 72–73
  creating 70–72
  DVR 95
  in Neutron 186–188
  listing 76, 190

S

SaaS (software as a service) 17
SAN (storage area network) 249, 287
screen console 32
SCSI (Small Computer System Interface) 208
SDN (software-defined networking) 18, 102, 105, 107
security
  passwords 286
  rules for VMs 39
SLAs (service level agreements) 252
Small Computer System Interface. *See* SCSI
SMP (symmetric multiprocessing) machines 4
snapshots
  images and 42–44
  LVM 202
software as a service. *See* SaaS
software-defined networking. *See* SDN
source command 129
ssh-copy-id tool 263
stacking 26
status command 319
storage
  backend 46
  OpenStack and
    block storage 15–16
    object storage 16–17
    overview 15
  replacing existing virtual server platforms 247–249
  types of 47
  vendor technologies
    how storage is used by VMs 96–98
    OpenStack support 98–100
    overview 96
storage area network. *See* SAN
storage node 115
storage transport protocol 97
storage_protocol statistic 100, 207
subnets
  defined 68
  external 75–78, 189–191
  internal 68–70, 184–186
sudo command 26
supporting projects 85
swap partition 340–342
Swift project 18, 86
switches 171
symmetric multiprocessing machines. *See* SMP machines
sysctl command 169–170, 223

T

tagged VLANs 279–280, 282
target framework. *See* tgt
Telemetry project 18, 86
Tempest
  defined 33
  DevStack and 34
  validation and 33
templates, Heat 304–306, 309
tenants 23
  assigning roles 64–65
  creating 62–63, 132
  creating users 63–64
  defined 6
  general discussion 59–78
  listing 63
  listing all users in 64
  listing roles for 63
  quotas for 79–80
  quotas for users 80–82
  tenant model 61–62
  tenant networks
    connecting router to public network 72–73
    creating external networks 73–74

tenants, tenant networks *(continued)*
creating external subnets 75–78
creating internal networks 67–68
creating internal subnets 68–70
creating routers 70–72
Neutron console 67
overview 65–67
text editors 29
tgt (target framework) 209
thin provisioning 202
total_capacity_gb statistic 100, 207
troubleshooting
checking for port listening 37
DevStack process failure 31
Trove project 86
type drivers 107
–type=compute parameter 155
–type=image parameter 137
–type=network parameter 150
–type=volume parameter 145

U

Ubuntu 116
disk configuration
overview 342–343
partitioning 337–339
root and swap partitions 340–342
images for 41
installing packages 343–346
network configuration
DNS server 334
host and domain names 334–335
host IP address 332
network gateway address 333
overview 330–332
subnet mask 333
preparing for installation 326–328
user configuration 335–336
Ubuntu Juju
deploying charms CLI 317–319
deploying GUI 319–324
installing 314–317
overview 88–90, 312
preparing OpenStack for 312–313
versions 314
udev manager 203
UEFI (Unified Extensible Firmware Interface) 286
unknown free space value 100
untagged VLANs 279–280
uplinks 281
usage-based pricing 253
user-role-add command 134
user-role-list command 134
users, Linux 335–336
users, tenant
assigning roles 64–65
creating 63–64, 133
listing all in tenant 64
quotas for 80–82
UTF8 error 140

V

V2V (Virtual to Virtual) 243
Vagrant 24
VDI (Virtual Disk Image) 42, 141
vendor technologies
network systems
examples of 107–108
how networking is used by VMs 102–105
OpenStack support 105–107
overview 101–102
storage systems
how storage is used by VMs 96–98
OpenStack support 98–100
overview 96
vendor neutrality of OpenStack 95–108
vendor_name statistic 100, 208
vgcreate command 204
vgdisplay command 206
VHD (Virtual Hard Disk) format 42, 140
Virtual Disk Image. *See* VDI
Virtual Local Area Networks. *See* VLANs
Virtual Machine Disk format. *See* VMDK format
virtual machine monitor. *See* VMM
virtual machines. *See* VMs
virtual private networks. *See* VPNs
Virtual to Virtual. *See* V2V
VirtualBox 24
virtualization
extensions for 227
nested 25
operating-system-level 14
VLANs (Virtual Local Area Networks) 168, 247, 279–280
VM interfaces 164
vm_state property 237
VMDK (Virtual Machine Disk) format 42, 141
VMM (virtual machine monitor) 11, 25
VMs (virtual machines)
companion for book 23–24
instances vs. 38
overview 5–6
PI 39
VMware 242–243, 250
volume groups 204–205

volume_backend_name statistic 100, 208
VOLUME_BACKING_FILE_SIZE value 46
Volumes screen, Dashboard 44–47
VPNs (virtual private networks) 173
vxlan module 173, 224

W

WordPress service, provisioning 318